RICE RESEARCH IN ASIA
Progress and Priorities

Edited by

R.E. Evenson
Economic Growth Center
Yale University, USA

R.W. Herdt
Director, Agricultural Sciences
Rockefeller Foundation
New York, USA

and

M. Hossain
International Rice Research Institute
Manila, Philippines

CAB INTERNATIONAL
in association with the
International Rice Research Institute

CAB INTERNATIONAL Tel: +44 (0)1491 832111
Wallingford Fax: +44 (0)1491 833508
Oxon OX10 8DE E-mail: cabi@cabi.org
UK Telex: 847964 (COMAGG G)

CAB INTERNATIONAL 1996. All rights reserved. No part of this publication may be reproduced in any form or by any means, electronically, mechanically, by photocopying, recording or otherwise, without the prior permission of the copyright owners.

Published in association with the:
International Rice Research Institute
PO Box 933
1099 Manila
The Philippines

A catalogue record for this book is available from the British Library.
ISBN 0 85198 997 7

Typeset in 10/12 Sabon by Techset Composition, Salisbury
Printed and bound in the UK at the University Press, Cambridge

Rice Research in Asia
Progress and Priorities

Contents

Contributors		vii
Preface		ix

PART I: BACKGROUND AND METHODS

1. Priorities for Rice Research: Introduction 3
   R.E. EVENSON, R.W. HERDT and M. HOSSAIN

2. Recent Developments in the Asian Rice Economy: Challenges for Rice Research 17
   M. HOSSAIN

3. Rice Ecosystems Analysis for Research Prioritization 35
   D.P. GARRITY, V.P. SINGH and M. HOSSAIN

4. Prospects of and Approaches to Increasing the Genetic Yield Potential of Rice 59
   G.S. KHUSH

5. The Economic Principles of Research Resource Allocation 73
   R.E. EVENSON

6. Priority-Setting Methods 91
   R.E. EVENSON

PART II: COUNTRY STUDIES

7. Prioritizing the Rice Research Agenda for Eastern India 109
   D.A. WIDAWSKY and J.C. O'TOOLE

8. Yield Gaps, Production Losses and Priority Research Problem Areas in West Bengal, India 131
   N.K. SAHA, S.K. BARDHAN ROY, S.K. GHOSH and M. HOSSAIN

9	Constraints to Higher Rice Yields in Different Rice Production Environments and Prioritization of Rice Research in Southern India C. RAMASAMY, T.R. SHANMUGAM and D. SURESH	145
10	Rice Production Constraints in China J.Y. LIN and M. SHEN	161
11	Rice Production Constraints in Bangladesh: Implications for Further Research Priorities M.M. DEY, M.N.I. MIAH, B.A.A. MUSTSFI and M. HOSSAIN	179
12	Rice Research in Nepal: Current State and Future Priorities H.K. UPADHYAYA	193
13	Rice Research Priorities in Thailand S. SETBOONSARNG	217
14	Constraints to Growth in Rice Production in the Philippines M. HOSSAIN, F.B. GASCON and I.M. REVILLA	239
15	Rice Production Losses from Pests in Indonesia T. JATILEKSONO	251

PART III: CROP-LOSS STUDIES

16	Technical Issues in Using Crop-Loss Data for Research Prioritization P.S. TENG and I.M. REVILLA	261
17	Priorities for Weed Science Research K. MOODY	277
18	Yield Loss Due to Drought, Cold and Submergence in Asia M.M. DEY and H.K. UPADHYAYA	291
19	Intercountry Comparison of Insect and Disease Losses C. RAMASAMY and T. JATILEKSONO	305
20	The Economic Impact of Rice Blast Disease in China M. SHEN and J.Y. LIN	317

PART IV: PRIORITY-SETTING APPLICATIONS

21	An Application of Priority-Setting Methods to the Rice Biotechnology Program R.E. EVENSON	327
22	Rice Research Priorities: an Application R.E. EVENSON, M.M. DEY and M. HOSSAIN	347
23	Summary, Conclusions and Implications R.W. HERDT	393

Index	407

Contributors

S.K. BARDHAN ROY, *Directorate of Agriculture, Government of West Bengal, Mission Compound, PO and District Midnapur, West Bengal, India.*

M.M. DEY, *International Center for Living Aquatic Resources Management (ICLARM), MC PO Box 2631, Makati Central, PO Box 0718 Makati, Metro Manila, Philippines.*

R.E. EVENSON, *Economic Growth Center, Yale University, 27 Hillhouse Avenue, New Haven, CT 06520, USA.*

D.P. GARRITY, *ICRAF Southeast Asia, Forest Research and Development Centre, Jalan Gunung Batu No. 5, PO Box 382, Bogor 16001, Indonesia.*

F.B. GASCON, *Social Sciences Division, International Rice Research Institute, PO Box 933, 1099 Manila, Philippines.*

S.K. GHOSH, *Director of Agriculture and Agricultural Economics, West Bengal, C/5, Shyamali Housing Estate, Block EA, Sector-1, Salt Lake, Calcutta 700 064, India.*

R.W. HERDT, *Director, Agricultural Sciences, Rockefeller Foundation, 420 Fifth Avenue, New York, NY 10018-2702, USA.*

M. HOSSAIN, *Social Sciences Division, International Rice Research Institute, PO Box 933, 1099 Manila, Philippines.*

T. JATILEKSONO, *Associate Professor of Agricultural Economics, Faculty of Agriculture, Gadja Mada University, Yogkarkarta, Indonesia.*

G.S. KHUSH, *Plant Breeding, Genetics and Biochemistry Division, International Rice Research Institute, PO Box 933, 1099 Manila, Philippines.*

JUSTIN YIFU LIN, *China Center for Economic Research, Peking University, Beijing 100871, People's Republic of China.*

M.N.I. MIAH, *Bangladesh Rice Research Institute, Gazipur 1701, Bangladesh.*

K. MOODY, *Agronomy, Plant Physiology and Agroecology Division, International Rice Research Institute, PO Box 933, 1099 Manila, Philippines.*

B.A.A. MUSTSFI, *Bangladesh Rice Research Institute, Gazipur 1701, Bangladesh.*

J.C. O'TOOLE, *Rockefeller Foundation, BB Building, Suite 1412, 54 Sukhumvit Soi 21 (Asoke), Bangkok 10110, Thailand.*

C. RAMASAMY, *Professor of Agricultural Economics, Tamil Nadu Agricultural University, Coimbatore 641 003, India.*

I.M. REVILLA, *Cross Ecosystem Program, Entomology and Plant Pathology Division, International Rice Research Institute, PO Box 933, 1099 Manila, Philippines.*

N.K. SAHA, *Additional Director of Agriculture, 17 S.P. Mukherjee Road, Calcutta, India.*

S. SETBOONSARNG, *ASEAN, Secretariat, Jakarta, Indonesia.*

T.R. SHANMUGAM, *Department of Agricultural Economics, Tamil Nadu Agricultural University, Coimbatore 641003, India.*

MINGAO SHEN, *Development Institute Research Center for Rural Development, 5 Liuliqiao Beili, Fengtaiqu, Beijing 100055, People's Republic of China.*

V.P. SINGH, *Agronomy, Plant Physiology and Agroecology Division, International Rice Research Institute, PO Box 933, 1099 Manila, Philippines.*

D. SURESH, *Department of Agricultural Economics, Tamil Nadu Agricultural University, Coimbatore 641 003, India.*

P.S. TENG, *Cross Ecosystem Program, Entomology and Plant Pathology Division, International Rice Research Institute, PO Box 933, 1099 Manila, Philippines.*

H.K. UPADHYAYA, *Center for Environmental and Agricultural Policy Research, Extension and Development (CEAPRED), PO Box 5752, Kathmandu, Nepal.*

D.A. WIDAWSKY, *Food Research Institute, Stanford University, Stanford, CA 94305, USA.*

Preface

Research organizations have a design structure that determines the conditions in which research activity takes place. Research design elements include physical infrastructure as well as human capital in the form of fields of science and the skill levels of staff scientists. These design elements constrain the actual research projects and programs that are undertaken in a given time period.

Research design elements can be changed, given time. Research project and program elements can be changed more readily, but are subject to design constraints in the short run. Decision makers determining both the long-run design elements and the short-run project and program elements generally accept the goal of economic efficiency. They seek to produce research goods or products in the most cost-efficient way. They also seek to produce the combination (or portfolio) of research goods or products that is of the highest value to their clients and supporters. By so doing they serve their clients and supporters best.

The problem of determining the most efficient combination of design elements and of project and program elements is a difficult one even in a simple static situation. It is confounded when research units in a research system have complex interunit relationships such as those reflected in the rice research system. The rice research system now encompasses international research centres (the International Rice Research Institute (IRRI), the West African Development Authority (WARDA) and the International Center for Tropical Agriculture (CIAT)), National Agricultural Research (NAR) programs and subnational rice research programs (e.g. the state programs in India). Each of these units has a 'comparative advantage' in the conduct of certain types of rice research.

The problem of determining (or planning) efficient system design and project program conduct is further compounded when the 'demand for research' conditions change. And it is even further and more profoundly compounded when the supply conditions (i.e. the methods of research and germplasm stocks) change.

Formal methods for allocating resources to agricultural research projects and programs have been in use for roughly three decades. These methods have evolved from early *ex ante* project and program evaluation techniques to *priority-setting* methods that have been applied in more recent years. Ironically, the early *ex ante* techniques were often more complex and more rigorous from a methodological perspective than the later priority-setting methods. Their very complexity, however, limited their actual use. (See Alston *et al.*, (1996) for treatments and applications of priority-setting methods.)

Informal planning and review procedures including peer reviews, administrative reviews and general program reviews have long been in use to plan and design research programs. Formal methods are not intended to supplement these informal procedures, but rather to complement them.

In this volume we develop and apply research priority-setting methods that differ from earlier methods in several ways. First we apply these methods to projects, programs and organizations directed toward research on a single commodity, rice. This single-commodity focus produces an emphasis on specific research problem areas (RPAs) that are usually not stressed in multiple-commodity research planning. Second, we stress the use of crop-loss data as economic unit measures to a greater degree than most other studies (although earlier rice research priority-setting studies pioneered this use, see Herdt and Rieley (1987)). Third, we have attempted to model the research discovery process and to utilize this model to guide our methods in a way that earlier studies have not. Fourth, we have attempted to address the comparative advantage relationship between different research units in this study. Finally, we have attempted to incorporate the latest advances in rice research techniques (i.e. the technology of rice research itself) into our analysis and to assess the expected contributions of alternative research techniques in this work.

In the first six chapters of the volume (Part I) we cover background (Chapters 1–4) and methods (Chapters 5, 6). Chapter 1 is a general introduction. Chapter 2 discusses changes in rice production and identifies some of the broad changes in the demand for rice (especially regarding grain quality) that are occurring. Chapter 3 discusses the eco-regional factors that influence rice research effectiveness and system design. Chapter 4 outlines recent developments on the supply side of rice research. It describes the evolution of the new plant type of rice, hybrids and new rice biotechnology research techniques.

Chapters 5 and 6 develop the methodology applied in later chapters. Chapter 5 develops the research model and derives allocative efficiency rules which, in turn, guide the procedures for assessing likely outcomes for RPA–research technique (RPA–RT) combinations. Chapter 6 develops the priority-setting procedure in more detail.

Part II of the volume includes eight country chapters (Chapters 7–15). Each of these chapters reports a yield-gap analysis and crop-loss assessments. The crop-loss assessments from these chapters constitute the most comprehensive evidence for crop losses available for any crop. Several of the country studies also report priority-setting exercises.

Part III of the volume (Chapters 16–20) covers comprehensive analysis of specific crop losses (rice blast, drought, cold and submergence) and of technical agronomic studies of crop losses.

Part IV of the volume (Chapters 21–23) includes three chapters applying the methods developed in Chapter 6 to rice research. Chapter 21 presents an assessment of rice biotechnology research where *ex post* estimates of rice-loss reduction are utilized in the exercise. Subjective probability estimates are confined to time-to-achievement estimates. Chapter 22 reports an application of our methods to a comprehensive set of RPA–RT combinations for different regions and ecosystems. It utilizes country study crop-loss data and is intended to be a generic study in that it illustrates the use of the methods to enable other more detailed studies. The final chapter provides an overview of the volume and relates it to earlier work on rice research priority setting.

References

Alston, J.M., Norton, G.W. and Pardey, G. (1996) *Science Under Scarcity*. Cambridge University Press, New York.

Herdt, R.W. and Rieley, F.Z. (1987) *International Rice Research Priorities: Implications for Biotechnology Initiatives*. Rockefeller Foundation Workshop in Allocating Resources for Developing Country Agricultural Research, Bellagio, Italy.

I Background and Methods

1 Priorities for Rice Research: Introduction

R.E. Evenson[1], R.W. Herdt[2] and M. Hossain[3]

[1] Economic Growth Center, Yale University, 27 Hillhouse Avenue, New Haven, CT 06520, USA; [2] Agricultural Sciences, Rockefeller Foundation, 420 Fifth Avenue, New York, NY 10018-2702, USA; [3] Social Sciences Division, International Rice Research Institute, PO Box 933, 1099 Manila, Philippines

I. Introduction

Research organizations, whether private or public, regional, national or international are not immune from the need to achieve economic efficiency. Shareholders, taxpayers and societies at large are served best by organizations that achieve economic efficiency. Economic efficiency can be divided into two parts. The first is technical or *cost efficiency*. Cost efficiency requires that resources are used so as to minimize the costs of producing a given product. The second is *allocative efficiency*. Allocative efficiency requires that the products or outputs of the organization are those of maximum value to the organization's clientele, given the organization's comparative advantage in a larger system of organizations.

For private sector firms, economic efficiency is required for economic survival in competitive markets. Private firms accordingly continuously engage in cost and management reviews and in product assessment. Achieving economic efficiency for a private sector firm is a dynamic process requiring continuous evaluation of resource use and of potential technological improvements in both process technology and product technology. The production of research products in the private sector is even more demanding of continuous cost and management review and product assessment than is the case for the production of normal goods and services because of the high level of uncertainty associated with research investment. Research programs in the private sector thus must also engage in continuous product assessment and be responsive to changes in clientele demand.

For public sector research organizations these same requirements hold. Serving taxpayers or citizens at large instead of shareholders does not alter the need for economic efficiency. Even if economic survival is not at stake, research organizations that are economically efficient serve their clientele best. For agricultural research organizations, this means that cost management reviews are continuously required. It also means that product assessment is continuously required by changes in the technology of research (i.e. changes in what products are possible) and by changes in clientele demand. These reviews and assessments need to consider the inherent uncertainty of research activities as well as the complexities of agricultural research systems and the comparative advantages of the relevant organizational units.

Agricultural research systems have evolved into complex forms over time. They have both ecosystem and hierarchical dimensions. The ecosystem dimension is required because of the high degree of sensitivity of plant and animal technology to ecosystem components (soil, water, temperature, etc.). One hierarchical dimension is the regional–national–international dimension. A second is the strategic (sometimes called basic or pre-invention science), applied (where the objective is technology invention), adaptive (where the objective is adaptive or sub-invention) dimension. These two dimensions are obviously closely related because of comparative advantages (see Chapter 5).

Agricultural research systems typically utilize the research *project* as the base unit for research performance. Projects are typically clustered into meaningful *programs* or research problem areas (RPAs). Allocative efficiency (and to a lesser degree cost efficiency) is typically achieved between projects by using *ex ante* (i.e. before the fact) project evaluation techniques to rank projects by potential economic pay-off. In principle, *ex ante* methods can be 'aggregated up' to allow comparisons between RPAs. But, in practice, this approach to achieving allocative efficiency is limited by the narrow vision of project developers and often fails to consider the broader issues of comparative advantage and ecosystem targeting. *Priority setting*, the topic of this volume, is the general approach applied to research planning to achieve allocative efficiency across RPAs.

Priority setting is not itself in conflict with *ex ante* project evaluation. Nor should it constrain and inhibit project generation activity within research organizations because it is largely at the project level (i.e. the bench scientist or field scientist level) that creative responses to changes in research opportunities (here termed research germplasm) are made. It is likewise at the project level that constructive formulations of the demand for germplasmic research are made. As a consequence, the technical or cost efficiency dimension of achieving economic efficiency is very much directed to project generation and project management activities. Research management, of course, is also effectively pursued at the RPA and higher levels. Priority setting does not and should not ignore cost management issues, but its principal use is in achieving allocative efficiency at the RPA level rather than cost or technical efficiency.

In this volume we discuss methods for priority setting and applications to rice research. Priority setting is related to project evaluation techniques. It is related to yield-accounting procedures and it is related to *ex post* studies. We discuss these relationships briefly in this introductory chapter and in more depth in subsequent chapters.

An important distinction may be made between priority setting for research programs directed toward the improvement of several commodities and priority setting for research on a single commodity. Most previous work on priority setting (or *ex ante* evaluations of agricultural research programs) has dealt with multiple-commodity applications. In this volume we deal with only one commodity – rice (although the different ecosystems might be considered to be multiple commodities). Accordingly the RPA definitions are directed to rice production problems.

Another feature of this volume is that it recognizes the fact that several research techniques (RTs) are available to address an RPA. For example, insect problems can be tackled through better management, through improved chemicals or through the incorporation of genetic host plant resistance. The latter may be achieved by conventional breeding programs or by wide crossing or transgenic breeding techniques. Thus, in this study we seek to establish research priorities not simply for RPAs but for RPA–RT combinations.

II. Priority Setting and Project Evaluation

It is natural that priority-setting methods are closely related to *ex ante* project evaluation methods (see Chapter 6 for specifics), but they are also amenable with, and draw policy insight from, more general economic models and economic principles. Chapter 5 of this volume develops some of these relationships and some of the policy insights for allocative efficiency from economic theory. The conditions under which research resource allocation is primarily governed by the 'demand' for research products or outcomes are outlined. The simple *congruence rule* (where research is allocated in proportion to the value of units affected) applies when specific research supply or cost conditions are met. Chapter 5 stresses the importance of first identifying the relatively easy to measure research demand variables and then identifying the more difficult to measure supply side variables (i.e. the research success and time-to-achievement probabilities) that would call for a departure from congruence.

Chapter 6 surveys the methods for priority setting. It begins with the basic methods for *ex ante* project evaluation and shows how economic pay-off to research is related to:

1. Project resources and design (costs).
2. Economic units affected by the project outcome (e.g. hectares of rice affected).
3. The economic contribution (e.g. cost reduction or yield increase) per economic unit.
4. The timing of research product effects.

The *ex ante* project analyst (and the priority setter as well) usually must estimate at least one of these elements by subjective probability estimation (SPE) methods. If one or two of the other elements can be based on outside evidence, or *ex post* evidence, the evaluation is much improved.

The option of relying on outside evidence is more applicable at the RPA level (the key focus of priority setting) than at the project level. By judicious definition of the RPA, this possibility can be enhanced and made more relevant. For example, in this volume, several country studies provide information as to the economic units (element 2 above) relevant to RPAs defined by crop-loss categories. These are the demand side variables referred to earlier.

A typical *ex ante* and priority-setting practice is then to fix (or assume a level for) element 1, the resources and design element, use outside information for (2), the demand side element, and use SPE methods for the two supply side probabilistic elements (3 and 4). In some cases *ex post* evidence may be used to fix element 3, leaving only one element to be estimated by SPE methods (see Chapter 21 for an application to rice biotechnology).

Priority setting thus is basically *ex ante* project evaluation 'writ large'. At the RPA level, the use of outside information is more feasible than at the project level. In some studies *ad hoc* 'scoring' methods are used (more often mis-used) in priority setting. These techniques are discussed in Chapter 6. Also discussed are the incorporation of special issues such as sustainability, equity-poverty alleviation and environmental effects into priority setting. The point is made (hopefully forcefully) that many of these issues do not belong in priority setting for rice research because rice research programs are not good instruments by which to address these important problems. Rice research is, after all, a very specialized activity and there is often little scope for achieving general economic outcomes without sacrificing research effectiveness.

III. Yield Gaps and Priority Setting

In the mid-1970s, after 5 or so years of broad-scale adoption of the modern rice varieties, two observations were made: not all farmers had adopted the first generation of modern varieties, and the yields obtained by farmers who had adopted them were significantly lower than yields on experiment stations. While experiment station yields were in the order of 10–12 tons ha^{-1}, farmers were averaging 3–5 tons ha^{-1}. The first issue formed the challenge for plant breeders to develop varieties that better suited the agroeconomical conditions of a broader range of farmers. The second was a challenge to agronomic researchers to understand why farmers' yields were low and what could be done to raise these yields. It led to the formulation of the 'yield gap' model, which defined the following concepts:

Yield A: actual farm yields – yields actually observed on farmers' fields.
Yield B: best practice yields (or potential farm yield) – the yield that could be obtained on farmers' fields if maximum yield practices were followed and no losses were experienced from pests.
Yield C: experiment station yield – the yield observed on experiment stations.

A series of on-farm experiments to measure these yields and attribute them to various factors were conducted in several Asian countries. These original constraints studies revealed two major limitations of this experimental approach to measuring constraints: lack of representativeness and the inherent limitations of experiments for the purpose.

The spatial variability and heterogeneity of rice production environments, even in a relatively small country, means that the range of conditions is quite large. In addition, the experiments would have to be spread so extensively over the landscape as to make it impractical for a single research team to conduct the experiments. The cost of doing an effective job was high. A second limitation was that the number of factors contributing to yield constraints are large and hence experiments to measure the contribution of each of the causal factors are very complex. Even partial factorial designs require a large number of treatments, while in practice it is difficult to conduct an experiment with over six or eight treatments in farmers' fields.

Even though the yield constraints experiments had their limitations, they generated interesting results and led researchers to define several gaps. Yield gap I was defined as the difference between the yield observed on experiment stations (yield C) and the best practice yield on farmers' fields (yield B). It was hypothesized that there are differences in the soil and water resources of experiment stations and farms that make the recovery of yield gap I quite difficult.

Yield gap II was defined as the difference between best practice and actual yields on farmers' fields. This gap could be overcome if farmers completely controlled pests, used maximum practices and the best varieties. It is a measure of the gap that can be overcome through the spread of the latest technology and hence is an indicator of potential for extension. It is also an indicator of the yield losses that are being experienced by farmers, largely because of uncontrolled pests and diseases, and has been the object of much of the research reported in the country studies.

While cost limitations made the continuation of the constraints experiments impractical, the appeal of the yield gap concept gave it life beyond the original project and it has been adapted to the problem of helping to determine research priorities. Stimulated by contemporary needs, two additional yield ideas were (implicitly) defined.

Over the 1980s, IRRI scientists gradually observed a downward drift in the maximum yields obtained in experiments designed to get maximum yields (Pingali et al., 1990). It was hypothesized that the production environment had changed in such a way that the full genetic potential of existing varieties was no longer being obtained, even on experiment stations. An alternative hypothesis, that the genetic potential of the latest varieties was lower, was generally rejected by tests with the early varieties that showed they also had lower yields in later years. It was clear that insect and disease pressures had built up at the IRRI experimental station over time, and probably at other similar stations, raising the intensity of such challenges far above what was experienced in farmers' fields. In addition, it was hypothesized that perhaps changes were taking place in the soil that changed the nutrient-supplying capacity or created toxicities, and affected the quality of irrigation water or other elements in the environment. With these ideas in mind another yield level was defined.

Yield D was defined as the yield that could be obtained on the experiment station, with the maximum-yield varieties, optimal soil conditions, optimal fertility control, optimal water quality and control, and complete absence of yield-reducing pests.

By the mid-1980s it also became obvious that the yield potential of the set of varieties created during the 1970s and 1980s were no higher than the first generation of modern varieties, and possibly even lower. The prospects for increasing actual national yields beyond the 6–7 tons ha^{-1} level seemed bleak, and the need to further raise the genetic yield potential of rice took on renewed urgency. One promising avenue was hybrid rice, which Chinese scientists had developed and which gave 20% or more higher yields under their conditions. Efforts were undertaken to develop hybrids for the tropics. In addition, efforts were made to conceptualize a plant type with a higher yield potential than the 1960s semidwarf. By 1989, IRRI scientists had, with the help of crop models, conceived of a new plant type with fewer tillers, more grains per tiller, a higher number of larger grains on each tiller and fewer non-productive tillers (see Chapter 4). Both hybrid rice and the new plant type reflect the idea that a higher yield potential is possible by changing the genetic composition of rice.

Hence we define Yield E as the yield that could be obtained by successfully pursuing presently conceived avenues to increasing yield potential. This yield is limited only by the imagination of scientists in thinking of possible avenues of improvement and convincing their peers of the realistic potential of such avenues. The new plant type presently being developed provides 20–25% increase in potential yield, and if it is made into a hybrid another 15–20% yield advantage might be obtained. Thus, yield E might be as much as 30–50% above the potential demonstrated by the present generation of best varieties (this assumes the solution of the agronomic challenges addressed by yield D).

The difference between yield C and yield E is defined as yield gap III. Yields D and E cannot be observed because each is defined as a level that could be obtained if a set of research activities were successful. However, it is important that these levels are estimated because they indicate the potential gains from research of quite distinct types. Yield D will be obtained largely with agronomic research that results in changed management, while yield E will be obtained largely with genetic research.

Yield gap III can be conceived of as a measure of the demand for yield-enhancing research. However, we have rather limited means for estimating its magnitude without considering the supply side of research. Yield potential from research per unit of rice (or per hectare) is really an expected research outcome from a research program and hence a supply side measure. The number of units (hectares) in the ecosystem is a measure of demand for yield-enhancement research. Few of the country studies actually estimated yield gap III, but it is important that we not reduce its importance in priority-setting exercises because of measurement difficulties.

Each of the country studies did estimate yield gap II (by ecosystem) and each provides detailed estimates of crop losses, the empirical centrepiece of this book. Recognizing that research may eliminate only a proportion of drought stress losses but may fully eliminate losses due to an insect pest, the estimated losses may be adjusted for the proportion subject to research impact and hence provide measures of the potential gains from research.

The yield gap accounting procedure imposes a consistency check on the crop-loss estimates and an allocation between the biotic and abiotic crop losses and unrealized production due to inadequate farm management, imperfect markets and poor infrastructure.

The country studies have attempted to provide these measures by ecosystem. They offer a relatively comprehensive set of measures of the scope or potential for a rice crop-loss reduction for which RPAs using different techniques (including biotechnology techniques) are meaningfully defined.

IV. *Ex post* Evaluation and Priority Setting

A considerable body of *ex post* evidence on the impact or economic contribution of rice (and other) research and extension programs now exists (Tables 1.1 and 1.2). These *ex post* studies are related to priority setting in at least two ways. First, the *ex post* perspective on the contribution of a research program is the same as the *ex ante* perspective except that the former is seen after the fact (and hence can be measured by evidence) and the latter is 'before the fact'. In each case a financial calculation can be made based on the following expressions:

$$PVB_0 = \sum_{t=0}^{n} b_t/(1+r)^t \qquad (1.1)$$

$$PVC_0 = \sum_{t=0}^{n} c_t/(1+r)^t \qquad (1.2)$$

PVB is the present value (at time = 0) of a benefit stream associated with the research project where b_t is the benefits in each period (from $t=1$ to $t=n$) and r is the discount rate. *PVC* is the present value of the costs (c) associated with the project.

Table 1.1. Summary of *ex post* imputation studies of rice research.

Study	Country	Time Period	Estimated AIRR
Hayami & Akino, 1977	Japan	1915–1950	25–27
Hayami & Akino, 1977	Japan	1930–1961	73–75
Hertford et al., 1977	Colombia	1937–1972	60–82
Avila, 1981	Brazil	1959–1978	87–119
Echeverria et al., 1988	Uruguay	1965–1985	52

AIRR, average internal rate of return.

Table 1.2. Summary of statistical studies of rice research.

Study	Country	Method	Time period	Estimated MIRR
Scobie & Posada, 1978	Bolivia	MPF	1957–1964	79–96
Pray, 1979	Bangladesh	MPF	1961–1977	30–35
Evenson & Flores, 1978	Asia (NAR Program)	TFP	1950–1965	32–39
Evenson & Flores, 1978	Asia (NAR Program)	TFP	1966–1975	73–78
Evenson & Flores, 1978	Asia (IRRI)	TFP	1966–1975	74–108
Flores et al., 1978	Philippines	D	1966–1978	75
Flores et al., 1978	Tropics	D	1966–1975	46–71
McKinsey & Evenson, 1991	India	D	1954–1984	65
Azam et al., 1991	Pakistan	D	1969–1988	84
Dey & Evenson, 1991	Bangladesh	TFP	1969–1989	165
Salmon, 1991	Indonesia	MFP	1969–1980	151
Setboonsarng & Evenson, 1991	Thailand	D	1967–1980	35

D, duality based; HYV, high-yielding variety; MFP, meta production function; MIRR, marginal internal rate of return; NAR, National Agricultural Research; TFP, total factor productivity decomposition.

In *ex post* studies actual estimates of b_t and c_t and their relationships are made. In *ex ante* or priority-setting studies one must estimate either b_t or c_t by SPEs. For example, a research project to be conducted at a certain level of effort over a certain period may be specified; this effectively fixes c_t. The problem then is to obtain the 'best' subjective judgements from scientists as to what b_t will be (see Chapters 6, 21 and 22 for applications). In practice, scientists base these judgements on past experience.

The second relationship between *ex post* and priority setting is that some *ex post* evidence can be formally utilized in priority-setting applications. This is typically confined to 'hedonic' *ex post* evidence (see below). For example, in Chapter 21, hedonic *ex post* estimates of the yield contribution (loss reduction) of several types of host plant resistance are used to estimate benefits streams (b_t) associated with biotechnology research. The timing dimension of benefits is then estimated by SPE methods.

There are basically three types of *ex post* studies of agricultural research.

1. Imputation studies where benefits estimates are imputed from experimental and field trial evidence (usually comparing improved or modern technology with previously available technology).
2. Statistical studies establishing a statistical link between research program costs (c_t) and benefits (b_t). These studies typically compile a 'research stock' variable by region and time period and relate this to measures of crop *productivity* or to crop *supply* and factor *demands*.
3. Statistical hedonic studies establishing a statistical link between technological characteristics (achieved via research) and benefits (b_t). These studies might relate host plant resistance to brown planthopper attacks to crop productivity (or to pesticide usage). They differ from the more general

statistical studies (eqn 1.2) in that they rely on technological characteristics measurement.

Table 1.1 provides an overview of five *ex post* rice research studies utilizing imputation methods. Typically these entailed field comparisons of improved and unimproved rice varieties. The measure of research effectiveness was the average internal rate of return (AIRR). This is defined as the discount rate, r, for which $PVB = PVC$ (see eqns 1.1 and 1.2). It represents a return on investment.

Table 1.2 provides an overview of 12 statistical studies. These are of three subtypes.

1. *Meta production function.* Statistical studies utilizing a production function model with meta variables (measuring research, extension, infrastructure and policy activities) to make a statistical inference as to the impact of research on production.
2. *Total factor productivity (TFP) decomposition.* Statistical studies in which TFP indexes are first computed, then statistically related to meta variables.
3. *Duality based.* Statistical studies utilizing the duality structure between rice supply and factor demand systems. In these studies meta variables are included in each equation in the system enabling direct statistical inference as to the impact of research on rice supply, labour demand, etc.

For these studies, the definition of the meta research variable is critical. The time dimension between spending and impact must be incorporated into the stock variable (and in some studies estimated – typically a period of 7–10 years after research conduct passes before the full impact is realized). In addition, the spatial dimension between research conducted in one location and impact in another location must also be specified. (Most studies have utilized geo-climate data in specifying this.

For statistically estimated coefficients one can calculate the marginal (incremental) contribution to benefits (b_t) from an increment to costs (c_t). Given the time and spatial dimensions in the specification this can be expressed as a marginal internal rate of return (MIRR). All statistical studies summarized in Table 1.2 reported statistically significant research impact and all reported high marginal (and average) rates of return to investment.

Table 1.3 summarizes findings from two *ex post* statistical hedonic studies. These studies attempted to determine whether the incorporation of host plant resistance to specific rice diseases and insect pests were associated with reduced crop losses and higher crop yields. The Indonesian study is the first to use actual crop-loss data to attempt to measure whether the incorporation of resistance in rice varieties planted in the region reduces these losses. Reference to Chapter 15 will show that the government-collected crop-loss data on which these estimates were based does not appear to be comparable to data from farmers. The Indian study was based on comparative yields in farmers' fields between varieties with specific characteristics and all varieties. These data show a considerable impact on yields for most of these host plant resistance traits. (See Chapter 21 for a further review and use of these hedonic estimates.)

Table 1.3. *Ex post* statistical hedonic studies of rice research.

Study	Country	Time period	Index	Technical characteristic	Estimate
Evenson, 1995	Indonesia (regions)	1971–1989	Crop loss	R: brown planthopper	0.021
				R: gall midge	0.019
				R: bacterial leaf blight	NS
				R: grassy stunt virus	NS
				R: rice tungro virus	0.003
				R: insects	0.11
				R: diseases	0.03
Evenson, 1995	India (state)	1972–1992	Comparative yields	R: blast	0.184
				R: bacterial leaf blight	Neg
				R: rice tungro virus	0.068
				R: sheath blight	0.108
				R: brown planthopper	0.033
				R: green leafhopper	0.123
				R: white-backed planthopper	0.377
				R: gall midge	0.174
				R: stemborer	0.141

R = resistance to.

V. Priority Setting and International Activity

Clearly, as discussed in the preceding section, investment in rice research has proved to have high rates of return in many locations. A question that remains, of special interest for international assistance donors, is whether there is any special value to international research apart from, or in addition to, national rice research programs.

Gollin and Evenson (1990) have examined the IRRI's international movements of rice germplasm, which are distinct from the IRRI's activities in breeding rice varieties. The former involves the collection, cataloguing, preservation and distribution of germplasm, and includes a large program in which varieties and breeding lines are packaged into sets and provided to cooperating researchers in rice-growing countries. The seeds in these rice 'nurseries' are planted and evaluated for performance by cooperating researchers. In addition, cooperating researchers may propose the lines for release as varieties in their own countries or, more frequently, the lines used as parents in making further crosses for local evaluation.

Gollin and Evenson found that International Network for Genetic Evaluation of Rice (INGER) significantly accelerated the exchange of rice-breeding materials and increased varietal release (Chapter 5 provides additional data). We also recognize that the IRRI has a comparative advantage not only in managing programs such as INGER, but in conducting strategic (basic) research that provides intellectual germplasm provided through INGER. Chapter 5 develops a full treatment of this issue.

VI. The Plan of this Volume

The plan of the volume is as follows. First, in Chapter 2 an overview of production and productivity trends as well as of consumption trends for rice is presented. This chapter also highlights the broad ecosystem dimensions of rice research. Chapter 3 deals further with the ecosystem element.

The importance of ecosystems in rice production and in rice research has long been recognized. Soil and climate factors affect rice performance and rice-improvement programs have responded to these factors. Modern rice-breeding programs have only partially overcome the Darwinian (natural selection) targeting of plants to ecological niches. The present-day system of national, regional and subregional research units each with some degree of 'ecosystem targeting' is still guided by Darwinian principles. The emerging biotechnology tools for rice research will not alter the ecosystem nature of rice research.

Chapter 4 describes the development of a new plant design (type) for rice in recent years. It is essential in priority setting that a means for recognizing the fundamental research problem areas such as plant design, photosynthetic efficiency and other factors determining biological efficiency be made. Chapter 4 is intended to illustrate the importance of this work.

Chapter 5 is directed toward setting the general economic framework for evaluating economic efficiency. It attempts to clarify the distinction between the demand side of research (i.e. the demand for research products) and the supply side of research (i.e. the probabilistic nature of research discoveries) and the relationship between expected discoveries and research resources. A model designed to deal with different research products (germplasm, inventions and adaptive or sub-inventions) that are typically produced by different research specializations (strategic, applied and adaptive research) is developed. Principles for efficient resource allocation are derived. The model is also used to address the comparative advantage question for different types of research organizations (international, national and regional). Chapter 5 also relates rice breeding to the economic model.

Chapter 6 provides a review of priority-setting methods and procedures. It relates these methods to more general product evaluation methods. Chapters 21 and 22 report applications of these methods.

Chapters 7–15 report country studies. Each of these studies provides historical background and ecosystem information. Each provides measures of crop losses from specific insect pests, rice diseases, other pests and abiotic stresses. Taken together they provide the first broad-scale comprehensive estimates of crop losses as estimated by farmers, scientists and extension agents for Asia. This evidence is important in its own right as well as being an integral part of priority setting. Each country study also reports a yield gap analysis and relates crop-loss data to yield gaps. Chapter 16 provides the experimental perspective on crop losses. International comparisons for rice blast, virus diseases, weeds, insect-related losses and abiotic stress losses are also reported in Chapters 17–20.

The final chapters develop the practical lessons for priority setting from the studies. Chapter 21 uses *ex post* evidence for trait values to illustrate the potential for using past experience for one of the two supply side elements (the economic contribution of research per unit) leaving only the time to achievement element for SPE estimation. Chapter 21 also applies this strategy to a priority-setting exercise for rice biotechnology research.

Chapter 22 develops a more comprehensive approach to rice research priority setting using recently elicited estimates of both rice supply components. A provisional set of priorities is developed and comparisons made with earlier estimates.

The final chapter attempts to summarize the state of the art for priority setting and to offer guidelines for further exercises.

References

Avila, A.F.D. (1981) Evaluation de la recherche agronomique au Bresil: Le ces de la recherche rizicole de l'IRGA ou Rio Grande do Sul. PhD dissertation, Faculté de Droit et des Sciences Economique, Montpellier.

Azam, Q.T., Bloom, E.A. and Evenson, R.E. (1991) *Agricultural Research Productivity in Pakistan*. Pakistan Agricultural Research Council, Islamabad.

Dey, M.M. and Evenson, R.E. (1991) The economic impact of rice research in Bangladesh. Unpublished manuscript, Bangladesh Rice Research Institute, Gazipur, Bangladesh/International Rice Research Institute, Los Baños, Philippines/Bangladesh Agricultural Research Council, Dhaka, Bangladesh.

Evenson, R.E. (1995) Science for Agriculture: International Dimensions. Economic Growth Center, Yale University (unpublished).

Evenson, R.E. and Flores, P. (1978) Economic consequences of new rice technology in Asia. International Rice Research Institute, Los Baños, Philippines.

Flores, R., Evenson, R.E. and Hayami, Y. (1978) Social returns to rice research in Philippines: domestic benefits and foreign spillover. *Economic Development and Cultural Change* 26, 591–607.

Gollin, D. and Evenson, R.E. (1990) Genetic resources and rice varietal improvement in India. Unpublished manuscript, Economic Growth Center, Yale University, New Haven.

Hayami, Y. and Akino, M. (1977) Organization and productivity of agricultural research systems in Japan. In: Thomas, M., Dalrymple, D.G. and Ruttan, V.W. (eds) *Research Allocation and Productivity in National and International Agricultural Research*. University of Minnesota Press, Minneapolis.

McKinsey, J. and Evenson, R.E. (1991) Research, extension, infrastructure and productivity change in Indian agriculture. In: Evenson, R.E. and Pray, C.E. (eds) *Research and Productivity in Asian Agriculture*. Cornell University Press, Ithaca.

Pingali, P.L., Moya, P.F. and Velasco, L.E. (1990) *The Post-Green Revolution Blues in Asia Rice Production – the Diminished Gap between Experiment Station and*

Farmer Yields. Social Science Division Paper No. 90-01. International Rice Research Institute, Los Baños, Philippines.

Pray, C.E. (1979) The economics of agricultural research in Bangladesh. *Bangladesh Journal of Agricultural Economics* 2, 1–36.

Salmon, D.C. (1991) Rice productivity and returns to rice research in Indonesia. In: Evenson, R.E. and Pray, C.E. (eds) *Research and Productivity in Asian Agriculture.* Cornell University Press, Ithaca.

Scobie, G.M. and Posada, R.T. (1978) The impact of technical change on income distribution: the case of rice in Colombia. *American Journal of Agricultural Economics* 60, 85–92.

Setboonsarng, S. and Evenson, R.E. (1991) Technology, infrastructure, output supply and factor demand in Thailand's agriculture. In: Evenson, R.E. and Pray, C.E. (eds) *Research and Productivity in Asian Agriculture.* Cornell University Press, Ithaca.

2 Recent Developments in the Asian Rice Economy: Challenges for Rice Research

M. Hossain
Social Sciences Division, International Rice Research Institute, PO Box 933, 1099 Manila, Philippines

I. Introduction

The 1960s were characterized by a prevailing mood of despair regarding the world's ability to cope with the food–population balance. The cultivation frontier was closing in most Asian countries, while population growth rates had accelerated due to rapidly declining mortality rates. International organizations and concerned professionals were busy organizing seminars and conferences to raise awareness regarding the ensuing food crisis and to mobilize global resources to address this emergency. In a book published in 1967, Paddock and Paddock predicted that ' ... Ten years from now, parts of the underdeveloped world will be suffering from famines. In fifteen years, the famines will be catastrophic, and revolutions and social turmoil and economic upheavals will sweep areas of Asia, Africa and Latin America ... ' (Paddock and Paddock, 1967).

The history of the green revolution in Asia over the last 30 years gives the impression that the said apprehension was unfounded. But this is primarily because those concerns regarding food shortages were able to mobilize financial and scientific resources for research on foodgrains that have succeeded in increasing productivity of the limited land resources. The population of already densely settled, rice-growing Asian countries has grown by another 70%, but Asian rice production has almost doubled since 1966 as a result of the spread of higher-yielding modern rice varieties. The average per capita consumption of rice today is about 15% higher than it was at that time. Many traditional rice-importing countries have achieved self-sufficiency in rice production and some are even struggling to deal with the issue of rice surpluses. Policymakers are now concerned with sustaining farmers' interest in rice production in view of the low and declining prices.

This chapter maintains that the race to stay ahead of the food crisis is not yet over in many parts of Asia. There are indications that the growth in the demand for rice will decelerate in the coming decades due to rising levels of income and declining population growth, but there are indications that the growth in rice supply may slacken even faster. The problem will be particularly serious in countries with unfavourable rice-growing environments. The focus of rice research needs to be adjusted from the past focus mostly on the irrigated ecosystems to addressing problems of the rainfed ecosystems, from increasing yield to improving the grain quality and sustaining the natural resource base, and from developing cultivars resistant to insects and diseases to those resistant to various abiotic stresses and problem soils.

II. Importance of Rice in the National Economy

Rice production and consumption are positively associated with low incomes and poverty. Of the 23 countries in the world that produce more than one million tons of rice, almost half have a per capita income of less than US$500. These are countries categorized by the World Bank as the least developed. Rice is one of their cheapest sources of food energy and their main source of protein. As incomes increase, people demand relatively higher-quality food, and resources are shifted from production of rice to production of other food and non-farm goods with high income elasticities of demand. The importance of rice in the national economy further dwindles as agriculture's share in the national income declines with faster growth of non-farm incomes. Increasing productivity of the rice sector, however, is an important means of raising the purchasing capacity of the poor and the alleviation of poverty in low-income countries.

Table 2.1 presents a comparative picture of the importance of rice in the national economies of the major rice-growing countries of Asia. The contribution of rice is inversely related to the level of per capita income of the country. In most countries with a per capita income of US$500 or less, rice accounts for one-third to one-half of the agricultural value added. China and India are exceptions, but in the major rice-growing regions (central and southeastern China and eastern India) the situation would be similar. In Japan, Korea and Malaysia, where income levels are high, rice is only a marginal sector of the economy but it occupies an important position in the foodgrain production sector. In the humid and subhumid tropics, rice is the primary source of human energy. In Bangladesh, Myanmar, Thailand, Laos, Cambodia, Vietnam and Indonesia, rice provides 50–80% of the calories consumed by the people. Even in Japan, where rice accounts for less than 3% of the agricultural incomes, nearly one-fourth of the human energy intake comes from rice. In most Asian countries, rice is still the prime mover of the food and economic conditions, particularly in rural areas.

Table 2.1. Contribution of rice to the national economy in Asia, 1990.

Country	Per capita income (US$)	Share of rice in total foodgrain production (%)	Share of rice in total calorie supply (%)	Gross value of rice production as percent of agricultural value added
Japan	25,430	48	24	2.8
Korea, Republic of	5,400	59	36	7.2
Malaysia	2,320	86	29	4.1
Thailand	1,420	95	55	39.3
Philippines	730	67	41	19.4
Indonesia	570	80	58	37.1
Myanmar	533	96	77	48.4
Sri Lanka	470	66	42	24.4
China	370	44	36	37.0
India	350	43	30	27.7
Vietnam	220	92	68	43.3
Bangladesh	210	84	75	62.4
Laos	200	93	70	46.3
Nepal	170	52	42	39.7

Source: *World Development Report* (World Bank, 1992); *Asian Development Outlook* (Asian Development Bank, 1992).

III. Recent Developments

A. Trends in rice production

Prior to the 1960s, the growth of rice production was slow and originated mostly from the expansion of cultivated land (Barker and Herdt, 1985). Yield growth was mainly limited to East Asian countries, where irrigation infrastructure was already developed and population pressure on limited land resources induced more intensive land cultivation and the development and spread of fertilizer-responsive, high-yielding japonica rice varieties. In South and Southeast Asia, rice yield was low and stagnant, and the increased demand for rice was met primarily by expanding the cultivated area. Rice research facilities were inadequate and irrigation and drainage facilities were poorly developed throughout much of the region. The lack of fertilizer-responsive indica varieties and the relatively poor market infrastructure contributed to the application of chemical fertilizers at low levels.

Dramatic changes in the rice production scene have, however, taken place throughout Asia since the mid-1960s with the introduction of IR8, a short-stemmed modern indica variety. IR8 was highly fertilizer-responsive and could be grown throughout the year. Scientists have incorporated many new traits in the modern rice varieties that followed IR8 – greater pest resistance, shorter crop duration and improved grain quality (IRRI, 1985). Farmers get two to three times higher yields from these varieties than from traditional cultivars.

Yields continued to increase as farmers gradually replaced traditional varieties by the modern ones. Over the last quarter century, rice production increased at 2.8% per year, enough to meet the population- and income growth-induced demand for rice in these countries. Nearly three-fifths of that growth came from the increase in crop yields

Table 2.2 provides information on the trends in rice production over the 1975–1991 period and compares it with the rate of population growth. Production growth surpassed the increase in population throughout Asia, except in Japan, the Republic of Korea, Malaysia and Pakistan. In the first three countries, per capita consumption has been declining because of changing food-consumption patterns with higher levels of income (see below). In Pakistan, rice is only a minor staple food. Rice yields increased faster than population in East Asia and kept balance with population growth in Southeast Asia. In South Asia, however, the long-term growth in yield had been slower than population growth and additional land had to be allocated to rice production to maintain the demand–supply balance.

Table 2.2. Trends in rice production in Asian countries, 1975–1991 (% per year).

Region/country	Area	Yield	Production	Population growth rate
East Asia	−0.78	3.02	2.23	1.3
China	−0.76	3.44	2.66	1.4
Japan	−1.87	0.43	−1.45	0.6
Korea, DPR	0.50	1.65	2.16	1.4
Korea, Republic of	0.18	0.53	0.71	1.1
Southeast Asia	0.82	2.81	3.63	2.0
Cambodia	2.85	1.01	3.87	2.6
Indonesia	1.54	3.46	5.00	1.8
Laos	−0.32	4.94	4.62	2.7
Malaysia	−0.88	−0.20	−1.08	2.6
Myanmar	−0.41	3.33	2.91	2.1
Philippines	−0.59	2.93	2.33	2.4
Thailand	1.19	−0.96	2.17	1.8
Vietnam	0.78	3.43	4.25	2.1
South Asia	0.42	2.58	3.02	2.3
Bangladesh	0.12	2.42	2.54	2.3
India	0.44	2.82	3.26	2.0
Nepal	1.10	1.51	2.61	2.6
Pakistan	0.90	0.10	1.00	3.1
Sri Lanka	0.38	3.31	3.69	1.4

Estimated by fitting semilogarithmic trend lines on 3-yearly moving averages of the time series for the respective variables.
Source: *Agrostat Database* (FAO, 1992).

Table 2.2 also shows an inverse relationship between the growth in rice yield and the increase in area under rice cultivation. In countries such as Thailand, Cambodia, Nepal and Pakistan (where growth in crop yield was slower than population growth), the area under rice cultivation was expanded to increase the supply of foodgrains. But in China, Myanmar and the Philippines (where yield growth was faster than population growth) there was an absolute decline in the area under rice cultivation. At the cross-country level, the following relationship is obtained between the expansion of rice area and the increase in rice yield:

$$d_A = 19.8 - 0.255\, d_Y - 0.0035\, PCI + 0.0041\, PLC; \quad R_2 = 0.49 \qquad (2.1)$$

$$(1.52)\ (-1.84)^*\ (-2.59)^{**} \qquad (1.07)$$

where d_A = percent increase in rice area during 1975–1990, d_Y = percent increase in rice yield, PCI = per capita income of the country (US$), and PCL = arable land per capita (m^2). (Figures within the parentheses are estimated 't' values of the regression coefficients, ** and * denote that the coefficient is statistically significant at the 5 and 10% level, respectively.)

These results show that the higher the level of income, the less the urge to extend the cultivation area under rice, presumably because of shifting demand to non-rice foods and the country's greater capacity to meet the demand through imports. Also, the higher the growth in rice yield, the less the expansion in rice area. The results likewise suggest that without impressive growth in rice productivity, the low-income Asian countries would have been forced to extend cultivation to marginal land, thus aggravating the problem of sustaining the natural resource base.

B. Agroclimatic factors and the diffusion of modern technologies

In spite of the impressive growth in rice yields all over Asia, the difference in yields across countries is still very large (Table 2.3). Yields vary from 1.4 tons in Cambodia to about 6.1 tons in Japan and South Korea. The difference in yields is largely due to variations in agroclimatic and socioeconomic factors which have influenced the adoption of modern varieties, the use of agrochemicals and the state of infrastructure development. China, Japan and South Korea have the highest rate of adoption of land-augmenting technology – irrigation, modern varieties and chemical fertilizers – and therefore have the highest yield levels. Yields are also high in these countries because of the low pressure of insects and diseases, longer day length, and favourable sunshine (the rice-growing area being located in the subtropical and temperate zones). On the other hand, adoption of modern varieties and rice yields are still low in the humid and subhumid tropics, such as eastern India, Bangladesh, Myanmar, Thailand and the Indochina countries. In Bangladesh and eastern India, the most densely populated regions in Asia, opportunities for increasing irrigated areas have been severely limited by the highly unfavourable physical conditions.

Table 2.3. Population pressure, use of modern inputs and rice yields in selected Asian countries.

Region/country	Rice yield, 1991–1993 (t ha^{-1})	Adoption of modern varieties, 1990 (%)	Percent of irrigated area, 1990	Fertilizer use, 1992 (kg NPK ha^{-1})*	Arable land per capita, 1992 (m^2)
East Asia					
China	5.8	100	93	313	723
Japan[†]	6.1	100	99	439	327
Korea, Republic of	6.1	100	99	507	430
Southeast Asia					
Cambodia	1.3	11	8	3	2678
Indonesia	4.4	77	72	157	858
Laos	2.4	2	10	4	1745
Malaysia	3.1	90	66	928	553
Myanmar	2.9	50	18	7	2183
Philippines	2.8	89	61	90	847
Thailand	2.1	18	27	64	3029
Vietnam	3.9	80	53	98	792
South Asia					
Bangladesh	2.6	51	22	114	738
India	2.6	66	44	74	1888
Nepal	2.3	36	23	40	1130
Pakistan	2.5	41	100	104	1655
Sri Lanka	3.1	91	77	197	526

*Per hectare of arable land. Information on crop-specific fertilizer use at the national level is not available.
[†] Excluding 1993, when the yield was severely reduced by early cold temperature.
Source: *Agrostat Database* (FAO, 1992).

For the cross-section of the major rice-growing countries, the following relationship is obtained between the level of rice yield and the use of modern inputs:

$$Y = 1.94 + 11.42\, FRT + 0.56 - MV - 1.09\, IRGN; \quad R_2 = 0.90 \qquad (2.2)$$
$$(4.08)\quad (5.69) \qquad (0.43) \quad (-0.70)$$

where Y is the yield of rice (tons ha^{-1}), FRT is the intensity of fertilizer consumption (NPK ha^{-1} of arable land), MV is the proportion of area under modern rices and $IRGN$ is the proportion of rice area with irrigation facilities. (The figures within the parentheses are estimated 't' values.) It appears that the difference in fertilizer use is the main determinant of variation in crop yield (Fig. 2.1). Modern rices increase crop yields by inducing farmers to use fertilizers in large amounts as these varieties are highly fertilizer responsive. The marginal

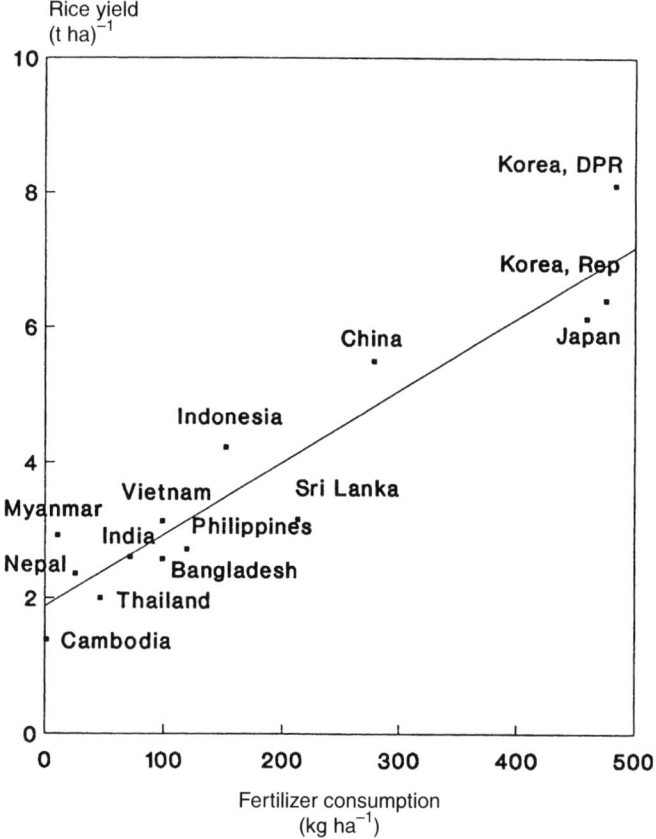

Fig. 2.1. The relationship between fertilizer consumption and rice yield, 1990–1992.

contribution of irrigation is probably higher at low levels of yield than at high levels, which may explain the negative regression coefficient for irrigation.

The relationship shows that 1 kg of NPK would increase rice (unhusked) yields on the margin by about 11.4 kg. At international prices, the fertilizer : paddy price ratio is about 2:1, so raising rice yields through greater use of chemical fertilizers is still highly economical. The difference in fertilizer consumption across Asian countries is substantial; from almost negligible in Myanmar, Cambodia and Laos, to less than 100 kg in Thailand, Philippines and India, and to more than 300 kg in East Asia (see Table 2.3). From the data, it appears that there is a vast potential for increasing rice production in South and Southeast Asia by increasing the consumption of chemical fertilizers.

Fertilizer application is, however, very risky in unfavourable production environments characterized by floods, droughts and variable rainfalls. Also, traditional varieties are much less responsive to chemical fertilizers than modern rices, so the scope of profitable use depends on the rate of adoption of modern varieties. Many farm household surveys in Asia note that the farmer who uses very little fertilizer in traditional varieties applies fertilizer in modern varieties at rates closer to the level recommended by extension agents if irrigation facilities are available. The potential for increasing rice production through the application of chemical fertilizer has remained unexploited in many parts of Asia, not only because of socioeconomic factors such as unfavourable input/output prices and the lack of purchasing capacity of the small farmer, but also due to agroclimatic factors such as waterlogging, droughts and climatic variations that constrain the adoption of modern varieties (David and Otsuka, 1994).

C. Economic prosperity and competitiveness of rice farming

Despite the impressive increase in land productivity, it has been difficult for the fast-growing Asian countries to sustain producers' interest in rice farming. First, many of the gains in productivity and efficiency have been reaped by consumers through the declining real rice prices. With the diffusion of modern varieties, the producers' surplus has been squeezed due to the agriculture's treadmill effect (Hayami and Ruttan, 1985). Second, and more important, the expansion of the non-farm sector and the rapidly rising labour productivity have pushed up non-farm wage rates which has attracted labour from rice farming and increased agricultural wages. Since traditional rice farming is a highly labour-intensive activity, the pressure from the non-farm sector has pushed up the cost of rice production and reduced profits and farmers' incomes. It is not only the wage labourers who are tempted to move to non-farm urban occupations, even the small-scale rice farmers find it more attractive to leave rice farming and join the industrial labour force. In Japan, Taiwan and South Korea, the constant outflow of the agricultural labour force has caused a continual decline in the farming population (Park, 1993). Ageing of the labour force and depopulation in remote areas has continued, making it difficult to sustain the existing rural communities in some areas. Malaysia, Thailand and China may soon follow this process.

The competitiveness of rice farming has sought to be maintained through: (i) improved farm management practices that increase efficiency in the use of non-land inputs and increase TFP; (ii) increased use of capital to replace labour through mechanization of farming operations so that labour productivity could be raised when no further increase in land productivity is possible; and (iii) using the price mechanism to transfer income from the relatively well-off rice consumers to the low-income rice producers so that the balance between the rural and urban incomes could be maintained. In addition, in Taiwan, the government developed infrastructure facilities in rural areas to promote the growth of rural non-farm activities that made possible involvement of rural households in both farm and non-farm activities. As part-time rice farming increased, the household could compensate the slow growth in farm incomes from the fast-growing incomes from non-farm sources, which partially checked the urge to migrate to urban areas.

An increase in yield potential of rice cultivars through technology development research could help improve the competitive strength of rice farming. Unfortunately, the yield potential of rice hardly increased after the introduction of IR8 in 1966. The later varieties incorporated new traits – resistance to insects and diseases, improved grain quality and shorter maturity period – but many of them had a lower yield potential than IR8. Once the potential yield under the optimum management condition is reached in the farmer's field, it will be difficult to increase the yield any further. In Japan and South Korea, rice yield has remained stagnant at around 6.0–6.5 tons ha^{-1} since the early 1970s. The results of long-term experiments at the IRRI farm and elsewhere show a tendency of rice yields to decline over time (Flinn and De Datta, 1984). Farmers may have been able to maintain yields of modern rices through the application of higher amounts of non-land inputs, which means a declining trend in TFP and profitability in irrigated rice farming (Cassman and Pingali, 1993).

In 1989, the IRRI began to design a new rice plant type, one that would make it possible to grow an irrigated rice crop with up to 30% higher yield mainly through increasing the harvest index from 0.5 to 0.6. It is designed to increase nutrient efficiency with fewer numbers of larger panicles per plant and erect and thick leaves, and will eliminate unproductive tillers (see Chapter 4). The field evaluations of breeding lines with the new plant characteristics has just begun, but the new architecture needs to be matched with agronomic practices – planting method, nitrogen application and weed control. Work is also needed to develop resistances to insects and diseases and to improve grain quality (see Chapter 4).

IV. Emerging Trends

A. Trends in demand for rice

East and Southeast Asia have already achieved high levels of income and food consumption due to rapid economic progress. Studies on consumption behaviour show that per capita rice intake largely depends on urbanization and changes in occupational structures. At low levels of income (below US$200 per capita), rice is considered a luxury commodity, and with increases in income people tend to substitute rice for low-cost sources of energy, such as coarse grains and sweet potatoes. But at high levels of income (above about US$1500 per capita), rice becomes an inferior good; as incomes rise further, consumers substitute rice for high-cost quality food, such as vegetables, bread, fish and meat. A recent study argued that rice has already become an inferior good in Asia (Ito *et al.*, 1989). The FAO food balance sheets, however, show that among Asian countries per capita rice consumption has declined substantially since the mid-1970s only in Japan, South Korea, Malaysia and Thailand; these are all middle- and high-income countries (Fig. 2.2). In Japan and Malaysia, the consumption in 1990 was about 30% lower than in 1975. But these countries account for less than 10% of Asian rice consumption.

A recent study conducted at the IRRI (Huang and David, 1992) showed that the threshold level of income at which consumers start substituting rice for other food has not yet been reached for the major rice-producing and rice-

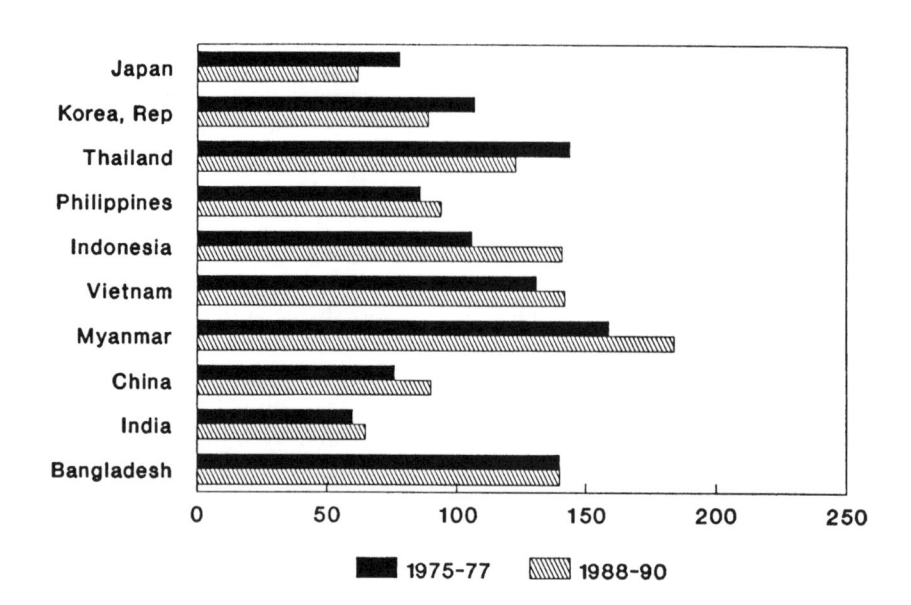

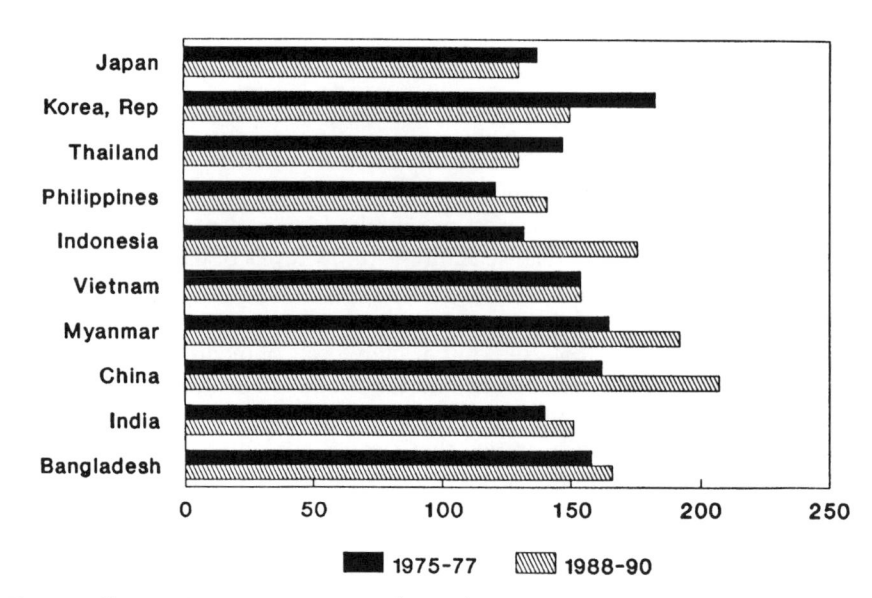

Fig. 2.2. Changes in per capita rice and cereal consumption, 1975–1990.

consuming countries, such as China, India, Indonesia and Bangladesh. These four countries account for more than 70% of the total rice consumption, and dominate the growth in demand for rice in the world.

Population growth is still a major force behind the increasing demand for rice. The population in Asia, whose main staple is rice, is growing by nearly 40 million each year. Population growth is not expected to level off until the middle of the 21st century. Asian population is expected to increase by 18% during the 1990s and by 53% in the next 35 years. Most of the additional population will be located in urban areas and marketed surpluses of rice have to increase to meet the urban demand. In South Asia alone, where unmet demand for food is still large due to widespread poverty, another 800 million people will be added over the next 35 years. Recent projections show that at prevailing price levels, the demand for rice may increase by 70% over the next three decades, most of it due to the need to feed a larger population (Agcaoili and Rosegrant, 1994).

B. Trends in rice supply

The rice supply has to increase by 1.7% per year to meet the growing demand. To many this may not appear to be a difficult challenge in view of the impressive growth in production over the last quarter century. But that historical growth will be difficult to maintain due to a number of factors mentioned below.

In the coming decades, there is going to be an absolute decline in area under rice. As an economy grows, prime rice land is lost to accommodate industrialization and urbanization. Water, a crucial input in rice cultivation, is becoming a scarce commodity as the demand in alternative uses is growing with economic prosperity. Salinization and degradation of irrigation and the scarcity of resources for maintaining public irrigation systems are reducing both the area under irrigation and the quality of irrigated land. Environmental concerns may induce policies to relieve marginal lands from rice cultivation and to move from intensive rice–rice to rice–non-rice cropping patterns. As rice area is expected to decline in the future, rice yield has to increase faster in order to meet a targeted increase in rice supply.

Recent developments suggest that the yield growth has started declining and reversing the trend will not be easy. In the irrigated rice ecosystem, which accounts for almost three-fourths of total rice supplies, most farmers have already planted high-yielding modern varieties and the best farmers' yields are approaching the potential that scientists are able to attain in the experimental fields with today's knowledge. The yield is still low in many regions in the humid tropics and subtropics due to natural factors such as floods, droughts, typhoons, temporary submergence from heavy rainfalls and tides, and salinity in coastal areas. But reducing the yield gap will be difficult as scientists have had limited success in developing varieties resistant to these abiotic stresses. Research for tackling these issues is underway but the outcome is uncertain.

Most of the increase in rice yields in favourable environments was achieved through intensive utilization of agrochemicals on genetically improved varieties that are responsive to these inputs. Undoubtedly, rice yield could be increased in many countries in humid and subhumid tropics through the increased use of fertilizers, but poor farmers may not undertake that investment unless the risk in rice cultivation is reduced. The concern for environmental protection may also discourage further intensive use of agrochemicals, thereby slowing down the growth in rice yields.

Another factor which may slow down the productivity growth is the rising demand for rice varieties that have better eating quality. In China, for example, rising incomes and reduced consumer subsidies for rice have contributed to an increased demand for high-quality rice, which faces premium prices in urban markets. In response to market signals, farmers are eager to grow quality varieties even if they are low yielding. The diffusion of hybrid rice has already slowed down in China because of the lower quality of grains. Rice scientists have so far had limited success in developing high-yielding cultivars which also have better eating quality. Also, tastes of consumers vary across countries and so the research on breeding for quality has to be country-specific which will increase the cost of such research. In Japan, consumers have started demanding environment-friendly rice produced under organic farming. The growing consciousness of consumers regarding eating quality and environment-friendly products may induce farmers to grow low-yielding traditional cultivars unless rice science develops technologies that reduce farmers' dependence on harmful agrochemicals.

The most recent trend in rice production shows a rapid deceleration in growth. Since 1989, global rice production remained almost stagnant at the level of 520 million tons. The annual growth in global rice production was only 1.8% per year during the 1985–1993 period, compared to 2.8% during 1975–1985, and 3.6% the decade earlier. Table 2.4 reports the most recent growth in rice production in the top ten rice growing countries in the world that account for 87% of global rice production. Since 1985, there has been a substantial deceleration in growth of production in these countries except in Bangladesh, Vietnam and India. And in most countries, the growth was lower than the rate of increase in population.

A small shortfall in rice production in a few major rice-consuming countries could have a dramatic effect on rice prices and food security for the poor. The volatility of the world market for rice is demonstrated by the recent surge in prices of quality rice in the world market in response to a 25% reduction in rice yield in Japan in 1993 due to abnormal weather. An increase in rice prices will have an adverse impact on the alleviation of poverty in the low-income Asian countries.

C. The road to the future

An important factor that would have significant effect on the future of rice economy is the agreement of the General Agreement on Tariffs and Trade (GATT) regarding the liberalization of agricultural trade. Numerous studies have been undertaken by economists to estimate the possible effects of trade liberalization on the world rice market (Childs, 1990). Most of the studies show

Table 2.4. Recent trends in growth of population and rice production in major rice-producing countries (% per year).

Countries	Rice production, 1993 (million tons)	Population growth, 1980–1993 (% per year)	Growth of rice production 1975–1985	Growth of rice production 1985–1993
China	187.2	1.5	2.9	0.3
India	111.0	2.0	2.2	2.9
Indonesia	47.9	2.0	5.7	2.9
Bangladesh	28.0	2.3	2.3	3.2
Vietnam	22.3	2.2	3.6	5.0
Thailand	19.1	1.5	2.8	−0.2
Myanmar	17.4	2.2	4.5	1.6
Japan	9.8	0.5	−1.6	−3.3
Brazil	10.2	2.0	2.8	0.5
Philippines	9.5	2.5	3.5	0.9
World	527.4	1.8	2.8	1.8

Source: For population: *World Development Report* (World Bank, 1992). For rice production: *Food Outlook* (FAO, December 1993). The growth of rice production has been estimated from the time-series in *Agrostat database* (FAO, 1992).

that liberalization will cause market prices of rice to increase, more for japonica rice than for indica rice, and more in the short run than in the long run. The volume of trade in the market is expected to increase with larger imports going to Japan, South Korea, Taiwan, the European Economic Community (EEC) and Brazil, and larger exports from China, Australia and the Indochina countries. The world rice market is, however, segmented by variety and quality and prices are often determined by tastes and preferences for particular varieties, which differ from country to country. The price elasticity of substitution among varieties may be quite low particularly at high levels of income, where the expenditure on rice accounts for a small fraction of total consumer expenditures. Also, the ability to increase the production of high-quality japonica rice in the major exporting countries – the USA, Australia, Thailand and Vietnam – may be limited at least in the short run. Northeastern China has a good potential to increase its production and export of japonica rice. But since the Chinese economy is also growing very fast, rice production for export may not remain competitive in the near future. Considering the above, and the scope of increasing farmers' incomes by consolidation into larger holdings, the fear of the extinction of rice farming in Japan, Korea and Taiwan may be misplaced.

The possibility of a re-emergence of a food crisis in land-scarce low- and middle-income countries, however, cannot be ruled out unless the trend toward the reduction of investment in irrigation infrastructure and in foodgrain research is reversed, and rice research succeeds in making breakthroughs in developing high-yielding varieties resistant to abiotic stresses. The apprehension that the countries which have achieved self-sufficiency in rice production may slip back to import dependence is demonstrated by the recent history of the Philippines and Indonesia. In the late 1970s, the Philippines became a minor exporter of rice from a major importer due to a rapid growth in rice production in 1970s. But the stagnation of yield since 1984 has turned it again into a rice-importing country in recent years. Indonesia became self-sufficient in rice production in 1984, but due to slow growth in rice production in the early 1990s it has started importing rice again. South Asia may face the same problem if the countries fail to sustain the growth in land productivity. Twenty-eight percent of the world's population and 50% of the world's poor are in South Asia. There is a large unmet demand for food and acceleration in the growth of per capita incomes will increase demand much above the level induced by population growth. The arable land per capita is low and farm size has been declining due to a rapid increase in population and slow absorption of the incremental labour force in non-farm activities. These areas have not benefited much from the green revolution as scientists have had limited success in developing varieties that can adapt to difficult natural and environmental conditions – drought, flood, temporary submergence, soil salinity – which are common in large parts of this region. The recent acceleration in rice yields may be short lived due to the high and rising cost of irrigation, the difficult physical conditions for irrigation development and the negative environmental effects of ground-water irrigation.

Thailand, Myanmar and Indochina have considerable excess capacity to meet potential shortages in other countries in South and Southeast Asia. But achieving food security through trade may not be possible due to the foreign exchange constraint. Also, politicians may not favour that option because of the risk of dependence on the volatile world market for rice which is a politically sensitive commodity. Since rice production is a major rural economic activity at low levels of income, and land cannot be easily diverted to other crops during the monsoon season, low-income households may not have enough productive employment and income to acquire food through trade. If the economic conditions of small farmers and landless workers fails to improve due to stagnant productivity of this most important economic activity in the humid tropics in Asia, there will be an increase in rice prices that may aggregate the already precarious poverty situation in South Asia.

V. Challenges to Rice Research

The rice research community is proud of the remarkable progress achieved in the global rice economy over the last quarter century. But there is no reason to be complacent, as the path ahead is not an easy one. Sustainability of productivity growth for the irrigated rice ecosystem, the development of high-yielding cultivars resistant to abiotic stresses in unfavourable environments, replenishing soil fertility, avoiding genetic vulnerability to pests and maintaining biodiversity, improving rice quality, increasing the value added and

Table 2.5. Population pressure, intensity of cereal production and satisfaction of energy needs in Asia, 1990.

Country	Person ha^{-1} of arable land	Cereal cropping intensity (%)	Per capita domestic production of cereals (kg person^{-1} year^{-1})	Per capita calorie intake (per day)
Group I: High income, self-reliant				
Japan	30	60	116	2926
Korea, Republic of	23	70	178	2840
Malaysia	4	65	101	2697
Group II: Excess rice production capacity				
Thailand	3	55	397	2271
Myanmar	5	55	328	2448
Cambodia	3	65	262	2114
Laos	5	70	349	2475
Group III: Risk of food insecurity				
Indonesia	12	88	291	2631
Southeast and central China	16	148	358	2647
Vietnam	13	134	319	2215
Philippines	14	146	211	2255
Eastern India	10	118	213	2243
Bangladesh	13	122	239	2019
Nepal	8	100	229	1957
Sri Lanka	19	89	129	2286

Source: *Agrostat Database* (FAO, 1992). For India and China: National Statistical Agency.

profitability through more efficient post-harvest operations, and many other emerging issues demand the attention of rice scientists. The central issue is how to balance the need for ever-greater food production at prices affordable by the urban poor and the rural landless, and at the same time profitable to farmers, against the very real concerns about protecting natural resources and the environment for the generations to come. This will require scientific breakthroughs that will allow farmers to produce more rice not only with less land but also with less labour, less water and less harmful chemicals.

The prioritization of rice research for different countries and ecosystems must take into account the level of economic development and the state of exploitation of natural resources. The Asian rice economies could be classified into three distinct groups for this purpose (Table 2.5).

At one end are the economically prosperous countries, Japan, South Korea, Taiwan and Malaysia who have the capacity to procure rice from the world market when there is a deficit. They have a high pressure of population on the land, but because of the impressive growth in both income and foodgrain production in the past, the energy needs of the population have already been met. The per capita consumption of rice has been declining and the population growth rate has reached low levels. The worry to sustain a high growth in rice productivity is over for these countries. They will rather pursue an increase in labour productivity and more efficient use of other inputs. The cost of producing rice in these countries is very high compared to the rice exporters in the world market (Yap, 1991), and even a substantial increase in rice yield would not enable them to produce surplus for the rice-deficient countries and compete in the world market. As land and labour generate higher returns in non-rice activities, rice cultivation will gradually decline with a growing trend in the liberalization of agricultural markets. The target for rice research should therefore be on reducing dependence on agrochemicals (environment-friendly rice), maintaining the resistance of modern cultivars to newly emerging insects and disease pressures, improving grain quality and maintaining efforts at shifting the yield potential.

At the other end are the land-scarce low-income countries of South Asia and Vietnam and the Philippines (possibly Indonesian Java) in Southeast Asia. Most of the land in these countries is already allocated to growing cereal grains, yet they are unable to meet the energy needs of the people. A major portion of the land is poorly drained and not suitable for other high-valued crops during the monsoon season. Labour engaged in rice farming cannot be easily diverted to other economic activities due to the small size and slow growth of the non-farm sector. The population is still growing at nearly 2% per year, and self-sufficiency in rice is partly due to the lack of purchasing capacity of a large section of the people living in poverty. The real challenge for the rice research community over the next few decades is to help this group achieve and sustain food security. A 2–3% per year increase in rice supplies has to be attained with less land to allow crop diversification, at the same time increasing the productivity of labour and maintaining the quality of natural resources. Rice research for this group of countries should focus on reducing the existing large yield gap in the rainfed lowlands, and shifting the yield potential for irrigated ecosystems to enhance productivity. This research must focus on understanding the processes and mechanisms that give traditional cultivars their capacity to withstand abiotic stresses, and use this knowledge to develop high-yielding cultivars with more stable yields. The crop management research will require a system approach to explore how non-rice crops could fit in the rice-based system to increase and stabilize farmers' income. An effective partnership between national and international scientists will be required to understand the problems of variable ecosystems and plant interactions for exploitation of the resource base in carefully selected key sites that represent large geographic areas.

In the middle are countries with substantial excess capacity in rice production, such as Myanmar, Thailand, Cambodia and Laos. The exploitation of this potential will depend more on the developments in the world rice market and the investment on rural infrastructure (the two are interdependent) than on the success of rice research in increasing yield rates. The target of rice research for this group should be improvement in grain quality to suit the needs of the major rice-importing countries and the development of technologies that save labour and agrochemicals.

References

Agcaoili, M. and Rosegrant, M.W. (1994) World rice trade: prospects and issues for the future. Presented at the Third Workshop of the Rice Supply and Demand Project, 24–26 January 1994, Bangkok, Thailand.

Asian Development Bank (1992) *Asian Development Outlook, 1992*. Asian Development Bank, Manila, Philippines.

Barker, R. and Herdt, R.W. (1985) *The Rice Economy of Asia*. Resources for the Future, Washington, DC.

Cassman, K.G. and Pingali, P.L. (1993) Extrapolating trends from long-term experiments to farmers fields: the case of irrigated rice systems in Asia. Paper presented at the Working Conference on Measuring Sustainability Using Long-Term Experiments, Rothamsted Experimental Station, 28–30 April 1993 (funded by the Agricultural Science Division, The Rockefeller Foundation).

Childs, N.W. (1990) *The World Rice Market: Government Intervention and Multilateral Policy Reform*. Economic Research Service, United States Department of Agriculture, Washington, DC.

David, C.C. and Otsuka, K. (1994) *Modern Rice Technology and Income Distribution in Asia*. International Rice Research Institute, Los Baños, Philippines.

Flinn, J.C. and De Datta, S.K. (1984) Trends in irrigated rice-yields under intensive cropping at Philippine research stations. *Field Crops Research* 9, 1–15.

Food and Agriculture Organization (FAO) (1992) *Agrostat Database*. FAO, Rome.

Food and Agriculture Organization (FAO) (1993) *Food Outlook*. FAO, Rome.

Hayami, Y. and Ruttan, V.W. (1985) *Agricultural Development: an International Perspective*, 2nd edn. Johns Hopkins University Press, Baltimore.

Huang, J. and David, C.C. (1992) *Demand for Cereal Grains in Asia: Effect of Urbanization*. Social Sciences Division, International Rice Research Institute, Manila, Philippines.

International Rice Research Institute (IRRI) (1985) *International Rice Research: 25 Years of Partnership*. IRRI, Manila, Philippines.

Ito, S., Peterson, W.W.F. and Grant, W.R. (1989) Rice in Asia: is it becoming an inferior good? *American Journal of Agricultural Economics*, 71, 32–42.

Paddock, W. and Paddock, P. (1967) *Time of Famines*. Little, Brown, Boston.

Park, J.K. (1993) Sustainability of Rice Farming in Korea. In: *Proceedings of the International Seminar on Recent Trends and Future Prospects of Rice Farming in Asia*. National Agricultural Cooperative Federation and Food and Fertilizer Technology Center for the Asia and Pacific Region. Seoul, Korea.

World Bank (1992) *World Development Report 1992: Development and the Environment*. Oxford University Press, Oxford, UK.

Yap, C.L. (1991) *A Comparison of the Cost of Producing Rice in Selected Countries*. FAO Economic and Social Development Paper No. 101. Food and Agriculture Organization, Rome.

3 Rice Ecosystems Analysis for Research Prioritization

D.P. Garrity[1], V.P. Singh[2] and M. Hossain[3]

[1] *ICRAF Southeast Asia, Forest Research and Development Centre, Jalan Gunung Batu No. 5, PO Box 382, Bogor 16001, Indonesia;* [2] *Agronomy, Plant Physiology and Agroecology Division, International Rice Research Institute, PO Box 933, 1099 Manila, Philippines;* [3] *Social Sciences Division, International Rice Research Institute, PO Box 933, 1099 Manila, Philippines*

I. Introduction

The prioritization of rice research among environments relies on reasonably useful and accurate information on what those environments are, how much rice area there is and how production is distributed among them. To date the rice resource allocation debate has been limited to relatively broad ecosystem distinctions, i.e. irrigated, rainfed lowland, upland and deepwater (Barker and Herdt, 1982; Herdt *et al.*, 1987). Consequently, data pertinent only to fairly general classes has been used. Detailed rice environmental datasets, with their inevitably more complex nature, have not been much employed.

There has been progress during the past decade in developing agroecological classifications at continental and national scales (Higgins *et al.*, 1987). In the rice world, a broadly applicable international ecosystems classification was derived (IRRI, 1984). Geographic databases and maps have been developed that are more or less consistent with these classes (e.g. Huke, 1982; Garrity *et al.* 1986; Jones and Garrity, 1986). Likewise, at the national level, massive agroecological classification efforts have been completed in several Asian countries, as in Bangladesh (FAO, 1988) and India. But the deployment of these classifications, and their rich datasets, in research prioritization analysis has been very limited.

Analyses at broad levels will continue to be needed, since major issues in research allocation are unresolved even at these levels. But as the formal tools of research prioritization are trained upon the rice sub-ecosystems, and the geographic scales of analysis are refined for national, regional and local levels, the need for the products of detailed ecosystems analysis is becoming more apparent.

This chapter reviews some of the databases that have been assembled through rice ecosystems analysis and their applicability to rice research prioritization. Section II explains the methodologies used in agroecological and ecosystem classification. Section III presents a review of existing databases and an examination of the issues of ecosystems data application at four scales: mega, macro, meso and micro. Section IV presents two applications of the ecosystem analysis, one for characterization of the agroecological zones in Asia and the other characterization of ecosystems at different scales in eastern India. The final section makes some concluding remarks regarding research priority setting in an agroecological framework *vis-à-vis* the conventional commodity-based approach. It examines the externalities imposed by a crop-production activity in one ecosystem on others.

II. Methodologies for Ecosystem Classification

A. Agroecological classification

To what extent can rice research prioritization rely upon the several important efforts at broad agroecological classification at the global level, specifically the classic work in climatic classification systems and the FAO agroecological zones (AEZ) studies? There are natural advantages in doing so, if this is feasible, since these efforts are generalized (with relatively fewer classes) and they are widely known.

Among the global climatic classifications, those of Koppen (1936), Thornthwaite (1948), Holdridge *et al.* (1971) and Papadakis (1975) have had wide currency. The Koppen system recognizes 13 tropical and subtropical climatic zones; the Thornthwaite system identifies 21. These relate locations to general climates, but are undeniably broad. Papadakis' work led to a much more complex classification system (with over 500 classes). It was a move in the direction of comprehensiveness but this sacrificed simplicity.

The FAO AEZ system (Kassam *et al.*, 1982) includes a broadly defined climate component based on temperature and length of the growing period, determined by the seasonal distribution of rainfall and evapotranspiration. The major climates can be used independently, or combined with growing period zones. This system of seven basic agroecological units (see below) was adopted for an analysis (TAC, 1990) of research priorities across commodities, and allocation of resources among the International Agricultural Research Centers (IARC). The major objective of the FAO system was to assess the suitability of land for different crops (Higgins *et al.*, 1987). When the climatic classification is combined with soils data, it yields a more comprehensive and complex global agroecological zones system. In addition to global and continental level studies, AEZ studies have been conducted in a number of countries, for example Bangladesh (FAO, 1988). A world databank on the AEZ system is maintained at the FAO headquarters in Rome.

The above review covers only some of the most well-known efforts. Young (1987) has reviewed a number of other systems and has discussed their relative utility for varied user purposes.

B. Rice ecosystems classification

Although the methods just discussed provide quite a range of flexibility in aggregation and data requirements, they have not been widely applied in characterizing and classifying rice environments. The main reason has been the uniqueness of rice's environmental situation compared to all other major crops: rice environments are dominated by surface flooding patterns. Therefore virtually all the rice classification systems (and there are at least 38) developed at national and international levels consider surface hydrology to be the dominant delineating variable (Garrity, 1984; Bowles and Garrity, 1988). Therefore, a meaningful classification of rice environments must proceed independently of the commonly known global systems.

The agroclimatic classification for rice and rice-based cropping systems that has been widely adopted (Oldeman, 1980; Oldeman and Frere, 1982) is based on the length of the rice-growing season. This is specified as the months in which surface flooding can be maintained (assumed to be the period with monthly rainfall exceeding 200 mm month^{-1}). National agroclimatic maps based on this system were derived for a number of countries, for example the Philippines, Bangladesh and Indonesia (Oldeman, 1980). Maps that uniformly classified all the countries of South and Southeast Asia in this system were compiled by Huke (1982).

The international *Terminology of Rice Growing Environments* (IRRI, 1984) established a standardized scheme of rice ecosystems. Its two-tiered structure subdivided the commonly accepted rice environments (i.e. irrigated, rainfed lowland, upland, deepwater and tidal wetlands) into varying numbers of sub-ecosystems, based on hydrological and (in some cases) soil factors (Table 3.1). The dominant emphasis in this definition was toward broad agronomic constraints, particularly those related to genetic improvement of the rice crop. The environmental characterization of the sub-ecosystems, and the rice area covered by each sub-ecosystem, remained uncertain. Subsequent efforts have attempted to sharpen the classes and provide better estimates of their overall extent.

III. Agroecosystem Databases: a Review

A. Mega-level analysis

When the IRRI restructured its research programs to explicitly address the rice ecosystems (IRRI, 1989b), decision criteria upon which to allocate funds on an ecosystem basis became more explicit. The rice area and production in each ecosystem were fundamental information in applying a resource-allocation

Table 3.1. Terminology for rice-growing environments.

Environment	Sub-ecosystem
Irrigated	Irrigated, with favourable temperature
	Irrigated, low temperature, tropical zone
	Irrigated, low temperature, temperate zone
Rainfed lowland	Rainfed shallow, favourable
	Rainfed shallow, drought-prone
	Rainfed shallow, drought- and submergence-prone
	Rainfed shallow, submergence-prone
	Rainfed medium deep, waterlogged
Deepwater	Deepwater
	Very deepwater
Upland	Favourable upland with long growing season (LF)
	Favourable upland with short growing season (SF)
	Unfavourable upland with long growing season (LU)
	Unfavourable upland with short growing season (SU)
Tidal wetlands	Tidal wetlands with perennially fresh water
	Tidal wetlands with seasonally or perennially saline water
	Tidal wetlands with acid sulphate soils
	Tidal wetlands with peat soils

Source: IRRI, 1984.

model. In congruence prioritization models, for example, in which the rice produced in each ecosystem is considered as a distinct commodity, the allocation of resources would be in direct proportion to the relative production of the ecosystem (Ruttan, 1982; Salmon, 1983).

Aggregate data and maps of the area of rice by cultural type have been standardized for more than a decade (Huke, 1982) on the basis of judicious estimates. The accuracy of the mega-level data on the amount of riceland in the five major ecosystems is still uncertain, however, because national-level statistics that distinguish the various types of non-irrigated riceland are generally unavailable. The IRRI *Rice Almanac* (IRRI, 1993), based on Hukes' work, contains a breakdown of estimated rice areas by cultural type in each country.

The work of Huke (1982) yielded standard maps of the regional allocation of rice land by ecosystem. The maps provided the basis for more comprehensive mega-level geographic databases to characterize the micro-regions and classify sub-ecosystems, particularly for the rainfed ecosystems.

The IRRI developed several mega-level geographic databases; the upland rice ecosystem geographic database was the initial product (Garrity, 1984; Garrity and Agustin, 1984; Jones and Garrity, 1986). It contains data for several agroclimatic and soil parameters for each of the approximately 4000 upland rice-growing locations designated on the Huke maps for South and Southeast Asia. The sites are uniformly classified according to a two-factor upland rice environmental classification based on the length of growing season and inherent soil fertility constraints, and also on a three-factor system which includes an estimate of seasonal moisture sufficiency (Garrity and Agustin, 1984). The two-factor classification conforms with the four broad upland subecosystems specified in the international terminology of rice environments (IRRI, 1984). The three-factor classification recognizes 12 major classes and, at a more detailed level, 72 classes.

The rainfed lowland rice ecosystem database (Garrity et al., 1986) was compiled using a similar methodology. Approximately 6300 rainfed sites were classified according to a number of agroclimatic and soil parameters. All of the rainfed lowland ricelands were then classified in a three-factor environmental classification that included the length of growing season, water balance and soil constraints as delimiters. The distribution of shallow rainfed environments for the region is shown in Fig. 3.1.

The Asian riceland soil constraints database covers all riceland in South and Southeast Asia, including irrigated and deepwater areas (IRRI, 1987). This database includes data from the FAO soils maps of the world (FAO, 1977, 1979), with soil constraints interpreted according to the fertility capability classification (Buol and Cuoto, 1981). It has enabled a breakdown of the rice area in each ecosystem by major soil constraints (Fig. 3.2).

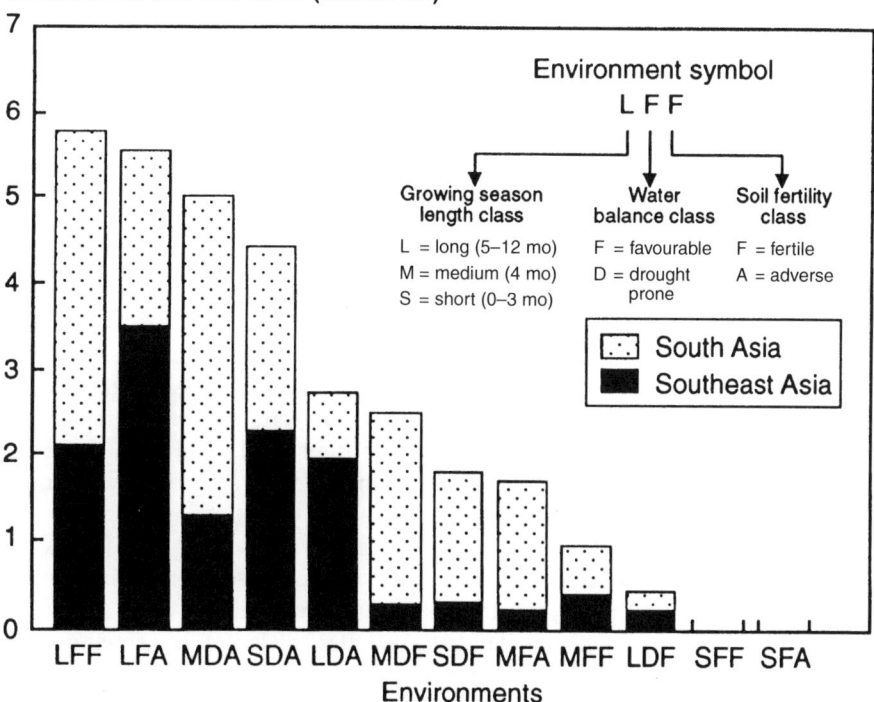

Fig. 3.1. Distribution of major shallow rainfed rice environments in South and Southeast Asia. Favourable water balance includes intermediate balances; drought-prone includes highly drought-prone.

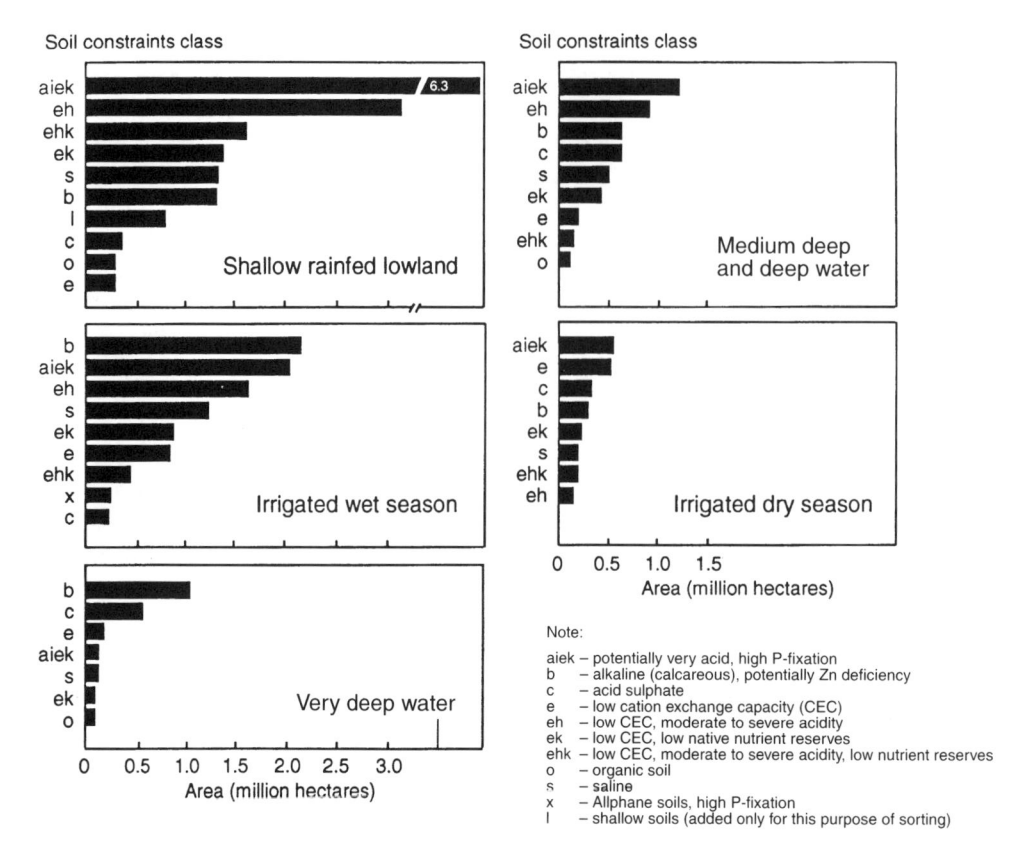

Fig. 3.2. Relative distribution of rice-cropped areas under various types of soil fertility constraints in the five lowland rice culture types for South and Southeast Asia.

These databases are useful in addressing general questions about the rice environments, particularly research priorities based on the congruence method. Their impact has been significant. Quantification of the dominant extent of upland rice area in the less favoured sub-ecosystems gave conclusive evidence that the IRRI's upland research was overly targeted to inherently fertile soils. The data helped initiate a major shift in upland breeding and agronomic research in the early 1980s, from young volcanic soils to acid upland soils, and from flat land to sloping lands.

The rainfed lowland database has had a lesser impact. This is most likely due to the limitation that it does not include data on the surface-water depth regime, particularly the frequency and duration of crop submergence and droughts. Although the database includes the length of growing season and a crude water-balance classification, data does not exist to classify the surface-water accumulation dynamics at the micro-region level. In its absence there is as yet no means to definitively classify and map the rainfed lowlands into the five sub-ecosystems. Confusion lingers as to the aggregate area and localities of 'favourable rainfed' and the four categories of 'unfavourable rainfed'. In lieu of a more precise delineation, it is often assumed that favourable rainfed corresponds to the area where semidwarf modern cultivars have been adopted.

The rice ecosystems geographic databases were originally developed as computerized databases linked to hand-drawn maps. The maps are now digitized and the entire database is being integrated into the Rice Ecosystem Geographic Information System (REGIS). The convenience and analytical power introduced by this transformation enhances the utility of the information. Recently, for example, the rice geographic databases and maps have seen application in an unanticipated research sphere: estimation of the impact of global climate change on rice production in Asia (Bachelet *et al.*, 1992). The project is modelling productivity changes on a spatial basis using the databases' climatic and soil information.

B. Macro-level analysis

The generalized nature and small scale of mega-level databases strongly limit their application beyond international issues. The nations are interested in research prioritization and extrapolation that typically require much more detailed information – at least at the macro (i.e. national) level. In a growing number of countries, some remarkably comprehensive and useful datasets have been developed; the challenge is to make better use of this data. An excellent example is the case of Bangladesh (Ahmed *et al.*, 1992). Extensive soil and land-use datasets and maps were developed over two decades by the Soils Resources and Development Institute in collaboration with the FAO (FAO, 1988). This included standard countrywide data on the surface flooding regime, a rare form of data in most countries. But in Bangladesh it was recognized early on to be critical to ecosystems analysis in this flood-prone country.

The Bangladesh Rice Research Institute perceived the value of this vast information pool, and began employing the data in research planning and extrapolation immediately after it was released. A national rice ecosystem map that conformed to the international terminology was prepared (see Fig. 3.3). It has aided significantly national research priority setting. The database and national maps have also sharpened target area delineation for the extrapolation of various institute technologies (Ahmed *et al.*, 1992).

C. Meso-level analysis

A major positive trend in many countries in recent years is the setting up of regional research stations for addressing local problems. The strengthening of regional universities and governmental research centres is enabling the development of institutions that strongly identify with the unique problems and research priorities of the specific areas where they are located. These institutions seek methods to establish priorities that are suited to these smaller

LEGEND

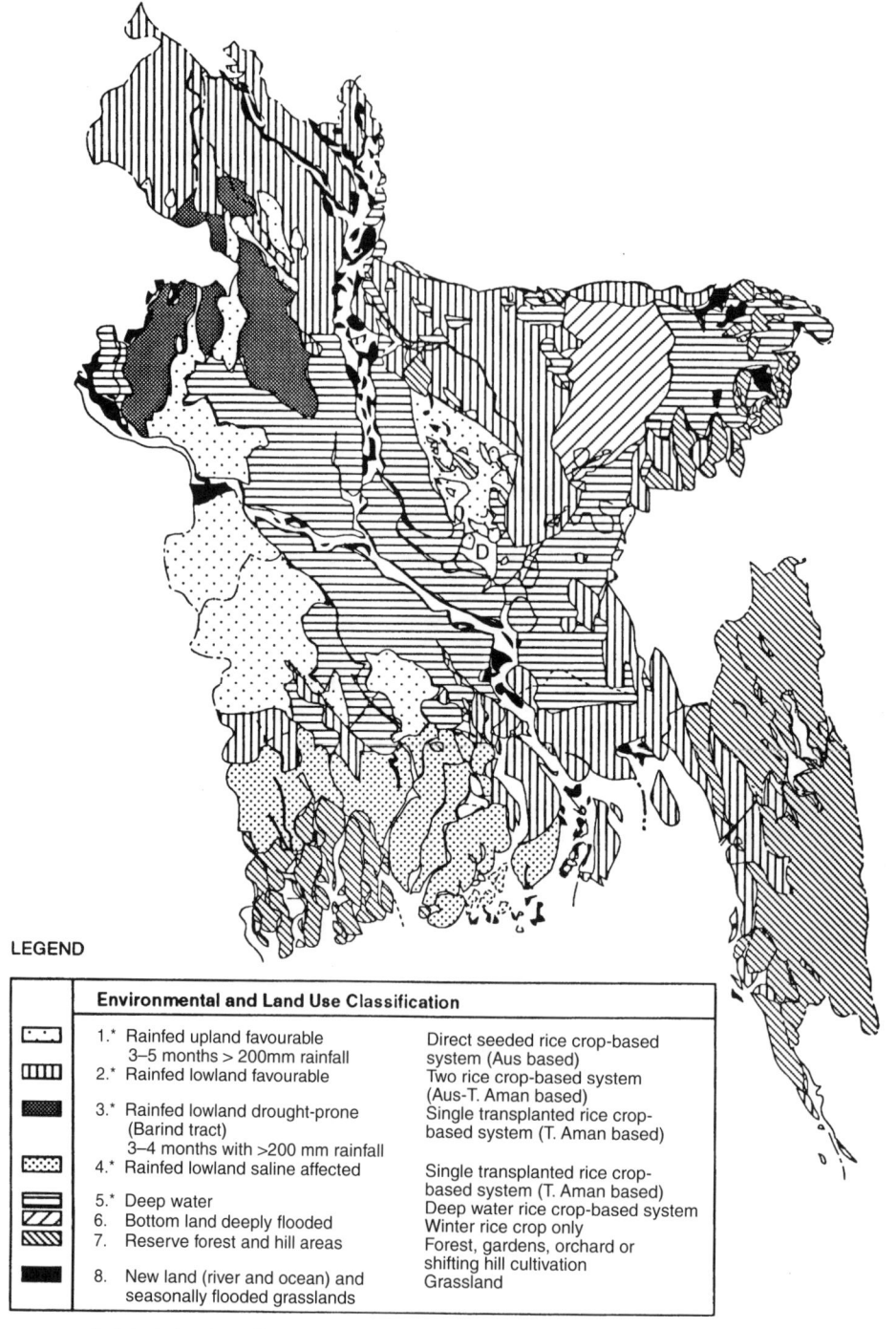

	Environmental and Land Use Classification	
▨	1.* Rainfed upland favourable 3–5 months > 200mm rainfall	Direct seeded rice crop-based system (Aus based)
▨	2.* Rainfed lowland favourable	Two rice crop-based system (Aus-T. Aman based)
▨	3.* Rainfed lowland drought-prone (Barind tract) 3–4 months with >200 mm rainfall	Single transplanted rice crop-based system (T. Aman based)
▨	4.* Rainfed lowland saline affected	Single transplanted rice crop-based system (T. Aman based)
▨	5.* Deep water	Deep water rice crop-based system
▨	6. Bottom land deeply flooded	Winter rice crop only
▨	7. Reserve forest and hill areas	Forest, gardens, orchard or shifting hill cultivation
▨	8. New land (river and ocean) and seasonally flooded grasslands	Grassland

* Irrigated systems dispersed throughout these environments and cover 20–25%.

Fig. 3.3. Environments and land-use associations in Bangladesh for rainfed systems.

geographic areas and larger mapping scales. The process is suited to more direct feedback from extension personnel and farm-level adaptive research.

Meso-level analysis is typically associated with a cultivated area in the range of about 100,000 ha or so, using a mapping scale of 1:25,000 to 1:100,000. Analysis at this scale can efficiently identify and delineate the rice ecosystems and sub-ecosystems in terms of surface hydrology, landform and soil classes. Associated with each rice ecosystem are the flood and drought frequencies and duration, prevalent cropping patterns and crop management practices.

An example of a useful meso-level analysis was that conducted for Bahraich district, Uttar Pradesh, India, to identify the problems causing low rice yields and to prepare the priority research agenda at a district level in eastern India (Singh and Pathak, 1990). The analysis determined the ecological variability in the district in terms of hydrology (rainfall pattern, water table depth, irrigation sources and drainage), landform and slope, length of growing season, frequencies and duration of floods and drought, and major insect, disease and weed pressures.

After characterization, the factors were combined to identify and delineate the area into homologous zones. They were manually mapped at a scale of 1:50,000 and classified under major rice cultural types (ecosystems). Combining the hydrological and topographic maps indicated that 70% of the rice area in this district was shallow rainfed (0–30 cm depth), 10% in medium-deep lowland (30–50 cm depth) and 15% in deepwater and very deepwater categories (50 to > 100 cm). Cropping pattern, variety use, crop management practices and input use and socioeconomic conditions were separately superimposed on each of the rice ecosystem maps. The rice ecosystems were then prioritized for research on the basis of the physical extent of area, number of affected households and potential possibilities of research success.

The IRRI and the Department of Agriculture Regional Office for the Cagayan Valley, Philippines (Region II) developed a meso-level classification of the valley's complex mosaic of rainfed ricelands (Garrity *et al.*, 1992). The exercise explored the utility of a computerized geographic database correlated with village-level maps of rainfed rice land types. The information was packaged as a field manual for extension personnel. Six rainfed rice sub-ecosystems were recognized on a hydrological basis. They were explicitly correlated with a range of associated information to specify their identification and the technology associated with them. The data on rice area, and the yield constraints associated with each rainfed riceland type, has facilitated rice research efforts, particularly the relative emphasis given to applied and adaptive research among land types.

D. Micro-level analysis

Agroecosystems analysis has become a very popular set of tools in micro-level prioritization (Conway, 1986; KEPAS, 1985). Fujisaka (1991) reviewed the use of micro-level diagnostic methods in greater detail. Micro-level analysis has been used extensively in the rainfed regions of eastern India, covering the states of Uttar Pradesh, Madhya Pradesh, Bihar, Assam, West Bengal and Orissa, to set research priorities within and among dominant rice-farming systems. Agroecosystems analysis techniques and rapid rural appraisal methodology were extensively employed. The methodology involved a two-tier training program for researchers on the methodology for setting research priorities by agroecosystems analysis with farmer participation. The analysis was carried out by 15 research centres in the region covering upland, rainfed lowland and deepwater rice ecosystems. The research diagnosis and prioritization at this level were conducted by multidisciplinary teams in the respective centres, with continuous involvement and interaction from groups of farmers.

The micro-level agroecosystem analyses (100 locations in eastern India) included detailed characterization and classification of the static and dynamic factors which differentiate agroecosystems in soils, hydrology, farming system practices and socioeconomic conditions (IRRI, 1989a). The sites were mapped on the scale of 1:2000 to 1:5000. At all sites the static factors studied were land types, land use, source of water supply and soil properties. The dynamic factors were field-water depth, rainfall land cropping patterns, crop yields, varieties and management practices, insects, diseases and weeds; landholding size, production costs and returns; labour supply pattern prices, assets and income distribution; and demography by social class.

The geographic area was zoned into agroecosystems and the problems and opportunities elucidated in each major agroecosystem. Among the different agroecosystems, highest priority was given to that with the largest extent. The research problems were then prioritized on the basis of the physical extent (coverage), number of affected households, complexity of the problem, severity of the problem (crop-loss estimates), frequency of problem occurrence and the importance of the affected enterprise in the farming system.

IV. Application of Ecosystem Analysis: Two Examples

A. *Socioeconomic characterization by agroecological zones*

AEZs are geographic mapping units developed by the FAO. They are based on climatic conditions and landforms that determine relatively homogeneous crop-growing environments. Characterization of AEZs permits a quantitative assessment of biophysical and socioeconomic constraints that agricultural research should address to improve the well-being of the people while preserving the resource base.

The Technical Advisory Committee (TAC) of the Consultative Group of International Agricultural Research (CGIAR) delineates the AEZs by applying simple crop-growth criteria on the FAOs agroecological database. These criteria are: (i) the moisture regime based on the length of growing period (LGP) under rainfed conditions, which uses data on the seasonal distribution of rainfall and evapotranspiration at different times of the year; and (ii) the thermal regime based on mean monthly temperature (MMT).

The AEZs are defined by a two-way matrix: one axis classifies the regions by LGP and the other by MMT. The definition of the terms used in AEZ classifications are give in Table 3.2. Data on socioeconomic variables are available only by political units, some at national and some at different subnational levels. Since many large countries cover several AEZs, we need to correlate the agroclimatic configuration on a physical geographic basis with available socioeconomic data from subnational political units.

For preparation of the Medium Term Plan 1994–1998, the IRRI developed a database by AEZs using subnational-level socioeconomic data for China and India and national-level data for other countries in Asia. The geographical boundaries used for AEZ delineation can be seen in Table 3.3. The delineation is far from perfect. Several countries have areas that do not belong to the AEZ into which it has been classified. For example, a part of central Myanmar is semi-arid, although all of the country is classified as warm subhumid tropics (AEZ 2), southern Thailand is humid, and central and western Madhya Pradesh (India) is semi-arid, but they are also classified as AEZ 2. Future ecoregional research should be to develop appropriate databases that allow more accurate analysis at the AEZ level.

The pattern of land use for the production of foodgrains in different agroecological zones can be reviewed from Table 3.4. In the semi-arid zones in the Asian tropics, the major crops are oilseeds, pulses and coarse grains such as sorghum and millets, which could be grown with little soil moisture. Wheat and maize are also grown in large parts of the semi-arid zones where irrigation facilities are developed. Wheat and maize are major foodgrain crops in the cool subtropics, and also in subhumid and semi-arid subtropics. Rice–wheat cropping systems predominate in the subhumid subtropics that benefit from adequate rain during summer and a relatively long and dry winter.

Rice is the predominant foodgrain crop in the humid tropics (AEZ 3) and subtropics (AEZ 7) due to abundant moisture. Rice accounts for three-fourths of total cropped land in these two AEZs. But rice is also the most important cereal crop in the subhumid subtropics (AEZ 2) where wheat cannot be grown

Table 3.2. Definition of terms used in the Technical Advisory Committee of the Consultative Group of International Agriculture Research's agroecological zones (AEZs).

AEZ term	Definition
Arid	LGP < 75 days
Semi-arid	LGP 75–180 days
Subhumid	LGP 180–270 days
Humid	LGP > 270 days
Tropics	MMT > 18 °C for all months
Subtropics	MMT 5–18 °C for 1 or more months
Temperate	MMT < 5 °C for 1 or more months
Warm	Daily mean temperature during the growing period of > 20 °C
Cool	Daily mean temperature during the growing period of 5–20 °C

LGP, length of growing period; MMT, mean monthly temperature.

Table 3.3. Geographical delineation of agroecological zones (AEZs) in Asia.*

Agroecological zones	Geographical boundaries
Warm semi-arid tropics (AEZ 1)	Southwestern India (Andhra Pradesh, Karnataka, Tamil Nadu, Maharashtra, Gujarat)
Warm subhumid tropics (AEZ 2)	Thailand, Myanmar, Eastern India (Madhya Pradesh, Orissa, Bihar)
Warm humid tropics (AEZ 3)	Indonesia, Malaysia, Philippines, Vietnam, Cambodia, Laos, Sri Lanka, Bangladesh, parts of India (Assam, Northeastern States, West Bengal, Kerala)
Warm semi-arid subtropics (AEZ 5)	Pakistan, parts of India (Rajasthan, Haryana), parts of China (Helong, Laioning, Jilin, Tianjin, Sandong, Hebei)
Warm subhumid subtropics (AEZ 6)	Northwestern India (Uttar Pradesh, Punjab), Nepal, parts of China (Jiangshu, Anhui, Hubei, Sichuan, Henan, Guizhou, Yunan), North Korea, South Korea
Warm humid subtropics (AEZ 7)	Parts of China (Shanghai, Zehjiang, Fujian, Jianxi, Hunan, Guandong, Guanxi) Taiwan
Cool subtropics temperate zone (AEZ 8)	Parts of India (Himachal Pradesh, Jammu and Kashmir), parts of China (Beijing, Shani, Inner Mongolia, Tibet, Gansu, Ningia, XinJiang, Quinghai)

*Excludes the Middle East and transitional economies in Central Asia.

because of the warm temperature during the dry season. Rice–pulses, rice–oilseeds or rice–fallow are the major cropping systems in this AEZ.

The interface between the rice ecosystems and the AEZs is shown in Table 3.5. The irrigated rice ecosystem accounts for 57% of the rice area and 76% of rice production in Asia. It is concentrated mostly in the subtropics and the semi-arid tropics. The irrigated ecosystem in the subtropics is gradually losing land to urbanization and industrialization. Farm yields are approaching the ceiling of average yields obtained in experimental stations. It is characterized by intensive use of agrochemicals with potential adverse effects on human health and the sustainability of the natural resource base. The unfavourable rice ecosystems (rainfed lowland, upland, deepwater and coastal wetlands) are predominant in the humid and subhumid tropics (AEZs 2 and 3). These are ecosystems where modern rice technologies have yet to make an impressive impact. Expansion to marginal land has been an important source of growth in rice production. Strategic research is needed for these regions to ease

Table 3.4. Relative importance of foodgrain crops by agroecological zones, 1991.

Agroecological zones	Percent of total area					
	Rice	Wheat	Maize	Coarse grains	Pulses	Oilseeds	Total
Warm tropics							
Semi-arid	18.5	3.4	2.0	36.1	16.2	23.8	100.0
Subhumid	51.3	9.3	6.2	7.6	15.4	10.2	100.0
Humid	75.4	1.7	12.9	0.3	3.7	6.0	100.0
Warm subtropics							
Semi-arid	9.4	38.0	15.6	12.7	17.9	6.4	100.0
Subhumid	35.7	29.4	10.8	4.8	10.0	9.3	100.0
Humid	75.6	3.6	3.3	0.1	7.3	10.2	100.0
Cool subtropics	3.1	45.3	20.5	6.1	14.0	10.8	100.0

Source: FAO agrostat database, and national statistical publications for China and India.

Table 3.5. The interface between agroecological zones and rice ecosystems, 1991.

Agroecological zones	Total rice cropped area (million ha)	Percent of total rice area in each rice ecosystem				
		Irrigated	Rainfed lowland	Rainfed upland	Flood-prone	Total
Warm tropics						
Semi-arid	9.68	75.0	12.4	10.8	1.8	100.0
Subhumid	28.94	23.3	53.9	10.6	12.1	100.0
Humid	44.52	42.2	32.0	10.3	15.5	100.0
Warm subtropics						
Semi-arid	7.47	99.7	0.0	0.3	0.0	100.0
Subhumid	23.91	76.6	13.8	5.2	4.4	100.0
Humid	18.35	92.1	6.4	1.5	0.0	100.0
Cool subtropics	0.4	100.0	0.0	0.0	0.0	100.0
Total	133.27	56.9	26.7	7.7	8.7	100.0

Source: IRRI rice statistics database.

constraints to growth in productivity imposed by abiotic stresses, drought, flooding, waterlogging and problem soils.

Population pressure and the natural resource balance for agricultural production in the four AEZs under study can be seen in Table 3.6. The most severe pressure of population on land resources is in the humid subtropics (AEZ 7), where 1 ha of arable land supported almost 11 persons as early as the beginning of the 1960s. With population growth at 2.0% per year over the last three decades, arable land per capita has now declined to less than 0.05 ha. The growing demand for food in this ecoregion has been met by expansion of irrigation facilities, which permitted multiple cropping with foodgrains, and adoption of the high-yielding modern varieties. In this ecoregion, the foodgrain area accounts for 158% of arable land; thus, a substantial portion of the area is multiple cropped under foodgrains.

The availability of land is somewhat better in the subhumid tropics, where at present four to five persons are supported per hectare of arable land. But the quality of this land is poor, and the development of irrigation has been limited. Less than one-third of the area under rice is currently irrigated. Due to dependence on erratic monsoon rains, farmers still grow mainly traditional varieties of grains, and agricultural productivity remains low. Data on the potential for irrigation development is not available by AEZs. But the fact that irrigation coverage remains lower in the tropics than in the subtropics in spite of the lower availability of foodgrain per capita (see Table 3.6) reveals greater economic and environmental constraints to the development of land under irrigation. Returns on investment in large-scale irrigation projects in the subhumid tropics may be low since these regions experience a monsoonal climate requiring only supplemental irrigation during the wet season (hence the capacity utilization of irrigation systems is low). During the dry season there is not enough surface water available for irrigating the rice crop. Abundant, yet poorly distributed, rainfall is the norm, and floods may be as severe a problem as drought.

AEZs 2 and 3 (the subhumid and humid tropics) have clearly less favourable grain-production environments than AEZs 6 and 7 (the subhumid and humid subtropics). Because of a more favourable land:person ratio, the production of foodgrains per capita was actually higher in the subhumid tropics than for other AEZs before the onset of the green revolution. But the subtropics benefited more from the diffusion of the green revolution technologies than did the tropics (Tables 3.7 and 3.8). In both the humid and subhumid subtropics, the productivity in foodgrain cultivation increased much faster than population. The impressive growth in rice yield has helped the humid subtropics to divert rice land to other crops since early 1980s, and meet the growing demand for non-grain foods without overexploitation of the limited land. Over the past three decades per capita production of foodgrains increased by about 100 kg year^{-1} for AEZs 6 and 7, and by about 53 kg year^{-1} in AEZ 3, but remained unchanged in AEZ 2. As a result, food insecurity and poverty in the subhumid and humid tropics increased more than in the subtropics. Compared to AEZ 7, which is in the most favourable position with respect to food security, the per capita energy intake is about

Table 3.6. Population pressure, foodgrain production and incidence of poverty in Asian agroecological zones, 1990.*

Agroecological zones	Share of arable land (%)	Share of population (%)	Arable land/ capita (ha)	Foodgrain production per ha of arable land (tons ha^{-1})	Foodgrain production/ capita (kg year^{-1})	Foodgrain cropped area as % of arable land	Calorie:poverty intake ratio (kcal % of capital)
Warm tropics							
Semi-arid	15.6	10.3	0.20	1.22	240	95.0	226,854.1
Subhumid	18.3	10.1	0.23	1.39	260	90.6	213,952.1
Humid	19.4	22.0	0.11	2.24	256	86.0	237,038.1
Warm subtropics							
Semi-arid	19.2	16.2	0.15	1.99	304	94.3	250,823.7
Subhumid	17.8	25.3	0.09	4.16	377	133.3	254,928.6
Humid	4.5	11.5	0.05	6.64	331	157.5	274,219.2
Cool subtropics	5.3	4.6	0.15	2.02	300	82.8	250,819.5
Asia	100.0	100.0	0.13	2.40	310	101.2	245,333.4

* Excludes middle east and transitional economies in Central Asia.
Source: FAO agrostat database, and national statistical publications.

Table 3.7. Sources of growth (% per year) in foodgrain production in Asia, 1961–1991.

Agroecological zones	Area	Yield	Production
Subhumid tropics	0.84 (0.06)	1.49 (0.13)	2.33 (0.15)
Humid tropics	0.84 (0.04)	2.34 (0.08)	3.18 (0.08)
Subhumid subtropics	1.45 (0.05)	3.43 (0.11)	4.88 (0.14)
Humid subtropics	0.07 (0.13)	3.08 (0.12)	3.15 (0.17)

Estimated by fitting semilogarithmic trend lines on time series data. Figures within parentheses are standard errors of estimated growth.

600 kcal lower in AEZ 2 and 372 kcal lower in AEZ 3. In 1990, nearly 52% of the people in AEZ 2 lived in poverty, compared to only 19% in AEZ 7 (see Table 3.8).

The overwhelming importance of rice in the humid and subhumid tropics and subtropics (see Table 3.4) clearly indicates that for ecoregional foodgrain research to have the desired impact, the primary focus must be on rice-based production systems. Rice will continue to dominate agriculture in Asia well into the next century. Efforts to recover degraded resources and increase sustainable productivity of the overall resource base in the humid and

Table 3.8. Sources of growth (% per year) in rice production, 1961–1991.

Agroecological zones	Area	Yield	Production
Subhumid tropics	0.56 (0.05)	1.38 (0.14)	1.94 (0.16)
Humid tropics	0.69 (0.03)	2.39 (0.08)	3.08 (0.08)
Subhumid subtropics	1.07 (0.11)	2.96 (0.11)	4.03 (0.16)
Humid subtropics	0.19 (0.13)	2.99 (0.11)	3.18 (0.17)

Estimated by fitting semilogarithmic trend lines on time series data. Figures within parentheses are standard errors of estimated growth.

subhumid tropics in Asia should give strong emphasis on rice-based ecosystems.

B. Ecosystem characterization from mega- to micro-level: an eastern India case study

An ecosystems analysis for research prioritization at all the levels throughout eastern India has been done in collaboration with the Indian Council for Agricultural Research, state agricultural universities and departments of agriculture (IRRI, 1992).

India has a 42 million ha rice-growing area. Mega-level analysis of the four regions of the country indicated that while rice yields in northern and southern India have increased rapidly in recent years, yields remained basically stagnant in eastern India (except in West Bengal which experienced a rapid growth in recent years). The eastern India region, comprising eastern Uttar Pradesh and Madhya Pradesh and the states of Assam, Bihar, West Bengal and Orissa, is the largest rice-growing region in the country and accounts for about 67% of India's rice area (26.8 million ha). In five of the six eastern states, average rice yield (1.8 tons ha^{-1}) is below the national average of 2.7 tons ha^{-1}. About 80% of rice farming in the region is rainfed. Rainfall is moderate to high, and is limited to a short period. This results in drought in the uplands and flooding in the lowlands.

Eastern India is the priority region for research because of its large rice area and low and stagnant rice yields. It is also a priority region because about half of the country's population of 960 million people live in the region and are largely dependent on rice farming.

The macro-level analysis of rice-growing ecosystems in eastern India revealed that only 21.2% (5.69 million ha) of the 26.80 million ha rice area is irrigated (Table 3.9). About 16.4% (4.38 million ha) is upland, 47.7% (12.78 million ha) is rainfed lowland (0–50 cm water depth), and the remaining 14.7% (3.95 million ha) is deep water or very deep water (50 to > 100 cm water depth). For the rainfed lowland ecosystem, about 83% (10.6 million ha) has shallow water depth (0–30 cm) and 17% (2.2 million ha) has medium water depth (30–50 cm).

Analysis of drought and flooding patterns, water balance, selected land characteristics and length of growing season in the shallow rainfed lowland ecosystem showed that 54.6% (5.7 million ha) is drought-prone, 25.5% (2.7 million ha) is drought- and submergence-prone, 10.3% (1.1 million ha) is submergence-prone and 9.6% (1.0 million ha) is favourable. The entire area in the medium-depth category of the rainfed lowlands is submergence-prone. The area of different rainfed lowland subecosystems in eastern India is given in Table 3.10. There is a wide fluctuation in the extent of the deepwater ecosystem and the medium-depth category of rainfed lowlands, depending on rainfall pattern and amount, and onset and cessation of the monsoon.

Rice yields in all the rainfed ecosystems are low and vary greatly from year to year. Yields in the irrigated areas average 3.2 tons ha^{-1}. Yield range is 0.6–

Table 3.9. Rice area in different ecosystems in eastern India.

State	Irrigated	Upland	Rainfed lowland 0–30 cm	Rainfed lowland 30–50 cm	Deepwater (50–100 cm)	Very deepwater (>100 cm)	Total
Assam	203	215	892	472	385	100	2,267
Bihar	1,512	531	1,698	465	382	672	5,260
Orissa	1,062	691	1,743	486	400	150	4,532
Madhya Pradesh	608	1,349	2,695	–	–	–	4,652
Uttar Pradesh	982	714	1,884	290	234	555	4,659
West Bengal	1,324	883	1,685	470	383	677	5,422
Total	5,691	4,383	10,597	2,183	1,784	2,154	26,792

Source: IRRI, 1992.

Table 3.10. Extent of the subecosystems of shallow rainfed lowland rice in eastern India.

	Shallow rainfed lowland subecosystems (ha × 10³)				
State	Drought-prone	Drought- and submergence-prone	Favourable	Submergence-prone	Total
Assam	–	–	450	442	892
Bihar	948	750	–	–	1,698
Orissa	700	500	275	268	1,743
Madhya Pradesh	2,695	–	–	–	2,695
Uttar Pradesh	784	1,100	–	–	1,884
West Bengal	660	350	300	375	1,685
Total	5,787	2,700	1,025	1,085	10,597

Source: IRRI, 1992.

1.5 tons ha^{-1} in the uplands, 0.9–2.4 tons ha^{-1} in the rainfed lowlands and 0.9–2.0 tons ha^{-1} in the deepwater and very deepwater areas. The first priority ecosystem in eastern India is the rainfed lowlands, because of its area, larger dependent population and potential for yield increase.

A meso-level analysis of rainfed rice ecosystems was conducted in the Bahraich and Faizabad districts of Uttar Pradesh, Hazaribagh district of Bihar and Raipur district of Madhya Pradesh. Characterization of the rice environments of the Faizabad district (total area of 451,100 ha and rice area of about 181,000 ha) was done with satellite remotely sensed data, selective field checks and auxiliary data. Maps (1:250,000 scale) were prepared to delineate physiographic units, land-utilization pattern, soils, flooding and drought. Information on climate, ground water, irrigation sources, landholding and input use was integrated with the maps.

The classification of rainfed rice environments showed that about 40% of the area in the Faizabad district is favourable rainfed lowland, 51% is drought-prone lowland, 2% is submergence-prone lowland and 4% is submergence- and drought-prone lowland. Apart from drought and submergence, soil sodicity was identified as the priority research area in the district.

Detailed meso-level analysis was done in the Masodha block of the Faizabad district. The block covers about 21,000 ha (total land area) and has about 8000 ha of rice area. Rice-growing environments in terms of physiography, land use, soils, flooding, drought, ground water and irrigation were studied in detail using remote sensing and conventional data. The major part of the block is classified as a shallow favourable rainfed rice subecosystem. Meso-level analysis showed that about 14% of the block area is affected by flooding, 10% by sodicity and 2% by waterlogging. Only 32% of the groundwater potential has been developed so far. Recharge-draft analysis showed that about 1600 ha m of ground water is still available for irrigation.

Within each of the upland, rainfed lowland and deepwater ecosystems in eastern India, target environments were characterized at the micro-level to set research priorities within and among dominant farming systems. Ninety sites were analysed.

Rapid rural appraisal techniques, which included agroecosystems mapping and diagnostic surveys, were employed at all sites. The analysis focused on spatial, temporal, resource flow and decision patterns. The methodology involved a two-tier training program for researchers on how to set research priorities using agroecosystems analysis. The research diagnosis and prioritization at this level were conducted by multidisciplinary teams, with continuous involvement and interactions from groups of farmers.

At all sites, the static factors studied were land types, land use, source of water supply and soil properties. The dynamic factors were field-water depth and rainfall; cropping pattern and crop calendars, crop yields, varieties and management practices; insects, diseases and weeds; production costs and returns; labour supply pattern, assets, income distribution, landholding size and demography by social class and gender.

The geographic area was zoned into agroecosystems and the problem and opportunities elucidated in each major agroecosystem. Highest priority was given to the agroecosystem with the largest rice area. The research problems were then prioritized on the basis of the physical extent (coverage), number of affected households, complexity, severity (crop-loss estimates), frequency of occurrence, importance of the affected enterprise in the farming system and the farmers' perceptions of the problem.

All site studies within each sub-ecosystem and ecosystem were pooled and compared to identify commonality of problems and opportunities. This provided an empirical picture of the entire ecosystem, which served as a basis in formulating a need-based research agenda and the allocation of resources at the national, regional and zonal level (IRRI, 1992).

V. Rice Research Prioritization: Beyond a Commodity Approach

The two most well-developed methods for rice research prioritization have examined the allocation of resources from a commodity approach. They have either attempted to estimate an optimal allocation among the rice ecosystems (Barker *et al.*, 1985) or among alternative rice-crop (i.e. genetic) improvement problems (Herdt *et al.*, 1987). Analyses of prospective allocation of research resources by rice ecosystem emphasized the direct marginal contributions to rice production that are likely to accrue to research effort expended in one or another ecosystem. They use congruence rules that are dominated by the value of current production.

Herdt *et al.* (1987) pointed out that research problems tend to fall into one of three types: (i) constraints to achieving currently possible yields (constraints research); (ii) increasing the level of potential yields (potential productivity); and (iii) maintaining current yields (maintenance research). Increasing awareness of environmentally damaging negative externalities in crop production has resulted in greater research resources directed toward technologies that will reduce these external costs, which is a form of maintenance research. The widespread interest in sustainability translates into greater relative emphasis on maintenance research to avoid negative external costs due to new technology, or due to the lack of it.

Environmental externalities have been embedded in the cost–benefit approach (Herdt et al., 1987). As ecosystems analysis improved our understanding of the rice environments, it became obvious that some major negative externalities are omitted from such exercises. Their inclusion would have significant effects on resource allocation, because the nature of the externalities, and their gross value differ among the ecosystems.

Negative externalities in the lowland rice ecosystem include such problems as ground-water pollution with high fertilization, pesticide residues and their danger to the health of human applicators. In the upland ecosystems, rice production on sloping lands with inappropriate farming practices results in accelerated soil erosion and soil degradation. In the upland ecosystem there are two major classes of negative externalities. First, massive sediment loads from the uplands devalue investments in irrigation systems in the lowlands, particularly storage reservoirs, and increase the maintenance costs of irrigation canals at primary and tertiary levels (P.L. Pingali, personal communication). The accelerated sediment deposition in river systems increases the flooding hazards and costs of flood prevention. Second, the degradation of the upland resource base, particularly soil quality and biological diversity, indicates a loss in the value of the land resource stocks over very large geographic areas. These stocks may be valued in terms of the current land productivity lost due to a degraded land base, and future costs in rehabilitating it to provide significant production and protection functions. These costs to society are large. Unlike for lowland rice, the aggregate land area on which upland rice is grown is many times larger than the amount of rice area in any given year.

Thus, the aerial extent of the negative externalities of upland rice is not adequately reflected in tallies of current crop area by ecosystem. The crop is shifted among fields in permanent as well as fallow rotation systems, and is therefore grown on a large proportion of the total area of acid tropical uplands. Thus, upland rice cultivation's effects on the lowland resource base greatly exceeds its effects on the land where it is grown in a given year.

Upland rice research in many environments is directed to reversing the process of land degradation due to inappropriate land use and and management. In such a framework it is important to recognize that the objectives of upland rice research differ from mainstream rice research in general. Given the impact of the negative externalities of upland rice cultivation on the lowlands, it is likely that a more definitive resource accounting would point toward greater relative attention to the uplands in research fund allocation than its share of global rice area and rice production suggests (P.L. Pingali, personal communication).

Adequate data do not exist to make reasonable estimates of the negative externalities. The relative gains to society from research that decreases the degradation of the land resource stocks can be explicitly embedded in these analyses. This also implies greater attention to AEZs as a basis for research prioritization; as the TAC of the CGIAR has attempted (TAC, 1990). Incorporation of these issues into formal research prioritization will undoubtedly stimulate further acceleration in ecosystem analysis work.

References

Ahmed, N.U., Magor, N.P., Islam, S., Khan, A.H. and Naseem, S.B. (1992) Extrapolation and testing of technologies in Bangladesh. In: *Proceedings of the 1990 Cagayan, Philippines Planning Workshop on Ecosystems Analysis for Extrapolation of Agricultural Technologies*. International Rice Research Institute, Los Baños, Philippines, pp. 26–59.

Bachelet, D., Brown, D., Bohm, M. and Russell, P. (1992) Climate change in Thailand and its potential impact on rice yield. *Climatic Change* 21, 347–366.

Barker, R. and Herdt, R.W. (1982) Setting priorities for rice research in Asia. In: Anderson, R.S., Brass, P.R., Levy, E. and Morrison, B.M. (eds) *Science, Politics and the Agricultural Revolution in Asia*. Westview Press, Boulder, Colorado, pp. 427–461.

Barker, R., Herdt, R.W. and Rose, B. (1985) *The Rice Economy of Asia*. Resources for the Future, Washington, DC.

Bowles, E.J. and Garrity, D.P. (1988) Development of a comprehensive classification for rice ecosystems. Paper presented at IRRI Special Seminar, 23 December. Multiple Cropping Department, International Rice Research Institute, Los Baños, Philippines.

Buol, S.W. and Cuoto, W. (1981) Soil fertility-capability assessment for use in the humid tropics. In: Greenland, D.J. (ed.) *Characterization of Soils*. Clarendon Press, Oxford, pp. 254–261.

Conway, G. (1986) *Agroecosystem Analysis for Research and Development*. Winrock International, Bangkok.

Food and Agricultural Organization (FAO) (1977) *FAO–Unesco Soil Map of the World*. Vol. VII. *South Asia*. Unesco, Paris.

Food and Agricultural Organization (FAO) (1979) *FAO–Unesco Soil Map of the World*. Vol. IX. *Southeast Asia*. Unesco, Paris.

Food and Agriculture Organization (FAO) (1988) *Land Resources Appraisal of Bangladesh for Agricultural Development*. BGD/81/035 Technical Reports 1–7. FAO, Rome.

Fujisaka, S. (1991) *A Set of Farmer-based Diagnostic Methods for Setting Post 'Green Revolution' Rice Research Priorities*. IRRI Social Science Division Paper 90-21. International Rice Research Institute, Los Baños, Philippines.

Garrity, D.P. (1984) Asian upland rice environments. In: *An Overview of Upland Rice Research. Proceedings of the 1982 Bouake, Ivory Coast, Upland Rice Workshop*. International Rice Research Institute, Los Baños, Philippines, pp. 161–183.

Garrity, D.P. and Agustin, P.C. (1984) A classification of Asian upland rice growing environments. Paper presented at the Workshop on Characterization and Classification of Upland Rice Environments, August 1984, CNPAF, EMBRAPA, Goiania.

Garrity, D.P., Oldeman, L.R., Morris, R.A. and Lenka, D. (1986) Rainfed lowland rice ecosystems: characterization and distribution. In: *Progress in Rainfed Lowland Rice*. International Rice Research Institute, Los Baños, Philippines, pp. 3–23.

Garrity, D.P., Agustin, P.C. Dacumos, R.Q. and Pernito, R.N. (1992) A method for extrapolating rainfed cropping systems by land type. In: *Proceedings of the 1990 Cagayan, Philippines Planning Workshop on Ecosystem Analysis for Extrapolation of Agricultural Technologies.* International Rice Research Institute, Los Baños, Philippines, pp. 10–25.

Herdt, R.W., Riely, R. Jr and Frank, Z. (1987) International rice research priorities: implications for biotechnology initiatives. Paper prepared for the Rockefeller Workshop on Allocating Resources for Developing Country Agricultural Research, Bellagio, Italy, 6–10 July 1987.

Higgins, G.M., Kassam, A.H., Van Velthuizen, H.T. and Purnell, M.F. (1987) In: Bunting, A.H. (ed.) *Agricultural Environments: Characterization, Classification and Mapping.* CAB International, Wallingford, UK, pp. 171–183.

Holdridge, L.R., Grenke, W.C., Hatheway, W.H., Liang, T. and Tosi, J.A. (1971) *Forest Environments in Tropical Life Zones. A Pilot Study.* Pergamon, Oxford.

Huke, R.E. (1982) *Rice Area by Type of Culture: South, Southeast and East Asia.* International Rice Research Institute, Los Baños, Philippines.

International Rice Research Institute (IRRI) (1984) *Terminology for Rice Growing Environments.* IRRI, Los Baños, Philippines.

International Rice Research Institute (IRRI) (1987) *Annual Report for 1986.* IRRI, Los Baños, Philippines.

International Rice Research Institute (IRRI) (1989a) *Program Report for 1989.* IRRI, Los Baños, Philippines.

International Rice Research Institute (IRRI) (1989b) *IRRI Toward 2000 and Beyond.* IRRI, Los Baños, Philippines.

International Rice Research Institute (IRRI) (1992) *Program Report for 1992.* IRRI, Los Baños, Philippines.

International Rice Research Institute (IRRI) (1993) *Rice Almanac.* IRRI, Los Baños, Philippines.

Jones, P.G. and Garrity, D.P. (1986) Toward a classification system for upland rice growing environments. In: *Progress in Upland Rice Research. Proceedings of the 1985 Jakarta Conference.* International Rice Research Institute, Los Baños, Philippines, pp. 107–116.

Kassam, A.H., van Velthuizen, H.T., Higgins, G.M., Christoforides, A., Voortman, R.L. and Spiers, B. (1982) *Assessment of Land Resources for Rainfed Crop Production in Mozambique.* FAO : AGOA : MOZ/75/011, Field Docs. 32–37. FAO, Rome.

Kelompok Penelitian Agro-Ekosistem (KEPAS) (1985) *The Critical Uplands of Eastern Java: An Agro-ecosystems Analysis.* KEPAS, Agency for Agricultural Research and Development, Indonesia.

Koppen, W. (1936) Gas geographische System der Klimate. In: *Handbuch der Climatologie.* Borntrager, Berlin.

Oldeman, L.R. (1980) The agroclimate classification of rice-growing environments in Indonesia. In: Cowell, R.L. (ed.) *Proceedings of a Symposium on the Agrometerology of the Rice Crop.* International Rice Research Institute, Los Baños, Philippines, pp. 47–55.

Oldeman, L.R. and Frere, M. (1982) *A Study of the Agroclimatology of Humid Tropics of Southeast Asia.* Food and Agriculture Organization, Rome.

Papadakis, J. (1975) *Climates of the World and their Agricultural Potentialities.* Published by the author, Buenos Aires.

Ruttan, V.W. (1982) *Agricultural Research Policy.* University of Minnesota Press, Minneapolis.

Salmon, D.C. (1983) *Congruence of Agricultural Research in Indonesia (1974–1978).*

Working Paper 83-1. Economic Development Center, University of Minnesota, Minneapolis.

Schgal, J.L., Mandal, D.K., Mandal, C. and Vadiucha, S. (1993) *Agro-ecological Regions of India*. Oxford and IBH Publications, New Delhi, 132 pp.

Singh, V.P. and Pathak, M.D. (1990) *Rice Growing Environments in Bahraich District of Uttar Pradesh*. Uttar Pradesh Council of Agricultural Research, Lucknow, India.

Technical Advisory Committee (TAC) (1990) *Towards a Review of CGIAR Priorities and Strategies*. Progress Report. TAC Consultative Group of International Agricultural Research, Rome.

Thornthwaite, C.W. (1948) An approach towards a rational classification of climate. *Geographical Review*, 38, 55–94.

Young, A. (1987) Methods developed outside the international agricultural research system. In: Bunting A.H. (ed.) *Agricultural Environments: Characterization, Classification and Mapping*. CAB International, Wallingford, UK, pp. 43–63.

4　Prospects of and Approaches to Increasing the Genetic Yield Potential of Rice

G.S. Khush

Plant Breeding, Genetics and Biochemistry Division, International Rice Research Institute, PO Box 933, 1099 Manila, Philippines

I. Introduction

Major increases in rice production have occurred during the last 25 years due to large-scale adoption of high-yielding semidwarf varieties and improved technology. World rice production doubled from 257 million tons in 1965 to 520 million tons in 1990. During this period, rice production increased at a slightly higher rate than population. However, the rate of increase of rice production is slowing down and if the trend is not reversed, severe food shortages will occur in the next century. The present world population of 5.6 billion is likely to reach 6 billion in the year 2000, 7 billion in 2010 and 8 billion in 2020. The population of rice-consuming countries is increasing at a faster rate than that of the rest of the world and the number of rice consumers will probably increase by 50% during the next 25 years. It is estimated that the demand for rice will exceed production by the end of this century (Pinstrup-Andersen, 1994).

Major increases in the area planted to rice are unlikely. The area planted to rice worldwide has remained stable since 1980. In fact it is likely to go down because of pressures on good rice lands from urbanization and industrialization. The increased demand for rice will have to be met from less land, with less water, less labour and less pesticides. Therefore, we need rice varieties with a higher yield potential and better management practices to meet the goals of increased rice production. In its strategy document for 2000 and beyond (IRRI, 1989), the IRRI accorded the highest priority to increasing the yield potential of rice.

II. Strategies for Increasing the Yield Potential

The yield potential of current high-yielding rice varieties in the tropics is 10 tons ha^{-1} during the dry season (DS) and 6.5 tons during the wet season

(WS). Plant physiologists have suggested that the physical environment in the tropics is not a limiting factor to increasing rice yields. Maximum yield potential was estimated to be 9.5 and 15.9 tons ha^{-1} during the WS and DS, respectively (Yoshida, 1981). Quantum jumps in the yield potential of crop plants have generally resulted from the modification of plant types. Thus, a new plant architecture permitted the yield potential of rice to be doubled in the mid-1960s. This new plant type was characterized by short stature, high tillering, sturdy stems and dark green, erect leaves. This plant type was extremely effective in increasing the productivity of rice lands and more than 60% of the world's rice area is now planted to varieties of the semidwarf plant type. During the 30 years after the development of this plant type, however, only marginal improvements have occurred in the yield potential of rice. This is because rice improvement efforts were directed towards incorporation of disease and insect resistance, shortening of growth duration and improving the grain quality. For another quantum jump in the yield potential, we must explore the possibility of further modifying the present high-yielding plant type. Another approach for increasing the yield potential of rice in the tropics is the exploitation of hybrid vigour or heterosis through hybrid rice breeding.

III. New Plant Type for Increased Yield Potential

Yield is a function of total dry matter and harvest index (HI). Therefore, yield can be increased either by enhancing the total dry matter (biomass) or the HI or both. The HI of modern high-yielding varieties is around 0.45–0.50. It should be possible to raise the HI to around 0.60. Possibilities for increasing the biomass by genetic as well as through management practices should also be explored. The following varietal characteristics may need improvement for increasing the harvest index and biomass production:

I. Increased HI:
  1. Increased sink size:
      (a) large spikelet number per panicle;
      (b) greater partition of assimilation in spikelet formation.
  2. Increased spikelet filling:
      (a) manipulation of canopy senescence;
      (b) higher proportion of high-density grains;
      (c) maintenance of healthy root system;
      (d) increased lodging resistance.

II. Increased biomass production:
  1. Establishment of desirable canopy structure:
      (a) rapid leaf area development;
      (b) rapid nutrient uptake.
  2. Reduced carbon consumption.

For achieving the above goals, a new plant type was conceived with the following attributes (Fig. 4.1):

Increasing the Genetic Yield Potential of Rice 61

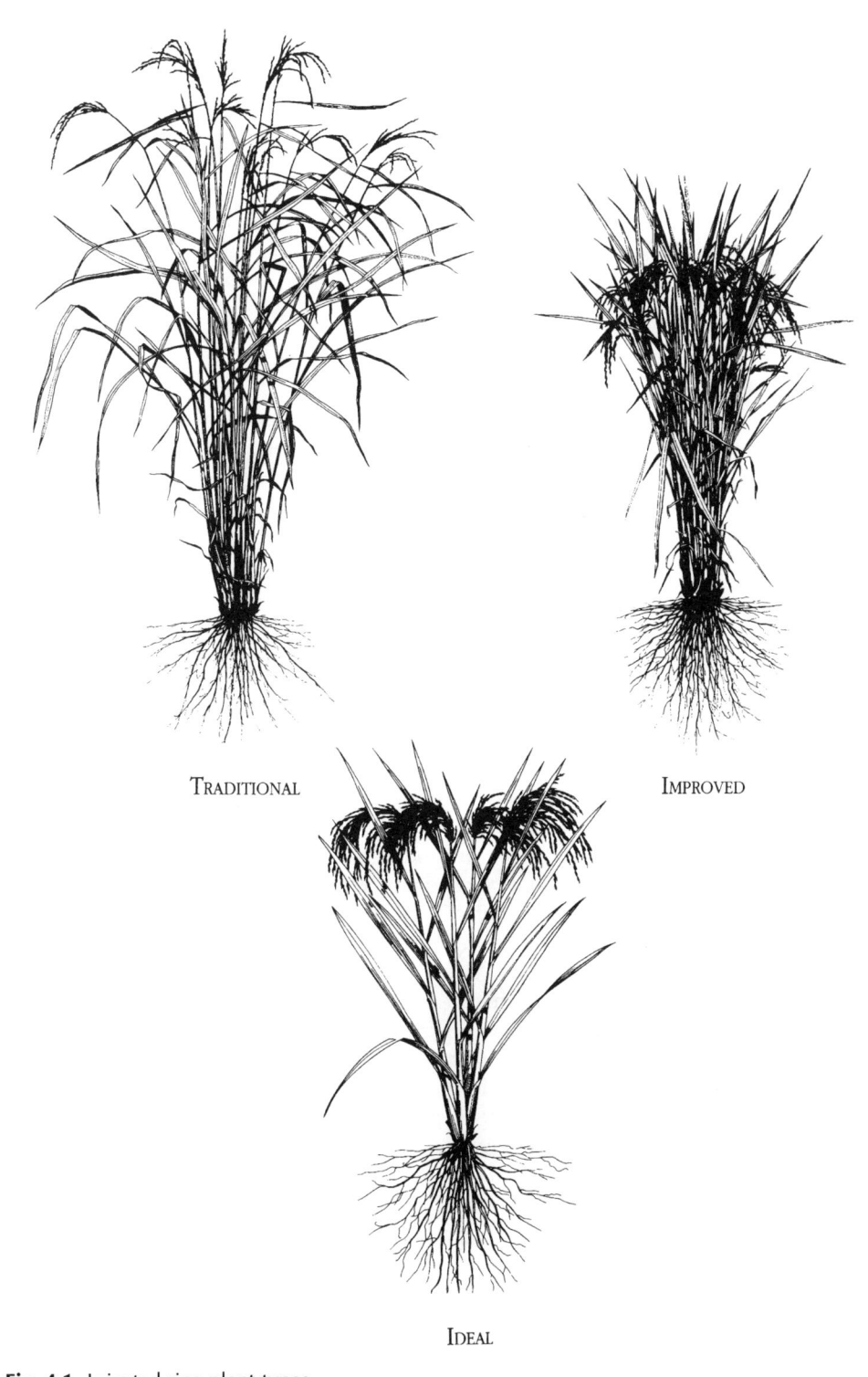

Fig. 4.1. Irrigated rice plant types.

- low tillering capacity (three to four tillers when direct seeded);
- no unproductive tillers;
- 200–250 grains per panicle;
- 90–100 cm tall;
- very sturdy stems;
- dark green, thick and erect leaves;
- vigorous root system;
- 100–130 days' growth duration;
- multiple disease and insect resistance;
- acceptable grain quality.

A. Reduced tillering and large panicles

Increases in the yield potential of other cereals such as corn and sorghum have resulted from increases in sink size. Selection and breeding for large sink size was accompanied by a decrease in tiller number. Modern corn (*Zea mays* L.) and sorghum (*Sorghum bicolor* L.) varieties are uniculm whereas primitive corn and sorghum had a large number of tillers and small cobs and heads (Khush, 1993). Teosinte, the ancestor of corn has 20–25 tillers and small cobs with a few grains. The agriculturists who domesticated the corn in the Americas continued to select corn with larger cobs and this selection resulted in a reduction in tiller number. By the 15th century when corn was introduced into Europe it had only four to five tillers. Further selection resulted in uniculm plants with very large cobs. Uniculm and short-statured sorghums were bred in the post-Mendelian era.

By contrast, modern rice varieties tiller profusely and have 20–25 tillers under favourable growth conditions. Only 14–15 tillers produce panicles and the rest remain unproductive. Unproductive tillers compete with productive tillers for assimilates, solar energy and mineral nutrients – particularly nitrogen. Elimination of the unproductive tillers could direct more nutrients to grain production, but the magnitude of the potential contribution to yield has not been quantified. Furthermore, the dense canopy that results from excess tiller production creates a humid micro-environment favourable for disease, especially the endogenous pathogens like sheath blight (*Rhizoctonia solani*) and stem rot (*Sclerotium oryzae*) that thrive in nitrogen-rich canopies (Mew, 1991).

Reduced tillering also facilitates synchronous flowering and maturity and more uniform panicle size. Low-tillering genotypes are also reported to have a larger proportion of high-density grains (Padmaja Rao, 1987b). The number of spikelets per unit land area is the primary determinant of grain yield in cereal crops grown in high-yield environments without stress (Takeda, 1984). The number of spikelets per unit land area can be increased by increasing the number of panicles or the number of spikelets per panicle. The modern high-yielding rice varieties have much greater panicle number than the traditional rice varieties they replaced. There is a limit, however, to how much the panicle number can be increased. Additional tillers become unproductive and lead to excessive leaf area index (LAI) and vegetative growth and a higher proportion of unfilled grains. Clearly, large panicle size is a prerequisite to compensate for the reduced panicle number in low-tillering plant types.

B. Grain size, grain density and grain-filling percentage

A thousand grain weight of about 25 g is considered ideal for rice. Larger grains tend to be chalky and thus have lower market value. Venkateswarlu *et al.* (1986) found a large variation in the weight of single rice grains within a panicle. Regardless of the growth duration of the varieties studied, the proportion of heavier, high-density grains was greatest at the top of the panicle (superior spikelet position) and smallest in the inferior spikelets of the lower panicles (Padmaja Rao, 1987a). High-density grains tend to occur on the primary branches of the panicle, while the spikelets of the secondary branches have lower grain weight (Ahn, 1986). The proportion of high-density grains was also greater on primary tillers than in secondary and tertiary tillers (Padmaja Rao, 1987b). The low-tillering varieties would have more primary tillers per unit land area and thus should produce a higher proportion of high-density grains and contribute to increased yield potential. The selection for high-density grains for increased yield potential assumes that there are sufficient assimilates or sources to make heavier grains. Thus, the role of source limitation in governing the proportion of high-density grains needs to be clarified.

Kato (1989) has emphasized the importance of three factors in the grain-filling process: (in) photosynthate production by the source; (ii) photosynthate translocation through the transport network; and (iii) photosynthate accumulation by the sink. The grain-filling process results from a balance between these three properties. For higher photosynthate production, dark green and thick leaves with slow senescence are desirable. Thick stems have more vascular bundles (Hayashi, 1980) which not only provide an efficient transport system but also contribute to lodging resistance.

C. Canopy and leaf characteristics

Erect leaf angle is a desirable trait for high-yielding varieties. Light is used more efficiently at a high LAI in an erect-leaved canopy (Yoshida, 1976). Carbon assimilation of a leaf exposed to light on only one side is lower than when the leaf is exposed on both sides and this difference is greatest when leaves have a high nitrogen content and greater thickness. Therefore, a plant community with vertically oriented leaves gives better light penetration and higher carbon assimilation per unit of leaf area (Tanaka, 1976). Droopy or horizontally oriented leaves increase the relative humidity and decrease temperature inside the canopy due to reduced light penetration and air movement. Although leaf thickness has not been associated with higher yield potential, it is positively correlated with leaf photosynthetic rate. Thicker leaves are therefore thought to be a desirable trait (Yoshida, 1972) and this provides a useful visual selection criterion for the new plant type.

D. Growth duration

The optimal growth duration for maximum rice yield in the tropics is thought to be about 120 days from seed to seed. Varieties of shorter growth duration usually give lower yields at conventional hill spacing in transplanted rice due to insufficient vegetative growth for maximum yield levels (Yoshida, 1976). A longer growth duration of about 120 days allows the plant to utilize more soil nitrogen and solar radiation and results in higher yields. Variation in growth duration largely reflects differences in the vegetative growth period (Vergara *et al.*, 1969). There is, however, a positive correlation between growth duration and the time from panicle initiation to heading. Thus an early-maturing rice crop has a relatively short period for panicle growth before heading and the spikelet number is positively correlated with crop growth rate during the period from panicle initiation to flowering. Shorter duration for panicle growth is often accompanied by decreased grain yield. The possibility of extending the duration of panicle growth and grain-filling duration independent of the total growth duration remains to be explored.

E. Plant height, stem thickness, biomass production and harvest index

Short stature reduces the susceptibility to lodging and leads to a higher harvest (HI). Shorter culms require less maintenance respiration and contribute to an improved photosynthesis–respiration balance (Tanaka *et al.*, 1966). A decrease in plant height for the present levels of current semidwarf varieties would reduce total dry matter. Thus, a plant height of 90–100 cm is considered ideal for maximum yield. It is estimated that stem reserves contribute about 2 tons ha^{-1} to grain yield. Thicker stems may serve to accumulate more assimilates.

Increased biomass production is not difficult to achieve when the rice crop is grown in a high solar radiation environment similar to the DS in the tropics, provided there is a luxuriant supply of nitrogen (Akita, 1989). Without a strong, thick culm and proper partitioning, however, increased biomass production results in lodging, mutual shading, increased disease incidence and decreased grain yield (Vergara, 1988). If lodging and disease problems can be solved, increased biomass production could contribute to increased yield potential in tropical environments.

For a specific environment, there is an optimum growth duration for high yield and high HI. The HI of modern high-yielding varieties varies from 0.45 to 0.53 depending upon the season. The HI is higher during the DS (Vergara and Visperas, 1977). Clearly there are limits to how far the HI can be increased in improved varieties. Austin *et al.* (1980) estimated that the maximum possible HI is 0.63 for wheat. If the same were true for rice, the present yield potential of

10 tons ha^{-1} with an HI of 0.53 would increase to 12 tons ha^{-1} assuming the same biomass and an HI of 0.63 for the new plant type (Evans, 1993).

F. Root system

Roots are the foundation of plants and yet they remain relatively unstudied compared to the rest of the plant. Rice varieties differ as much in plant parts below the soil surface as in parts above the ground. For example, cultivar differences are known in root length, degree of branching, root volume and thickness. Thicker and deeper roots provide better anchorage and lodging resistance. Healthy roots are beneficial for nutrient supply, particularly during grain-filling period.

G. Disease and insect resistance

For the full expression of yield potential, genetic resistance to diseases and insects is essential. Resistance is a must under tropical conditions. The major diseases of rice to consider are blast, bacterial blight, sheath blight and several virus diseases; the major insects are the leaf and planthoppers and stem borers.

IV. Present Status of Breeding for the New Plant Type

Breeding work on the new plant type was started in 1989 when about 2000 entries from the IRRI germplasm bank were grown during dry and wet seasons to identify donors for various traits. Donors with low tillering, large panicles, thick stems, vigorous root system and short stature were identified (Table 4.1). Hybridization work was undertaken in the 1990 DS and F$_1$ progenies were grown for the first time in the 1990 WS, F$_2$ progenies in the 1991 DS and a pedigree nursery in the 1991 WS. Since then 1350 crosses have been made, and 60,500 pedigree nursery rows have been grown. Breeding lines with targeted traits of the proposed ideotype have been selected. These were grown in the

Table 4.1. Donors for various traits being used for developing the new plant type.

Trait	Donors
Short stature	MD2, Sheng-Nung 89-366
Low tillering	Merim, Gaok, Gendjah Gempol, Gendjah Wangkal
Large panicles	Daringan, Djawa Serang, Ketan Gubat
Thick stems	Sengkeu, Sipapak, Sirah Bareh
Grain quality	Jhum Paddy, WRC4, Azucena, Turpan4
Donor for resistance	
Bacterial blight	Ketan Lumbu, Laos Gedjah, Tulak Bala
Blast	Moroberekan, Pring, Ketan Aram, Mauni
Tungro	Gundil Kuning, Djawa Serut, Jimbrug, Lembang
Green leafhopper	Pulut Cenrana, Pulut Senteus, Tua Dikin

observational trials for the first time in the 1993 WS. Their yield potential is being evaluated in replicated field plots under various management practices.

On the basis of observations during the 1993 WS and 1994 DS field trials, the following characteristics of new plant type lines have been observed.

- The new plant-type lines produced less tillers than improved indicas. The tillering response to spacing and nitrogen levels was similar for both types.
- The spikelet number per panicle of the new plant types was two to three times greater than IR72, the improved indica variety.
- Some of the new plant-type lines have about 20% more spikelets per square metre and thus 20% larger potential sink size than IR72. The percent filled spikelets and 1000 grain weight was similar in both types.
- Single-leaf photosynthesis per unit leaf area of the new plant-type line was 10–15% higher than that of IR72 at the vegetative and reproductive stages. This advantage was mainly due to higher leaf nitrogen concentration in the new plant-type lines.
- The new plant-type lines had greener, thicker and more erect leaves than those of IR72. They also had one to two more functional leaves at flowering as compared to IR72.
- Functional duration of flag leaves of the new plant-type lines appears to be longer than those of IR72. This may result in longer grain-filling duration and contribute to increased yield potential.
- The new plant-type lines have thicker and sturdier stems and much greater lodging resistance.
- The growth duration of the new plant-type lines is similar (115–120 days).

Further improvements are aimed at improving grain quality and incorporation of disease and insect resistance. Donors for disease and insect resistance and for better grain quality have been identified (see Table 4.1) and have been used in the hybridization program.

A. Germplasm for the new plant type

When we started looking for donors for various characteristics to breed the new plant type, we first examined the germplasm classified as bulus or javanicas from Indonesia. Bulus are known for low tillering, large panicles and sturdy stems. The javanica rices are genetically very close to the japonicas grown in temperate areas. On the basis of allelic constitution at 15 isozyme loci, it was found that javanicas and japonicas belong to the same varietal group. We therefore now refer to the javanicas as tropical japonicas. Crosses between tropical and temperate japonicas are fully fertile and there are no barriers to recombination. On the other hand, crosses between indicas and japonicas have varying levels of sterility and give poor recombinant progenies as restrictions to recombination exist in such crosses.

A decision was made to limit the hybridization work for the new plant type within the tropical japonica germplasm with selective introduction of genes from temperate japonicas and indicas. The reasons for this approach are threefold.

1. Intercrosses within the japonica germplasm would not encounter problems of sterility and restriction to recombination.

2. After the major breakthrough in raising the yield potential of indicas through the introduction of genes for short stature in mid-1960s, significant increase in their yield potential has not occurred in spite of efforts by international and national rice improvement programs. We therefore decided to work with entirely different germplasm to explore the possibility of raising the yield potential.

3. F_1 hybrids between indica and japonica rices are expected to have higher heterosis, but the temperate japonicas are not adapted to tropical conditions and thus cannot be used for producing F_1 hybrids. Improved tropical japonicas with genes for wide compatibility, short stature, disease and insect resistance and long slender grains would be most suitable for this purpose (Khush and Aquino, 1994).

To identify additional donors for the tropical japonica improvement program, we are systematically classifying the germplasm form the Southeast Asian countries. In addition to bulus from Indonesia, many tropical japonica varieties have been identified in the germplasm from Malaysia, Thailand, Myanmar, Laos, Vietnam and the Philippines. Tropical japonicas are grown both under upland as well as lowland conditions in these countries. Upland rices in Thailand and Laos, for example, are primarily japonicas. Some of the upland rices in the Philippines and Malaysia are japonicas and other are indicas. Tropical japonicas are grown on lowlands in the Philippines and Indonesia. Tinawon rices, for example, grown in the irrigated rice terraces in the Philippines and the lowland bulu rices of Java and Bali in Indonesia are japonicas.

Upland japonicas from Southeast Asia were introduced into Guinea and other West African countries by Portuguese and Spanish priests, starting in the 16th century. Thus, many of the upland rices in West Africa such as Moroberekan, OS4 and OS6 are japonicas. Upland japonicas from West Africa were introduced by the Portuguese to Brazil and spread to other Latin American countries and many of the upland rices in Latin America are japonicas. We have crossed several long grain upland rices from Brazil and the International Center for Tropical Agriculture (CIAT) with our tropical japonica lines and we did not encounter any problems of sterility and barriers to recombination.

V. Hybrid Rice for Increasing the Yield Potential

Hybrid rices have been grown in China since 1976 and on average have a yield advantage of about 15% over the best inbred varieties. Approximately 50% of China's rice area is now planted to rice hybrids (Yuan *et al.*, 1992). These hybrids were evaluated in tropical countries and were found to be unadapted.

Hybrid rice research was initiated at the IRRI in 1978. Selected hybrids also show a yield advantage of about 15% under tropical conditions. Increased yield of tropical rice hybrids is due to increased total biomass, higher spikelet number and to some extent higher 1000 grain weight (Ponnuthurai et al., 1984).

Rice hybrids in China are based on cytoplasmic male sterility (CMS) and fertility restoration systems. Hundreds of CMS lines have been bred in China for hybrid seed production. These CMS lines could not be used as such to develop rice hybrids for the tropics because of their susceptibility to diseases and insects, poor adaptability and poor grain quality. Therefore, new CMS lines were bred at the IRRI and by the national programs using the wild abortive (WA) cytoplasmic male sterility system from China. There is no dearth of restorers among the elite indica rice germplasm in the tropics and subtropics. New sources of CMS have also been identified at the IRRI.

The thermosensitive genetic male sterility (TGMS) system and photoperiod-sensitive genetic male sterility (PGMS) system simplifies the hybrid rice seed production. Several sources of TGMS and PGMS have been reported from China (Sun et al., 1989; Wu et al., 1991). A few mutants have been identified in Japan (Maruyama et al., 1991) and at the IRRI (Virmani and Voc, 1991). TGMS and PGMS systems do not require maintainer lines for multiplication and hybrids can be developed by using only two lines (instead of three in the case of the CMS system), e.g. TGMS and pollen parent. The latter does not have to be a restorer. Two-line hybrids are likely to show higher heterosis because there are less restrictions on the choice of parents in comparison to the CMS system. It should be possible to use the TGMS system to develop rice hybrids in the tropics by utilizing the temperature differences at different altitudes or in different rice-growing seasons. The TGMS system should make a good alternative to the complex and cumbersome CMS system.

Technology for producing hybrid rice seeds in the tropics has been outlined (Virmani and Sharma, 1993). Using this technology, yields of 1–2 tons ha^{-1} of hybrid seed and 1–2 tons ha^{-1} of pollen parent have been obtained by the IRRI and at some national programs. Seed yields can be increased further by improving the outcrossing potential of the parental lines and fine tuning of technology by the prospective seed growers and in the national programs.

A. Indica/japonica hybrids

The magnitude of heterosis (hybrid vigour) depends upon the genetic diversity between the two parents of the hybrids. The greater the genetic difference in the parents, the higher the heterosis. During the past 30 years, the genetic diversity among the improved indica rices has narrowed down due to massive international exchange of germplasm (Khush and Aquino, 1994). Indica and japonica germplasm have, however, remained distinct as there has been very little gene flow between these two varietal groups. As expected, the hybrids between indica and japonica parents showed higher heterosis for yield (Yuan et al., 1989). Therefore, as discussed earlier, the new plant-type development program was based on tropical japonica germplasm so that this improved germplasm would also be utilized for producing hybrids with higher heterosis. Our preliminary results show that the level of heterosis in the indica/tropical japonica hybrids is higher than that of indica/indica hybrids.

Major constraints in utilizing hybrid rice technology for increasing rice production are; (i) the need to buy fresh hybrid seeds for every planting season; (ii) expensive hybrid seeds; and (iii) the need to establish a seed-production infrastructure in developing countries. Farmers would be willing to buy fresh seeds at a price higher than inbred rice provided there is a cost:benefit ratio of 1:4. With such a cost:benefit ratio, national programs would also invest to strengthen or establish the seed industry in the public/private and/or cooperative sectors. These constraints can also be overcome if true-breeding hybrids with permanently fixed heterosis are developed by using apomixis. Search for apomixis in rice or inducing it through mutagenesis is underway at the IRRI and in China.

VI. Summary

The genetic yield potential of tropical rice was doubled in the 1960s by modification of the plant type. However, significant increases in the yield potential of rice have not occurred during the last 30 years. For a further quantum improvement in the yield potential of rice, a new plant type was conceptualized and donors for developing this plant type were identified. The breeding program was started in 1989. Prototype breeding lines for the new plant type have been developed. These new plant-type lines have short stature (90–100 cm), low tillering (six to 10 tillers under transplanting), no unproductive tillers, large panicles (200–250 grains per panicle), thick and sturdy stems and thicker, dark green and erect leaves. The germplasm used in developing these plant types belongs to the tropical japonica group. These new plant types are likely to have 20% higher yield potential than the existing high-yielding indicas. The new plant types will be employed in developing indica/japonica hybrids which may have a yield advantage of 20–25% over the best inbred lines. A combination of the two approaches may raise the yield potential of tropical rice by 50%.

References

Ahn, J.K. (1986) Physiological factors affecting grain filling in rice. PhD thesis, University of Philippines, Los Baños.

Akita, S. (1989) Improving yield potential in tropical rice. In: *Progress in Irrigated Rice Research. Proceedings of the International Rice Research Conference 21–25 September 1987, Hangzhou, China.* International Rice Research Institute, Los Baños, Philippines, pp. 41–73.

Austin, R.B., Bingham, J., Blackwell, R.D., Evans, L.T., Ford, M.A., Morgan, C.L. and Taylor, M. (1980) Genetic improvements in winter wheat yields since 1990 and associated physiological changes. *Journal of Agricultural Science (Cambridge)* 94, 675–689.

Evans, L.T. (1993) Raising the ceiling to yield: the key role of synergism between agronomy and plant breeding. In: Muraldiharan and Siddiq, E.A. (eds.) *New Frontiers in Rice Research*. Directorate of Rice Research, Hyderabad, India, pp. 103–107.

Hayashi, H. (1980) Studies on large vascular bundles in paddy rice plant and panicle formation. IV–V. *Bulletin of the Fukui Agriculture Experimental Station* 17, 31–35 (in Japanese with English summary).

International Rice Research Institute (IRRI) (1989) *IRRI Towards 2000 and Beyond*. IRRI, Los Baños, Philippines.

Kato, T. (1989) Relationship between grain filling process and sink capacity in rice (*Oryza sativa* L.) *Japanese Journal of Breeding* 39, 431–438.

Khush, G.S. (1993) Varietal needs for different environments and breeding strategies. In: Muralidharan, K. and Siddiq, E.A. (eds.) *New Frontiers in Rice Research*. Directorate of Rice Research, Hyderabad, India, pp. 68–75.

Khush, G.S. and Aquino, R. (1994) Breeding tropical japonicas for hybrid rice production. In: *Hybrid Rice Technology, New Development and Future Prospects*. International Rice Research Institute, Los Baños, Philippines, pp. 33–36.

Maruyama, K., Araki, H. and Kato, H. (1991) Thermosensitive genetic male sterility induction by irradiation. In: *Rice Genetics*. International Rice Research Institute, Los Baños, Philippines, pp. 227–232.

Mew, T. (1991) Disease management in rice. In: Pimentel, D. (ed.) *CRC Handbook of Pest Management in Agriculture*, Vol. 111, 2nd edn. CRC Press, Boston, pp. 279–299.

Padmaja Rao, S. (1987a) Panicle type: few structural considerations for higher yield potential in rice. *Indian Journal of Plant Physiology* 30(1), 87–90.

Padmaja Rao, S. (1987b) High density grain among primary and secondary tillers of short- and long-duration rices. *International Rice Research Newsletter* 12(4), 12.

Pinstrup-Andersen, P. (1994) *World Food Trends and Future Food Security*. Food Policy Report. International Food Policy Research Institute, Washington, DC.

Ponnuthurai, S., Virmani, S.S. and Vergara, B.S.(1984) Comparative studies on the growth and grain yield of some F_1 rice (*Oryza sativa* L.) hybrids. *Philippine Journal of Crop Science* 9(3), 183–193.

Sun, Z.X., Xiong, Z.M., Min, S.K. and Si, H.M. (1989) Identification of the temperature sensitive male sterile rice. *Chinese Journal of Rice Science*, 3(2), 49–55.

Takeda, T. (1984) Physiological and ecological characteristics of high yielding varieties of lowland rice. In: *Proceedings of the International Crop Science Symposium, 17—20 October*, Kyushu University, Fukuoka, Japan.

Tanaka, T. (1976) Regulation of plant type and carbon assimilation of rice. *Japan Agricultural Research Quarterly* 10(4), 161–167.

Tanaka, A., Kawano, K. and Yamaguchi, J. (1966) *Photosynthesis, Respiration, and Plant Type of the Tropical Rice Plant*. IRRI Technical Bulletin No. 7. International Rice Research Institute, Los Baños, Philippines.

Venkateswarlu, B., Parao, F.T., Visperas, R.M. and Verara, B.S. (1986) Screening quality grains of rice with a seed blower. *Society for the Advancement of Breeding Researchers in Asia and Oceania Journal* 18(1), 19–24.

Vergara, B.S. (1988) Raising the yield potential of rice. *Philippine Technical Journal* 13, 3–9.

Vergara, B.S. and Visperas, R.M. (1977) Harvest index: criterion for selecting rice plants with high yielding ability. IRRI Saturday Seminar, 10 September 1977. International Rice Research Instituted, Los Baños, Philippines.

Vergara, B.S., Chang, T.T. and Lilis, R. (1969) *The Flowering Response of the Rice*

Plant to Photoperiod. IRRI Technical Bulletin No. 8. International Rice Research Institute, Los Baños, Philippines.

Virmani, S.S. and Sharma, H.L.(1993) *Manual for Hybrid Rice Seed Production.* International Rice Research Institute, Los Baños, Philippines.

Virmani, S.S. and Voc, P.C. (1991) Induction of photo and thermo-sensitive male sterility in indica rice. *Agronomy Abstracts* 119.

Wu, X.J., Yin, H.Q. and Yin, H. (1991) Preliminary study of the temperature effect on Annong S-1 and W 6154S. *Crop Research in China* 5(2), 4–6.

Yoshida, S. (1972) Physiological aspects of grain yield. *Annual Review of Plant Physiology* 23, 437–464.

Yoshida, S. (1976) Physiological consequences of altering plant type and maturity. In: *Proceedings of the International Rice Research Conference.* International Rice Research Institute, Los Baños, Philippines, p. 268.

Yoshida, S. (1981) *Fundamentals of Rice Crop Science.* International Rice Research Institute, Philippines.

Yuan, L.P., Virmani, S.S. and Mao, C.X. (1989) Hybrid rice – achievements and future outlook. In: *Progress in Irrigated Rice Research.* International Rice Research Institute, Los Baños, Philippines, pp. 219–223.

Yuan, L.P., Yang, Z.Y. and Yang, J.B. (1992) Hybrid rice in China. Paper presented to the Second International Symposium on Hybrid Rice, IRRI, Los Baños, 21–25 April 1992.

5 The Economic Principles of Research Resource Allocation

R.E. Evenson
Economic Growth Center, Yale University, 27 Hillhouse Avenue, New Haven, CT 06520, USA

I. Introduction

In this chapter we state some of the principles of resource allocation derived from economic theory to guide priority-setting exercises. Chapter 6 is more specific in translating these principles into practices. Both Chapters 5 and 6 and the applications reported in Chapters 21 and 22 are motivated by the importance of economic efficiency in the allocation of rice research resources.

The 'economic theory of the firm' provides the basic guidelines for achieving economic efficiency. This theory is useful in terms of providing conditions both for *cost efficiency*, i.e. the rules for cost-minimizing input or resource choice, and for *allocative efficiency*, i.e. the rules for choosing the value-maximizing set of outputs of the firm. It is also possible to consider interlinkages between firms and to consider the rules for *comparative advantage* as well. This chapter seeks to develop these rules and discuss their application to rice research prioritization.

Although research organizations are not firms in the ordinary sense, standard economic rules can describe how they should behave if they are to make best use of the resources available to them. There are two contentions raised to distinguish between research organizations and ordinary firms. The first is that most rice research organizations are not profit-making organizations and that economic rules derived for profit-maximizing conditions are thus not applicable. The second is that the activities of a research organization are subject to such a high level of uncertainty as to success in achieving an outcome, that standard production ideas do not apply.

The first of these objections to the application of economic efficiency rules is easily dealt with. Economic efficiency, as noted above, has two parts. The first part is cost efficiency and cost-minimization rules govern this part. All producing organizations, whether public or private and whether they are profit-making or not, have sound reasons to minimize cost, i.e. to produce whatever they produce using 'best practices' or the best technology and the most efficient mix of resources or inputs. The second part is allocative efficiency – choosing the products. Economic rules for choosing products are not confined to profit-maximizing rules. A research program may chose to maximize its contribution to the utility of a particular client group, e.g. small farmers or poor consumers, or both. There are economic rules for these cases just as there are for profit maximization (and they are often the same rules).

The second objection is that research is different from producing ordinary goods and services. Much research is a trial and error process of searching for particular outcomes. The basic objective of many research projects is to produce an 'invention'. Inventions are differentiated from normal discoveries in that they entail an inventive step. The legal definition of an inventive step is that it is 'unobvious' to a practitioner of the art. It is true that research activities are subject to inventiveness or unobviousness and thus have uncertain outcomes. But do these properties remove them from the pale of economic analysis?

Uncertainty itself is not a basis for repealing the rules of economic efficiency. Many economic models deal with uncertainty and with risk-related behaviour by individuals. Inventiveness (unobviousness) is also not beyond the pale of economics, at least if it can be meaningfully captured in the conceptualization of search (including sophisticated search) behaviour. It is argued in this chapter (and elsewhere in the economic literature) that a search foundation of the research production process is a meaningful way to model the research organization's economic problem. If so, we can apply economic principles to the research priority-setting process.

Part II of this chapter reviews the basic rules for economic efficiency. Part III develops a search model to characterize research activities and to clarify interorganizational germplasmic exchanges. Part IV develops a model of three hierarchical research organizations and shows how one can derive resource-allocation rules and how comparative-advantage principles apply to research organizations. Part V discusses international dimensions of rice breeding to illustrate the relevance of the analysis to rice research.

II. The Basic Economic Rules

A. *Simple cost minimization*

First, consider the rules for cost minimization for a producing firm. Cost-minimizing rules hold for any level or mix of production or outputs. Consider the simplest case of one product produced by two factors (or inputs). Since this is a research firm, consider the product to be inventions of improved products or processes for rice farmers. For the moment, consider this to be characterized by function 5.1 (in part III this function is derived from a search process):

$$I = F(R, G). \tag{5.1}$$

This describes a production process where inventions (I) are produced over a specified time period with research effort (R) and a stock of 'invention germplasm' (G). The invention germplasm includes genetic resources (see part V) as well as 'intellectual germplasm' in the form of methodologies, scientific equipment and experimental plots. Note that G may be produced in other organizations and, as will be noted later, may actually be a by-product of past invention in another research organization.

The firm's costs, given that it uses 'best practice' technology (i.e. actually produces according to eqn 5.1) are:*

$$C = P_r R + P_g G \qquad (5.2)$$

where P_r and P_g are prices per unit of R and G. (In practice, G may not be marketed, but for present purposes, suppose it is.[†]) Then for any level of inventions output, I (say I_o), the firm will wish to minimize cost (C). The mathematics of this simple problem require a Lagrangian set-up:

$$L = P_r R + P_g G - \lambda(I_o - F(R, G)) \qquad (5.3)$$

where the term in parentheses is the output constant constraint. The first order conditions for a minimum are:[‡]

$$\delta L/\delta R = 0 = P_r - \lambda F_r$$
$$\delta L/\delta G = 0 = P_g - \lambda F_g \qquad (5.4)$$

where $F_r = \delta F/\delta R$ and $F_g = \delta F/\delta G$ are the marginal products of R and G. This implies the basic rule of cost minimization:

$$\lambda = P_r/F_r = P_g/F_g \qquad (5.5)$$

and this implies that

$$P_r/P_g = F_r/F_g. \qquad (5.6)$$

Cost-minimizing firms then use inputs or resources in such a way as to meet this rule. Since there are diminishing returns to each factor, by choosing the appropriate ratio of R to G they can meet this condition. When G is abundant, for example, P_g will be low. This will induce entrepreneurs to raise the ratio of F_r/F_g by using less R per unit of G.

---

* Note that we are ignoring costs of the externality type, i.e. costs borne by others.
† Intellectual Property Rights (see below) provided a means by which G and I may be marketed.
‡ The necessary condition for a minimum is that:

$$dL = \frac{\delta L}{\delta R} dR + \frac{\delta L}{\delta G} dG = 0.$$

This is satisfied for dR or dG not equal to zero only when:

$$\frac{\delta L}{\delta R} = \frac{\delta L}{\delta G} = 0.$$

The sufficient condition is $d^2 L > 0$.

The term λ in eqn 5.3 (the Lagrangian multiplier) is actually the marginal cost of producing I, given that the enterprise is both technically and cost-efficient.*

B. Profit maximization

So far we have not developed a rule for the optimal production of the product, I. The cost-minimization rules hold for any level of I.

But now suppose the firm is maximizing profits where profits are:

$$\Pi = P_I I - P_r R - P_G G = P_I F(R, G) - P_r R - P_g G. \tag{5.7}$$

The conditions for maximizing are (note that there are no constraints on I, R and G here, hence no Lagrangian expression is needed):

$$\begin{aligned} \delta\Pi/\delta R &= 0 = P_I F_r - P_r = 0 \\ \delta\Pi/\delta G &= 0 = P_I F_g - P_g = 0. \end{aligned} \tag{5.8}$$

These conditions imply:

$$\begin{aligned} P_r &= P_I F_r \\ P_g &= P_I F_g \\ P_r/P_g &= F_r/F_g. \end{aligned} \tag{5.9}$$

Thus profit maximization implies cost minimization plus the condition that the value of the marginal product (VMP) for each factor is equal to the price of the factor. It is also readily obvious by comparing eqns 5.6 and 5.9 that $\lambda = P_I$, i.e. that the price of output is equal to the marginal cost of producing that output.

These are elementary and basic economic ideas. But they do have implications for research prioritization. It first should be noted that a research organization, if it is to make best use of the resources at its disposal, typically maximizes something like profit even if it is not set up to make profits or to sell its product. It produces inventions, I, and these have some value, V_I, to the clientele that the organization serves. The organization faces prices P_r and perhaps also P_g, and wishes to maximize the value of its product as judged by its clientele, even if there is no market for I; hence it will face the same economic problems as the profit-maximizing firm.

---

* We can see this by substitution of the first order conditions (eqn 5.4) into the differential of eqn 5.1 with respect to I:

$$\frac{dI}{dI} = 1 = F_R \frac{dR}{dI} + F_G \frac{dG}{dI},$$

substituting we obtain:

$$\lambda = P_R \frac{dR}{dI} + P_G \frac{dG}{dI} = \frac{dC}{dI}.$$

C. Aggregate conditions for economic efficiency

The general conditions for economic efficiency for an economy that wishes to achieve the maximum value (as expressed by its consumers, given their incomes) from the resources in the economy are straightforward extensions of the above rules for individual enterprises. In fact we need only add markets to the system. Markets ensure that all purchasers of the resources R and G will pay a common market price for them. They also ensure that all producers of a good will receive a common price for the goods. And, in the long run, price relationships between goods will be governed by costs of production.

Perhaps the most important proposition to come from these conditions is the value of marginal product rule. Suppose a second research product is produced by another enterprise (say I_1 and I_2). Both enterprises will meet the VMP rule for research:

$$P_{r1} = P_{I1}F_{r1} = VMP_{r1}$$
$$P_{r2} = P_{I2}F_{r2} = VMP_{r2}$$
(5.10)

and if $P_{r1} = P_{r2}$ then $VMP_{r1} = VMP_{r2}$.

This principle provides simple rules for allocation of research between two research problems areas (RPAs), as shown below. Indeed it holds for any pair of RPAs in any arbitrary number of research programs. But to be meaningful, a convincing invention production function must be developed. Fortunately such a function can be developed and it has remarkably simple properties that enable us to develop simple research allocation rules.

III. Invention Production Functions

In this section, we develop an invention production function of the following general form:

$$E(I) = \alpha(G) + \lambda \ln(R).$$
(5.11)

This expression states that expected inventions, $E(I)$, are simply related to G and $\ln(R)$. This function can be derived from simple probability principles or from more sophisticated 'order statistics' principles. Note that it incorporates germplasm (G) into the expression in a simple way (this will be modified shortly). Note also that it has a specific functional form with a simple marginal product:

$$\delta E(I)/\delta R = \lambda/R.$$
(5.12)

There are two approaches to deriving eqn 5.11. The first relies on simple probability logic in a search process. We begin by noting that search is an activity that is consistent with both uncertainty and inventions (unobvious discovery).

Consider the following logic. A 'pool' of potential inventions exist. Suppose each invention has a value index, X_i. The inventor draws X_is randomly from a distribution $f(x)$.

Define the variable Y_i as equal to one if the draw X_i exceeds prior draws and zero if it does not. This is the basic definition of a real invention, i.e. it must be an improvement over prior art. Because of independent sampling, each of the random variables, $X_1, X_2, \ldots X_i$, is equally likely to be the largest.

Thus the probability that the ith draw will be a successful invention is a Bernouli trial:

$$E(Y_i) = \text{Prob}(Y_i = 1) = 1/i. \tag{5.13}$$

The X_i draws are a sequence of independent 'Bernouli' trials. The probability that $Y_n = 1$ depends on the absolute value of previous X_i, but knowledge of the ordinal ranking provides no information about the likely value of X_n. Accordingly, Y_i and Y_j are uncorrelated. Now define:

$$Z_n = \sum_{i=1}^{n} Y_i \tag{5.14}$$

Where Z_n is the cumulated inventions realized by the nth draw or search.

It follows that:

$$E(Z_n) = \sum_{i=1}^{n} \frac{1}{i}$$
$$\text{Var}(Z_n) = \sum_{i=1}^{n} \frac{1}{i} - \sum i = 1^n \frac{1}{i^2}. \tag{5.15}$$

This then is the invention function for a given pool of potential inventions. And it is the appropriate invention function for any distribution $f(x)$.

The expressions in eqn 5.15 can be approximated by:

$$E(Z_n) = \ln(n) + 0.5772$$
$$\text{Var}(Z_n) = \ln(n) - \frac{\Pi^2}{6} = \ln(n) - 1.0677. \tag{5.16}$$

This approximation is accurate even when n is small.

Evenson and Kislev (1975) derived a similar expression utilizing the statistical theory of extreme values (order statistics) applied to an exponential function, as follows. The statistical theory of extreme values states that the expected maximum value of a sample of size n drawn from a distribution $f(x)$ is:

$$H_n(z) = n F^{n-1}(z) f(z). \tag{5.17}$$

Applied to the exponential function $(\lambda e^{-\lambda(X-G)})$ the probability density function of extreme values is:

$$h_n(z) = \lambda n \left[1 - e^{-\lambda(z-)}\right]^{n-1} e^{-\lambda(z-)} \tag{5.18}$$

and the expected value and variance are:

$$E_n(Z) \approx E(Z_n) = \theta + \frac{1}{\lambda}\sum_{i=1}^{n}\frac{1}{i} \approx \theta' + \frac{1}{\lambda}\ln(n) \quad (5.19)$$

$$\text{Var}_n(Z) = \text{Var}(Z_n) = \frac{1}{\lambda^2}\sum_{i=1}^{n}\frac{1}{i^2}. \quad (5.20)$$

The similarities with eqn 5.16 are obvious. These results hold the pool, $f(x)$, constant. Germplasm, G, effectively determines $f(x)$, thus changes in germplasm will shift $f(x)$ Germplasm can be defined as being of two types. One type produces adaptations of inventions made elsewhere. This is *adaptive* germplasm and is linked to past inventions in other pools. A second type is more fundamental; this is *pool-expanding* germplasm. This is the germplasm from 'upstream' scientific research producing new tools and procedures, as well as research leading to expanding the raw materials – the building blocks of invention.* Access to a larger pool of genetic resources, to better techniques for genetic transformation, better tests for genetic resistance, molecular markers and gene maps are also pool-expanding germplasm or lead directly to pool-expanding germplasm.

IV. A Multi-Research Problem Area, Multi-Organization Model

We now have the relevant analytics in hand to develop economic efficiency (hence research priority) rules in a reasonably realistic multi-RPA, multi-organization model. This model will illuminate the nature of comparative advantage as well as the priority-setting rules (guidelines).

We will have three research institutions; each with two RPAs. The first research institution is an international institute (e.g. the IRRI). This organization is primarily a germplasm supplier and it produces two types of pre-invention germplasm, each targeted to an RPA.† Its discovery functions can be described as:

$$\begin{aligned} G_1 &= \Theta_1 + \gamma_{12}G_2 + \lambda_1 \ln(R_{g1}) \\ G_2 &= \Theta_2 + \gamma_{21}G_1 + \lambda_2 \ln(R_{g2}). \end{aligned} \quad (5.21)$$

Note here that we are allowing for cross-program germplasmic effects through the γ_{12} and γ_{21} terms. That is, germplasm for one program may expand the pool of the other. The Θ_1, Θ_2 terms reflect the contributions of the higher sciences.

---

* One could think of the adaptive germplasm as 'raw materials' used in production, while pool-expanding germplasm is a type of capital good (see Chapters 21 and 22).
† Germplasmic research is sometimes termed 'strategic' research.

The second research institution is a national research institute. Its chief objective is to produce inventions, I_1 and I_2, for each of the RPAs. Its invention functions can be stated as:

$$I_1 = \phi_1 + (\alpha_1 + \alpha_{12} I_2 + \delta_1 G_1) \ln(R_{I1})$$
$$I_2 = \phi_2 + (\alpha_2 + \alpha_{21} I_1 + \delta_2 G_2) \ln(R_{I2}). \tag{5.22}$$

Note here that both forms of germplasm enter these invention functions. One is the pre-invention and pool-expanding germplasm from above, i.e. from the first institution. G_1 and G_2 are specified to shift the search distributions for I_1 and I_2. The second form is the invention 'disclosure' germplasm reflected in α_{12} and α_{21} and this is specified to affect discovery through adaptive invention. Thus inventions in RPA$_1$ may be adapted by or stimulate inventions in RPA$_2$ through this mechanism.

The third research institution is an institution that is subordinate to the second (which, in turn, is subordinate to the first). It seeks only adaptive sub-inventions. Its invention functions can be stated as:

$$A_1 = \beta_1 + \Omega_1 I_1 \ln(R_{A1})$$
$$A_2 = \beta_2 + \Omega_2 I_2 \ln(R_{A2}). \tag{5.23}$$

These sub-inventions are tailored or 'fine-tuned' to specific ecological and economic conditions in the subordinate region. Hence, they are a type of 'end of the chain' invention that does not have a further by-product germplasm effect. Inventions from the higher level institutions provide the only germplasm here. These institutions benefit only indirectly from the G_1 and G_2 germplasm.*

We can now note that we have depicted a hierarchy of research programs with the third type being subordinate to the second type which is subordinate to the first type. This is a reasonable approximation of the international rice research system. There is one international institute (the IRRI), many national agriculture research (NAR) programs and, subordinate to them, many regional and substation programs. The factors determining the number of subordinate programs include both political (national boundaries) and ecosystem factors.

We can now proceed to examine the value of marginal product rules. The value of marginal product rules are derived in the Appendix. Here we will examine the ratio equations for each of the three types of research programs using the property that the prices of research resources in RPA$_1$ and RPA$_2$ in each program are equal. This means that the value of marginal products can be set to equal each other. From this condition the ratios of optimal research can be derived.

---

* This structure of research institutions is 'top–down' in one sense, but not in another. Most adaptive inventions are valuable in a limited environmental setting. They thus are usually not valuable in the more aggregate environments for which the adapted inventions were suited.

Begin with the adaptive invention programs. The ratio rule from this program is:

$$\frac{R_{A1}}{R_{A2}} = \frac{V_{A1}\Omega_1 I_1}{V_{A2}\Omega_2 I_2}. \tag{5.24}$$

This rule tells us that the optimal allocation of research resources in subordinate units will be proportional to the economic value of the units affected, if $\Omega_1 I_1 = \Omega_2 I_2$. The economic value is $V_{A1} = A_1 U_1 P_1$, where A_1 is the number of inventions, U_1 the number of units affected per invention and P_1 the price of each unit (similarly for V_{A2}).

The units, U_1 and U_2, can be affected by the location of subordinate programs. If it were the case that the ratio of subordinate units was proportional to the base invention units for each, i.e.

$$\frac{V_{A1}}{V_1} = \frac{V_{A2}}{V_2} \text{ or } \frac{V_{A1}}{V_{A2}} = \frac{U_1}{U_2} = \frac{V_1}{V_2} \text{ and } \frac{V_1}{V_2} = \frac{I_1}{I_2}, \tag{5.25}$$

then optimal research resource allocation will be proportional to economic units. This is the *congruence rule*, i.e. the rule that research resources should be congruent with economic value (as long as sub-invention productivity is equal (i.e. $\Omega_1 = \Omega_2$)).

We now turn to the research resource-allocation rule at the base invention level. Setting $\text{VMPR}_{I1} = \text{VMPR}_{I2}$ we obtain:

$$\frac{R_{I1}}{R_{I2}} = \frac{(\alpha_1 + \alpha_{12} I_2 + \delta_1 G_1)(V_1(1 + \Omega_1 R^*_{A1}) + \alpha_{21} \ln(R_{I2})V_2}{(\alpha_2 + \alpha_{21} I_1 + \delta_2 G_2)(V_2(1 + \Omega_2 R^*_{A2}) + \alpha_{12} \ln(R_{I1})V_1} \tag{5.26}$$

where $V_1 = \Sigma_A V_{A1}$, $V_2 = \Sigma_A V_{A2}$, $R^*_{A1} = \Sigma + A \ln(R_{A1})$ and $R^*_{A2} = \Sigma_A \ln(R_{A2})$.

Inspection of eqn 5.26 will show that the congruence rule applies if the following conditions are met:

1. $\alpha_{12} = \alpha_{21} = 0$, i.e. no cross-program adaptive disclosure invention effects are present,
2. congruence also holds at the sub-invention level,
3. $\alpha_1 = \alpha_2$, i.e. the basic productivity of research in the two programs is equal, and
4. $\delta_1 G_1 = \delta_2 G_2$, the germplasm flows and effects are equal.

We have already noted that congruence holds at the sub-invention level if it does at the base invention level and if ecosystem distribution is similar. Thus, as a working proposition, the congruence rule is a good place to start the consideration of research resource allocation.*

Now consider departures from the congruence rule. They include:

1. Cross-RPA invention disclosure effects, α_{12} and α_{21}.
2. Sub-invention ecosystem differences, R^*_{A1} and R^*_{A2}.

---

* Note that the congruence rule is not an *ad hoc* rule (a step above rules of thumb); it is a rigorously derived rule.

3. Research productivity effects, α_1 and α_2.
4. Scale economies.
5. Upstream germplasm effects, $\delta_1 G_1$ and $\delta_2 G_2$.

Sub-invention differences are likely to be relatively small. This is because, even if there are ecosystem distribution differences, they largely impact on the resources allocated to the subregions.

The validity of the condition that there are no cross-field invention disclosure effects will depend on the nature of the two research programs and on institutions to facilitate such effects. For some RPA comparisons these effects will be small. For others, they will be considerable and they will be facilitated by international programs. For example, if RP_1 and RP_2 are two national rice-breeding programs, the evidence (see below) is quite strong that inventions in one program can 'parent' inventions in another. It is also the case that the IRRI's research program has increased these flows (i.e. it has caused an increase in α_{21} and α_{12}).

Inspection of eqn 5.26 also shows that if $\alpha_{21} = \alpha_{12}$ and both are greater than zero, optimal research allocation will move in the direction of equality between the programs. If $V_1 > V_2$ as α_{21} ($=\alpha_{12}$) rises, R_{I1}/V_1 will fall relative to R_{I2}/V_2.

On the other hand, if these effects are asymmetric ($\alpha_{21} > \alpha_{12}$) the optimal allocation between the two will move in the direction of the 'stronger' program. These effects will be strengthened; the more important are the subordinate programs.

Perhaps the most valid reason for departure from congruence, however, is that the germplasm bases are not equal, i.e. $\delta_1 G_1 \neq \delta_2 G_2$ (condition 1). A reading of rice-breeding history indicates that the germplasm base differs considerably by ecosystem focus. With the development of the semidwarf plant type for rice and with the development of genetic resource stocks the germplasm base for rice in irrigated conditions was substantially better than for other ecosystem conditions (especially upland and deepwater). It remains so today.

Expression 5.26 also tells us that unless the cross-program invention by-product germplasm is significant, one will not solve a low germplasm situation through invention research. More germplasm must be produced.

The scale economies effect may be important here for the research unit. The marginal product rule applies to the marginal search. There are fixed costs and fixed resources in research programs – and these will have some effect on resource allocation. The more units that they can be spread over, the more efficient the program. But this does not mean that at the margin more should be spent on a research program that is efficient in that sense. In fact, average spending per economic unit should be lower as a result. Accordingly, a scale parameter could be incorporated into the resource allocation rule.*

---

* There may be a minimum scale for efficient research programs.

Now consider the rule for the allocation of germplasm research:

$$\frac{R_{G1}}{R_{G2}} = \frac{\lambda_1 \delta_1 V_1^* + \lambda_1 \gamma_{12} \delta_2 V_2^*}{\lambda_2 \delta_2 V_2^* + \lambda_2 \gamma_{21} \delta_1 V_1^*} \tag{5.27}$$

where $V_1^* = \Sigma_K V_{1K}$ and $V_2^* = \Sigma_K V_{2K}$, i.e. where these values are the values of all programs to which germplasm flows.

We note again that if the following conditions are met we have the congruence rule at this level:

1. $\lambda_1 = \lambda_2$,
2. $\delta_1 = \delta_2$, and
3. $\gamma_{12} = \gamma_{21} = 0$.

Expression 5.27 thus has implications, both for the international centre (the IRRI) and for its targeting to applied research clients. Its major message is similar to that for eqn 5.25. Under certain circumstances it should be guided by target congruence – both in terms of its aggregate program emphasis and in terms of targeting its germplasm effects to clients.

Departure from congruence is associated with the γ_{21} and γ_{12} parameters, the λ_1 and λ_2 parameters and the δ_1 and δ_2 parameters. As with α_{12} and α_{21} in eqn 5.22, if γ_{21} and γ_{12} are high, this will move research allocation away from congruence toward equality of resource allocation. If one program (say a biotechnology tool program) is particularly strong ($\gamma_{12} > \gamma_{21}$), that program should receive more emphasis.

The parameters, λ_1 and λ_2, matter because they govern the basic productivity of germplasm research. They can be regarded to be given by higher order science and by nature. Biotechnology tools and instruments are part of λ_1 and λ_2. Germplasm-producing institutes sometimes have to produce their own λs. Their absorptive capacity is important, just as is the absorptive capacity of their clients. Clearly, if one field of science moves ahead of another, these germplasm producers should be investing more in the field with the better λs.*

The absorptive capacities of national program clients (i.e. δ_1 and δ_2) matter in terms of regional emphasis. It may seem counterintuitive to some, but the model shows that research impact will be maximized by favouring the programs with the best absorptive capacity. But this depends on whether the international centre sees its comparative advantage strictly as a germplasm supplier (see below) or as a producer of inventions.

To summarize then, the allocative efficiency rules described here should guide thinking about resource allocation. The congruence conditions are a useful starting point for priority setting. These depend on good measures of the units (V_1 and V_2) and these are related to the demand side of the model (see Chapter 4 for modifications about demand for equity, sustainability, etc.). But then the priority setter has to recognize that the supply side parameters discussed in this section are the basis for departures from congruence.

---

* As noted in several chapters, rice research for irrigated ecosystems has had superior λs.

A. A note on comparative advantage

The model developed here presumed a hierarchy from international to national to regional and on associated specialization, germplasm inventions to adaptive inventions. This implied specialization in strategic research (R_{G1}, R_{G2}), applied research (R_{I1}, R_{I2}) and adaptive research (R_{A1}, R_{A2}); and this implies comparative advantage. Is this justified? Particularly in the light of the 'mandate' that many International Agricultural Research Centers (IARCs) (including the IRRI) have?

The comparative advantage implied by the model is essentially based on the 'summation signs' in the model. A research unit which serves other units serves from above, i.e. it does not compete with its clients, it complements them. And to do so, it must provide germplasm in one form or another. Its comparative advantage is based on scale. Its research affects multiple clients. Thus, the model clearly indicates that the IRRI (and other IARCs) has a comparative advantage in, and a clear responsibility for, the provision of germplasm (in many forms) to national programs. It has a comparative-advantage responsibility as the gatekeeper for the higher sciences (θ_1 and θ_2) and for developing tools and methods.

This conclusion is not inconsistent with a mission of producing inventions in the IARCs because inventions can serve as germplasm. There may be a case for some compensatory behaviour on the part of IARCs, i.e. providing inventions for weak (or non-existing) national programs. But these activities are secondary to the larger comparative-advantage responsibilities of the IARCs.

It is probably the case that most IARCs do not see the comparative advantage responsibilities outlined here. They were as Alex McCalla has noted 'set up to use science, not to do science' (A. McCalla, unpublished comment) and all have a strong applied-research direct-service culture. There is also a strong compensatory motive alive in the IARCs.

But the fact that the designers of the system (and its donors) did not fully appreciate the nature of comparative advantage or the fact that the 'culture' on these systems is not fully consistent with the comparative advantage does not mean that it does not exist. The overall contribution of the international agricultural research system is maximized when economic efficiency, including realizing comparative advantage, is realized.

V. Rice Research and Economic Modelling

Is the economic model sketched out here relevant for rice research?

- Do hierarchical organizers exist?
- Is there evidence for germplasmic effects?
- Does the IRRI exhibit comparative advantages?
- Are RPAs related to economic units (V_1 and V_2)?

Clearly hierarchical organizations exist. The IRRI has an international mandate and NARs are perceived to be the IRRI's 'clientele' organizations. NARs differ in strength from country to country. India clearly exhibits a hierarchy between national programs and state and regional programs. And this is the case for China, Pakistan and a number of other countries as well.

The evidence for germplasmic effects can be found in a number of places. But, it is nowhere so clear as it is for rice-breeding programs. Table 5.1 illustrates the point. The table reports the number of released varieties from national rice-breeding programs for 1965–1992 by origin of the cross from which the variety was selected. Of the 1709 varieties released by these programs, 294 (17%) originated from the IRRI and an additional 96 (6%) originated from other national programs.

Of perhaps more relevance are the data for varietal parents. Nearly three-quarters of the released varieties have at least one parent that was the product of a cross made outside the releasing countries. Eight hundred and ten (47%) had a parent from the IRRI and 519 (36%) had a parent from another national program. For all varietal ancestors the proportions were higher. Nearly three-quarters of all successful varietal development since 1965 have at least one ancestor contributed by the IRRI.

This clearly attests to the germplasmic role of the IRRI and, for that matter, of inter-program invention by-product germplasm. In this case the germplasm is literally genetic resource germplasm, but this same pattern would hold for other forms of germplasm as well.

Table 5.2 provides additional insight into institutions that facilitate germplasmic exchange in rice research. In 1975, the IRRI established the International Rice Testing Program, later renamed the International Network for the Genetic Evaluation of Rice (INGER). Table 5.2 reports the frequency of inclusion of released varieties in the various INGER nurseries. In addition, the incidence of variety (or lines) as ancestors is noted.

These data show that while a number of released varieties have appeared in yield nurseries, relatively few ancestors of released varieties have appeared in yield nurseries. Screening nurseries and observational nurseries for particular traits on the other hand have been important sources of ancestral material.

INGER has played an important role in facilitating the international exchange of cultivars that ultimately served as ancestors to released varieties. Annual varietal releases in the 1971–1975 period were roughly 60 per year from the 40 releasing countries. After 1975, this rose to 80 per year. Of these 80, 10–15 had appeared in INGER (10 or so were IRRI crosses first placed in INGER). But INGER was also the source for 210 parental and grandparental cultivars by the late 1980s. On the basis of a statistical model (Gollin and Evenson, 1996) it has been concluded that the INGER system had increased the annual flow of released varieties in the late 1980s by approximately 25% over what it would have been in the absence of INGER. This evidence shows how international programs such as INGER have increased the germplasm flows from the IRRI (G_1 and G_2), but perhaps more importantly, have increased the flows of invention by-product germplasm (the α_{12} and α_{21} terms) when RPAs are treated as different national breeding programs.

Table 5.1. International flows of rice varieties and parents.

Origin	IRRI	CIAT	IITA	Brazil	Mexico	Other Latin American countries	Africa	Bangladesh	India	Sri Lanka	Nepal
International centres											
IRRI	7(8)								0(3)	0(1)	
CIAT	1(18)	13(3)		2(0)		1			0(2)	0(1)	
IITA	5(13)		32(6)	0(1)					0(33)	1(3)	
USA	2(6)			1(3)			1(5)		0(15)	0(1)	
Oceania	5(5)								0(2)		
Latin America											
Brazil	12(45)	2(7)	0(1)	74(29)		1(2)			2(16)	0(4)	
Mexico	8(12)	0(2)			27(5)		0(3)		0(49)	0(1)	
Other	19(89)	3(11)		4(6)		79(10)			0(13)	1(0)	
Africa	22(25)	1(0)	1(0)	1(2)	1(2)	1(2)	48(15)	48(15)	2(39)	2(39)	2(39)
South Asia											
Bangladesh	11(34)							17(1)	4(18)		
India	55(363)		0(1)			0(1)	0(3)	0(3)	623(711)	5(10)	
Sri Lanka	2(23)								0(4)	55(52)	
Nepal	5(10)									1(2)	10(1)
Pakistan	8(13)						0(1)	0(1)			
Southeast Asia									2(9)		
Burma	21(40)								0(12)		
Indonesia	19(47)							1(0)	1(72)	2(0)	
Philippines	26(47)								0(27)		
Thailand	8(8)								0(11)		
Vietnam	44(64)								0(14)		
Other	7(19)								0(36)	0(1)	0(1)
East Asia									1(13)		
China	13(76)								1(46)		
Japan	0(0)									2(2)	2(2)
South Korea	1(54)								0(7)		
Taiwan	0(5)								0(3)		

Table 5.1 (continued)

Origin	Pakistan	Burma	Indonesia	Philippines	Thailand	Vietnam	Other Southeast Asian countries	China	Japan	Korea	Taiwan
International centres											
IRRI										1(0)	
CIAT											
IITA											
USA				0(1)	0(2)	0(2)	0(2)			0(1)	0(1)
Oceania					0(2)						0(1)
Latin America											
Brazil				0(6)	0(3)	0(1)					0(3)
Mexico			0(1)		0(6)						0(2)
Other			0(6)		0(2)	0(3)					0(5)
Africa				0(2)	0(5)	0(2)				0(3)	0(10)
			0(6)	1(3)			0(4)	0(2)			
South Asia											
Bangladesh			0(2)	0(2)		0(1)	1(2)	0(1)			0(1)
India			1(15)	0(17)	1(7)	1(0)	0(26)	1(10)	0(2)		2(59)
Sri Lanka			0(3)						0(1)		0(1)
Nepal			0(2)	0(2)			0(2)	0(1)			
Pakistan	8(0)			0(2)	0(2)			0(2)			
Southeast Asia											
Burma		34(2)	1(5)	1(7)	12(0)	2(3)	0(1)	0(1)			0(2)
Indonesia			37(23)		0(2)		0(1)	0(1)			
Philippines			1(5)	27(18)	25(11)	1(0)	0(2)		0(1)		
Thailand				0(1)							
Vietnam			0(3)	0(1)		25(7)	0(2)	0(2)			0(1)
Other			0(5)	0(3)	0(2)		27(13)				0(2)
East Asia											
China			0(2)					87(26)	26(37)	2(1)	0(1)
Japan									1(13)	0(1)	0(1)
South Korea								0(1)	0(1)	117(76)	
Taiwan											19(50)

Number of parents by origin are in parentheses.
Source: IRRI varietal release data; David and Evenson, 1993.

Table 5.2. International Network for the Genetic Evaluation of Rice (INGER) nurseries as indicators of characteristics of varieties in datasets.

Nursery name			Presence in entry	Presence in ancestry
IRYN-M	Rice yield – medium	Yield	44	5
IRYN-L	Rice yield – late	Yield	74	6
IURYN	Upland rice yield	Yield	9	2
IRYN-VE	Rice yield – very early	Yield	27	2
IRRESWYN	Rainfed shallow water	Yield	21	–
IURYN-E	Upland yield – early	Yield	18	–
IURYN-M	Upland yield – medium	Yield	14	–
IRDYN	Deepwater yield	Yield	14	–
IRON	Rice observational nursery	Obs.	12	–
IRGON	Rice grain observational	Obs.	231	87
IRDTN	Drought tolerance	Obs.	8	–
IRSATON	Salinity and alkalinity tolerance	Obs.	25	–
IRCTN	Cold tolerance	Obs.	118	5
IURON	Upland observational	Obs.	75	131
ORDON	Rice deepwater observational	Obs.	164	36
IRRSWON	Rainfed shallow water	Obs.	31	18
IRARON	Arid region observational	Obs.	134	34
IRBN	Blast	Obs.	46	28
IRTN	Tungro	Screen	281	152
ISHBN	Sheath blight	Screen	165	56
IRBPHN	Brown planthopper	Screen	47	260
IRSBN	Stem borer	Screen	168	107
IRGMN	Gall midge	Screen	82	157
IRWBPHN	Whitebacked planthopper	Screen	67	7
FLDTOL	Flood tolerance	Screen	99	7
MEDDEEP	Medium deepwater	Screen	12	2
IFRON	Floating rice observational	Obs.	11	–
ITPRON	Tidal prone observational	Obs.	11	–
IRONTOX	Iron toxicity	Screen	29	–
ACDSUF	Acid sulphate	Screen	35	1
ACDUPL	Acid uplands	Screen	45	1
PEAT SOILS	Peat soils	Screen	67	–
ACID LOW	Acid lowlands	Screen	21	4
BTRSPOT	Brown spot	Screen	68	1
CERCOSP	Cercospora	Screen	25	–
SHROT	Sheath rot	Screen	7	–
IRBBN	Bacterial blight	Screen	7	–
RAG STUNT	Ragged stunt	Screen	161	6
LEAF FOLD	Leaf folder	Screen	7	45
LEAF SCALD	Leaf scald	Screen	3	41
YSB	Yellow stemborer	Screen	20	1
THRIPS	Thrips	Screen	9	54
GLH	Green leafhopper	Screen	19	16
SSB	Striped stemborer	Screen	–	–
SBPH	Small brown planthopper	Screen	–	–
STINK	Stink bug	Screen	–	–
ZIGZAG	Zigzag leafhopper	Screen	–	–
IRRSWYN-E	Rainfed shallow water – early	Yield	–	–
IRRSWYN-M	Rainfed shallow water – medium	Yield	9	–
IRSTYN	Salinity tolerance	Obs.	12	–

Obs., observational.
Source: Gollin and Evenson (1996).

The development of biotechnology techniques has not materially altered the nature of RPAs and rice research objectives (see Chapter 21). Genetic improvement continues to seek objectives as reflected in Table 5.2. Indeed Table 5.2 offers a useful guide to the definition of RPAs in rice (see Chapters 7–14 on the measurement of crop losses from RPA-associated stresses).

Throughout the chapters in this book, the contributions from economic theory will be apparent in the interpretation of data. Theory has also influenced the practical problem of priority setting to which chapter 6 is addressed.

References

David, C. and Evenson, R.E. (1993) *Adjustment and Technology: The Case of Rice.* OECD, Paris.

Evenson, R.E. and Kislev, Y. (1975) *Agricultural Research and Productivity.* Yale University Press, New Haven.

Gollin, D. and Evenson, R.E. (1996) Genetic resources, international organization and rice varietal improvement. *Economic Development and Cultural Change* Vol. 44, No. 4.

APPENDIX 5.1: Computation of Marginal Products

From eqn 5.23:

$$VMPR_{A1} = V_{A1}(\delta A_1/\delta R_{A1}) = V_{A1}\Omega_1 I_1/R_{A1}$$
$$VMPR_{A2} = V_{A2}(\delta A_1/\delta R_{A2}) = V_{A2}\Omega_2 I_2/R_{A2}.$$

This provides the first rule (eqn 5.24):

$$\frac{R_{A1}}{R_{A2}} = \frac{V_{A1}\Omega_1 I_1}{V_{A2}\Omega_2 I_2}.$$

Next from eqn 5.20:

$$VMPR_{I1} = V_1(\delta I_1/\delta R_{A1}) + V_2(\delta I_2/\delta I_1)(\delta I_1/\delta R_{I1}) + \Sigma V_{A1}(\delta A_1/\delta I_1)(\delta I_1/\delta R_{I1})$$
$$= [(\alpha_1 + \alpha_{12}I_2 + \delta_1 G_1)/R_{I1}][V_1 + V_2\alpha_{21}(\ln R_{I2}) + \Sigma V_{A1}\Omega_1 R_{A1}]$$
$$VMPR_{I2} = V_2(\delta I_2/\delta R_{A2}) + V_1(\delta I_1/\delta I_2)(\delta I_2/\delta R_{I2}) + \Sigma V_{A2}(\delta A_2/\delta I_2)(\delta I_2/\delta R_{I2})$$
$$= [(\alpha_2 + \alpha_{12}I_2 + \delta_2 G_2)/R_{I2}][V_2 + V_1\alpha_{12}(\ln R_{I2}) + \Sigma V_{A2}\Omega_2 R_{A2}].$$

This provides rule 5.26:
$$\frac{R_{I1}}{R_{I2}} = \frac{(\alpha_1 + \alpha_{12}I_2 + \delta_1 G_1)(V_1(1+\Omega_1 R_{A1}^*) + \alpha_{21}\ln(R_{I2})V_2}{(\alpha_2 + \alpha_{12}I_1 + \delta_2 G_2)(V_2(1+\Omega_2 R_{A2}^*) + \alpha_{12}\ln(R_{I1})V_1}.$$

From eqn 5.19:
$$\begin{aligned}VMPR_{g1} &= \Sigma V_1(\delta I_1/\delta G_1)(\delta G_1/\delta R_{G1}) + \Sigma V_2(\delta I_2/\delta G_2)(\delta G_2/\delta G_1)(\delta G_1/\delta R_{G1}) \\ &= [\Sigma V_1\delta_1 \ln(R_{I1}) + \gamma_{12}\Sigma V_2\delta_2 \ln(R_{I1})]\lambda_1/R_{G1}\end{aligned}$$

$$\begin{aligned}VMPR_{g2} &= \Sigma V_2(\delta I_2/\delta G_2)(\delta G_2/\delta R_{G2}) + \Sigma V_1(\delta I_1/\delta G_1)(\delta G_1/\delta G_2)(\delta G_1/\delta R_{G1}) \\ &= [\Sigma V_2\delta_2 \ln(R_{I2}) + \gamma_{21}\Sigma V_1\delta_1 \ln(R_{I1})]\lambda_2/R_{G2}\end{aligned}$$

From eqn 5.19:
$$\frac{R_{G1}}{R_{G2}} = \frac{\lambda_1\delta_1\Sigma V_1 \ln(R_{I1}) + \lambda_1\gamma_{12}\delta_2\Sigma V_2 \ln(R_{I2})}{\lambda_2\delta_2\Sigma V_2 \ln(R_{I2}) + \lambda_2\gamma_{21}\delta_1\Sigma V_1 \ln(R_{I1})}.$$

6 Priority-Setting Methods

R.E. Evenson
Economic Growth Center, Yale University, 27 Hillhouse Avenue, New Haven, CT 06520, USA

I. Introduction

Research systems typically are organized with the research *project* as the basic performing unit. A set of projects is typically aggregated to form a research problem area (RPA). Projects may also be aggregated to program, departmental or divisional levels. RPAs may, and typically do, cross departmental or divisional lines.

Priority setting may be undertaken for RPAs and, if so, the appropriate definition of RPAs is critical. The actual performance of research itself may be governed by the project, department, administrative and management structure and these, of course, are important. But priority setting is not fundamentally a management or administrative mechanism. It is designated to inform and guide the administration, management and performance of research. It is not a substitute for administration and management reviews. Its chief purpose is to allocate resources, not to manage them.

Priority setting, like administration and management review is a never-ending process. It need not be undertaken on a continuous basis but it does require a periodic process because both the supply side of research (see Chapter 3) and the demand side for research are constantly changing. Not only should priority setting be a periodic dynamic exercise, but special precautions should be taken to ensure that it does not become a 'curtailing' activity by impeding the process of creative project design in research organizations.

This chapter discusses formal and less formal methods for priority setting. Since priority setting is closely related to *ex ante* project evaluation (designed for application at the project level, but sometimes applied to higher levels) much of the basic methodology is derived from project evaluation methods. Section II of this chapter is directed to a discussion of these methods. Section III deals with the appropriate definition of RPAs, including conformity to policy issues as well as to the use of *ex post* evidence in priority setting. It also discusses measurement problems regarding the 'demand side' of priority setting (this is particularly critical in view of the stress in Chapter 5 on congruency with demand factors).

Part IV discusses the use of crop-loss data and yield gap evidence as measures of research potential. Parts V and VI deal with the difficult supply side parameters that are critical to determining departures from congruence (see Chapter 5). Section V address this use of *ex post* evidence from supply parameters. Section VI addresses subjective probability estimates of supply side parameters. The final section of the chapter deals with the modification of the demand side variables to consider (or not to consider) such issues as sustainability, equity, gender and other development questions as well as the general issue of compensatory research *vis-à-vis* weak national programs and pioneering research in the spirit of producing germplasm for NARs.

II. *Ex Ante* Project Evaluation Principles

It is useful to begin this discussion of methods by reference to the most micro-resource allocation decision, i.e. the decision to invest in a specific project. We can think of this as a priority-setting decision even though we usually apply priority setting to a more aggregate level.

The steps in *ex ante* project evaluations are:

1. To specify the project in terms of objectives, techniques, inputs, costs and time.
2. To estimate the *benefits stream* associated with the project.
3. To make economic calculations: present value of benefits (PVB), present value of costs (PVC), benefit/cost (PVB/PVC) and internal rate of return (IRR).

Many projects are such that the component estimates can be based entirely on past experience. The technology of converting project inputs into project outputs on a given time schedule for some projects are well known (e.g. the building of a bridge). But for research projects this is not the case. One of the properties of research is that research projects are not repetitive. A research project seeks to discover or to invent something new, and this activity is subject to uncertainty. A research project builds on past success and failures and on discoveries elsewhere (see Chapter 5 for a discussion of by-product germplasm). It has a germplasm base and good research always involves creativity and inventiveness. (See Chapter 5 for a fuller discussion.)

Some forms of uncertainty can be treated as random elements and *expected values* can be based on past experience. But for most research projects, some form of *subjective* estimation as to the project benefits is required. The detailed knowledge required to make informed judgements as to project outcomes is typically not widely held. And, unfortunately, the holders of this knowledge are not always free of bias. The proponents of a particular line of research are usually self-selected into the conduct of that line of research. They may know a good deal about the likelihood of success of that particular line of research, and they may be the right people to pursue the research, but they are not always unbiased. Most research managers and administrators have difficulty with the task of cutting off funding for research programs that have not produced according to the expectations of those who conduct the research.

Priority-Setting Methods

Because of this estimation problem, it is generally regarded to be best to use as much *ex post* evidence as possible in estimating benefit streams. Typically, the *ex ante* evaluator pre-specifies the cost (resources) and timing side of the problem. (This is also possible with priority setting.)

The basic components of the present value of benefits (PVB_0) calculation for a given project specified as to methods, resources used and timing is:

$$PVB_0 = \sum_{t=0}^{n} B_t/(1+r)^t$$
$$= \sum_{t=0}^{n} (b/u)_t U_t/(1+r)^t \quad (6.1)$$
$$= \sum_{t=0}^{n} (b/u)^* U^*(W_t)/(1+r)^t.$$

This expression is written in three ways. In the first, B_t is simply expressed as the value of benefits in each future period, with the term $(1+r)^t$ as the discounting factor to convert the future benefits stream into present values. In the second, the B_t value is broken into two parts, benefits per unit, $(b/u)_t$, and units benefited, U_t, in each future period. In the third, the benefits per unit and units benefited are expressed as the level achieved after n periods (the economic life of the project) and a set of time weights, W_t, associated with the timing of benefits (W_t ranges from 0 to 1).

Costs can be handled in several ways. The first is simply to subtract them from the benefits of each period. This produces a net present value (NPV_0) of the project:

$$NPV_0 = \sum_{t=0}^{n} (B_t - C_t)/(1+r). \quad (6.2)$$

Alternatively the PVC may be calculated:

$$PVC_0 = \sum_{t=0}^{n} C_t/(1+r)^t \quad (6.3)$$

and the benefit/cost ratio (B/C) computed:

$$B/C = PVB_0/PVC_0. \quad (6.4)$$

Finally, one may compute the IRR for which $NPV_0 = 0$:

$$\sum_{t=0}^{n} (B_t - C_t)/(1+IRR)^t = 0. \quad (6.5)$$

These are the standard economic calculations for project evaluation. They can be applied to *ex ante* or *ex post* calculations.

Consider the components of PVB_0 which must be estimated or calculated. They are the third version of eqn 6.1.

1. $(b/u)^*$, the potential benefits per unit benefited (for example, this might be the reduction in costs per unit produced, or the reduction in crop losses per hectare or per unit produced),
2. U^*, the economic units (price × quantity) that are potentially affected,
3. W_t, the timing of benefit achievement, and
4. r, the discount rate.

For many projects, like constructing a building or a road, all four components can be computed from past evidence. The value per unit benefited can be computed from other similar projects and units benefited can be computed based on past usage. The discount rate can be a standard rate, or two or three alternatives can be used.

For *ex ante* evaluation of research projects using past evidence is not generally possible. Some components typically must be set using subjective probability estimation (SPE) techniques. Setting aside r, the discount rate, as not requiring estimation, the evaluator has four components to work with: C_t (the cost stream), $(b/u)^*$, (U^*) and W_t. The evaluator can set or 'fix' one or more of the components and estimate the remaining using *ex post* evidence or SPE methods.

As noted, traditional project evaluations fix C_t (i.e. the project cost) and calculate or estimate the remaining components. For research projects it is not unreasonable to fix $(b/u)^*$ and (U^*) or (b/u), U and W_t and estimate C_t (e.g. by asking how much would it cost to achieve specified results in a specified time frame). These options are also available to the priority setter.

More typically though, *ex ante* research project (and RPA) evaluation proceeds in the normal way, i.e. by fixing C_t. Options are then to:

1. Estimate B_t directly by SPE.
2. Estimate $(b/u)_t$ and U_t by SPE.
3. Estimate U_t from past experience and $(b/u)_t$ by SPE.
4. Estimate $(b/u)^*$, U^* and W_t by SPE.
5. Estimate $(b/u)^*$ from past experience and U^* and W_t by SPE.
6. Estimate U^* from past experience and $(b/u)^*$ and W_t by SPE.
7. Estimate U^* and $(b/u)^*$ from past experience and W_t by SPE.

In practice all the options have probably been attempted. We believe the quality of the estimates tends to improve as one moves from 1 to 7. This is, in part, because the quality of SPE techniques improves. (The low-quality SPE scoring methods tend to be applied to 1, 2 and 3, and timing elements are usually not well handled by SPE methods unless specifically addressed as in options 4–7.)

As noted earlier, priority setting is an extension of *ex ante* evaluation. When well done it should preserve the strengths of *ex ante* project evaluation. (In other words it should not be used as an excuse to downgrade SPE methods.) The country studies in Part II of this volume provide measures of U^* and $(b/u)^*$ for many RPAs and enable the use of options 6 and 7 (option 5 tends to be dominated by option 6 for priority setting). This is the demand side of priority setting discussed further in Section III below. (Chapters 7–15 provide evidence for U^*. Chapter 21 reports a priority-setting exercise based on option 6 for rice biotechnology RPAs. Chapter 22 reports a broader (option 7) priority-setting exercise for rice research.)

III. Research Program Areas and Units (U)*; the Demand Side of Priority Setting

RPAs are the basic units for priority setting. It is critical that they be defined so as to: (i) be conformable to the use of *ex post* evidence to the greatest extent possible; (ii) be conformable to research project structures; and (iii) allow meaningful comparisons of projects. This means that attention must be paid in defining RPAs to:

1. Economic data (e.g. on crop losses).
2. Ecosystem dimensions.
3. Research objectives.
4. Technique or research tool (germplasmic) dimensions.

Ideally, consistent accounting systems by RPAs should be developed. RPAs should be similarly defined and used for different research organizations dealing with similar problems. In Chapter 5 relevant comparisons for priority setting for similar RPAs at different positions in the hierarchy were discussed.

RPAs clearly have an ecosystem dimension; and so does the measurement of units (V_1 and V_2 in Chapter 5). One wishes to compare (i.e. set priorities for) projects in different ecosystems. The international rice research system is strengthening its ecosystem structure and data are more readily becoming available by ecosystem (see Chapter 3). RPAs should be consistent with perceived research problems. In rice improvement, there has long been a focus on directing research toward:

1. Rice yield or productivity enhancement.
2. Rice quality enhancement.
3. Reducing constraints to economic and technical production efficiency.
4. Reducing constraints associated with crop losses. These include losses due to:
    (a) insect pests;
    (b) diseases;
    (c) other biotic stresses (weeds, rodents, birds, etc.);
    (d) abiotic stresses (drought, submergence, extreme temperature, etc.).

The IRRI has conducted several multicountry projects identifying constraints generally and crop-loss constraints specifically (the project underlying this volume). Because of these studies, there is some concern in this priority-setting endeavour that the basic yield enhancement RPAs are under-weighted. (The China study in Chapter 10 addresses this specifically). It is important that they are not, particularly as they have large germplasmic dimensions (and this is IRRI's comparative advantage.) There is also concern that the natural conformity of many rice biotechnology research objectives to crop-loss-based RPAs will further tilt priority setting toward these RPAs. It is important that priority-setting RPAs are comprehensive (see section IV below).

Since most research problems, particularly with the development of biotechnology research techniques, can be addressed by one or more techniques, it is important that RPAs are either defined as technique-specific or, more properly, cross compared by technique (see below and Chapters 21 and 22 for examples in the context of obtaining SPEs).

The natural units (U^*) of RPAs are usually land units (although they could be rice output units). Crop-loss RPAs have the desirable property that they inherently offer a scope dimension, in that the upper limit of a research program directed toward reducing or eliminating a crop loss covering X ha in ecosystem 1 is the estimated value of the loss. This is an advantage over RPAs where there is no natural upper limit. Of course, for some crop-loss RPAs, for example, drought stress or submergence stress, research cannot, even if fully successful, eliminate the loss. And, even for RPAs where loss elimination is feasible (i.e. for a specific insect pest), the dynamics of pests and pathogens are likely to produce an alternative loss.

There is also an important issue in the units and benefits measurement associated with pricing, variable costs and land rents. Rice production (as with other production) has three classes of factors:

1. Variable in the short run (labour, fuel, etc.).
2. Quasi-fixed (fixed in the short run, variable in the long run).
3. Fixed in the short and long run. These factors, such as sunshine, land, etc. obtain a *rent* related to the difference between price and variable costs.

In some project evaluations, product price is taken to be the key valuation procedure. A project is specified to produce more product (rice) holding constant factor use, and this product is valued at a distortion-free price. Rice prices and agricultural prices generally have been steadily falling over time (see Chapter 2). This decline has, in fact, been the result of efficiency gains. But cost components 1 and 2 above have not declined. It is the third component that has declined. Product price is not the relevant measure for projects with large rent components.

To illustrate this point, consider the basic cost–supply relationship, in Fig. 6.1. The initial equilibrium has two types of farms. Type 1 farms are small and inefficient, i.e. they have high average variable costs ($AVCs$). Type 2 farms are large and efficient. Average total costs are equal to price (P) because the rents to fixed factors are simply capitalizations of the $P - AVC$ difference. The supply function to the market is the summation of the marginal cost (mc) curves above the minimum point on the AVC curves. (Supply curves do not go through the origin as depicted in many treatments of the value of research.) Now consider a decrease in AVC for type 1 farms to AVC. This shifts the supply curve as depicted. Note that the shift in the supply curve is parallel – this is also the case as long as the new MC curves have the same slope as the original MC curves.

The benefits from this technology are measured as the area B_1. They are well approximated by $K_1 \times P_0 Q_0$, where K_1 is the decline in AVC_1 and Q_1, the

Priority-Setting Methods 97

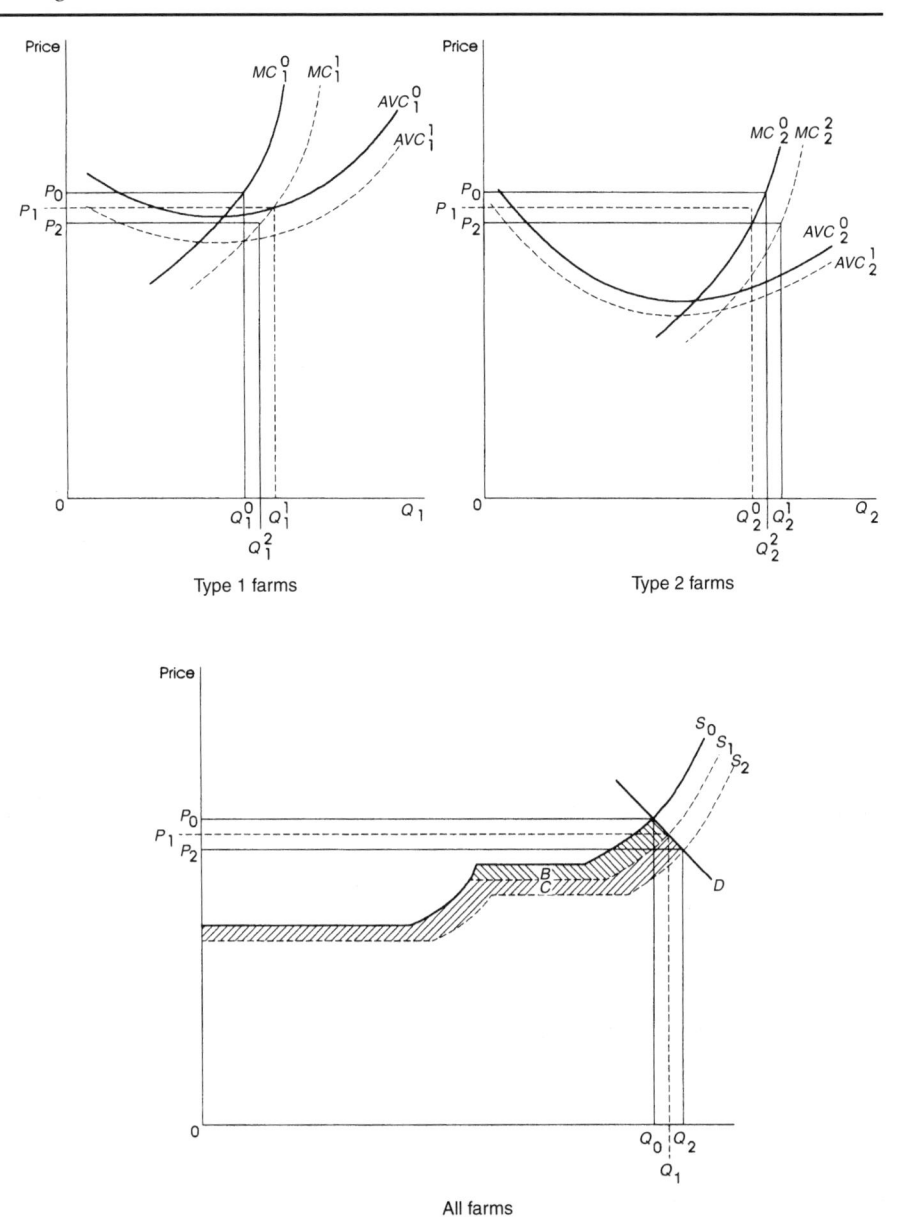

Fig. 6.1. Benefits (losses) from productivity change by farm type.

production of type 1 farms. These benefits are distributed between producers and consumers. In this case, prices fall, so consumers gain. This reduces producer rents to type 2 farms so they suffer a capital loss $(P_1 - AVC_2)$. Type 1 farms will experience a gain because this $P_1 - AVC_1$ term has risen. Note that price does not enter into the calculation of the benefits from research except to a minor extent (the size of the triangle is affected but this is a second order effect). Price does determine the distribution of benefits.

Now consider a second research program that lowers the AVC of type 2 farms – but does not affect type 1 farms. This provides a further *parallel* shift in the supply curve and, a further set of benefits, B_2. Price declines further, consumers benefit further. Producer rents (surplus) to type 1 farms are now reduced. Producer rents to type 2 farms are increased.

Now note that the size of the research benefits for the second cost reduction are not lower because the first had occurred. Prices are lower because the first reduction occurred but the benefits are not lower. Given the linear demand curve, the benefits to the two research projects are the same, independent of the order in which they occurred and independent of the initial price. Benefits are measured as the change in variable costs and they depend on the prices of variable factors, not on the price of the product.

It is sometimes alleged that with international trade, prices matter. But this is not so! Consider Fig. 6.1 where the two types of farms are now two different countries. A downward (rightward) shift in supply in country 1 shifts world supply and causes a reduction in price. It may seem that this reduces the area B_1 by B_1^* but this is not the case because it is global welfare that should be measured. Consumers in other countries have benefited, so the analyses of benefits is properly made in the world market. The implications for this market are the same as those for the farm types. (See section VI below for more regional analysis of distributional impacts.)

IV. Measuring Crop Losses and Yield Gap Consistency

Crop-loss RPAs are important for rice research. They meet the requirements that they are real research invention targets. Research programs using different techniques are directed toward their reduction. Measures of crop losses thus provide natural measures of target units (U) and to some degree of target impacts (benefits) per unit (b/u). There are three areas of caution needed when examining crop-loss-based estimates of (U) and (b/u).

1. Reported losses are usually losses in the presence of some forms of protection.
2. Losses should be consistent with more general measures of production.
3. Crop losses cover only part of the crop-improvement scope. Yield enhancement through higher biological efficiency is also important and typically not covered by crop losses.

The first point regarding losses under protection is important because the scope for research benefits may actually be greater than the reported losses. Incorporating insect pest resistance genetically into rice plants may reduce actual losses due to insects and reduce protection (pesticide use) costs (see Chapter 19 for evidence on this.) Ideally protection cost measures should be considered to be part of the crop-loss estimates.

The second point is one of consistency. Crop losses, when added up, should be consistent with evidence for actual and loss-free crop yields. Yield gap analysis helps to achieve more realistic estimates of crop losses. In this volume and in prior studies at the IRRI, two yield gaps have been identified. Yield gap I, the difference between (i) research potential yields under best practices, infrastructure, experiment station environments and loss-free conditions and (ii) yields under best practices, infrastructure, farmer's environment and loss-free conditions represents yield-enhancement research opportunities. Yield gap II, the difference between (i) best practices, farmer's environment and loss-free yields and (ii) actual farmer yields can be divided into crop losses and a residual due to the difference between best and actual practices and best and actual infrastructure and best and actual market conditions. The crop-loss components should not exceed reasonable independent estimates of yield gap II.

Yield gap I is not easily measured. The difference between experiment station and best farmer yields is not necessarily a good measure of the scope for yield-enhancing research. It really reflects environmental conditions. Some yield-enhancement research may be guided by a strategy of bringing poor environment yields up to best environment yields, but in general yield enhancement applies to all environments.

V. The Use of *ex post* Evidence in Priority Setting

The crop-loss and economic units data discussed in the previous section is a type of *ex post* evidence in that it incorporates past experience as to the spread and diffusion patterns of improved varieties and the effect of ecosystem differences. *Ex post* evidence is also useful for estimating timing (W_t) or impact parameters (b/u) in cases where continuity or regularity might reasonably be expected.

A case in point is illustrated in Chapter 21 where 'trait values', which are a type of (b/u) index in that they measure how much yield increase and pesticide use decrease is associated with the achievement of insect-resistance traits in rice varieties, are estimated. It is then argued that biotechnology will allow more traits and 'stronger' traits than realizable by conventional breeding methods but that the economic value of these traits is likely to be comparable to those achieved in the past.

This may not always be a good argument, but one has to weigh its use against the use of SPEs (see below) and these too are subject to error. In practice it may be useful to use both types of measures. Good SPEs are based on *ex post* evidence to a considerable degree. Many research objectives remain constant over time even though the degree of achievement varies.

The more aggregate the RPA, in general, the more viable is *ex post* evidence. For some broad purposes such as comparing rice research priorities for broad ecoregions or for other RPAs, *ex post* evidence for research impacts from statistical studies may be the best guide. Consider the evidence for irrigated and upland rice in the tropical climate regions of the world. *Ex post* studies clearly show that irrigated rice research programs (in the aggregate) have had high impacts. Few, if any, upland rice research programs have shown impacts. It is quite possible that research events in the past decade are providing more potential (more germplasm) for upland research. If so, this is not apparent from *ex post* evidence and thus *ex post* estimates of research impact will be wrong. But if it is not so, the experience reflected in the *ex post* experience will continue.

The reason that we need to turn to SPE estimation is to incorporate information and evidence that is not available in the *ex post* evidence.

VI. Subjective Probability Estimates

SPEs are notoriously difficult to obtain. They are subjective assessments and the provider of such estimates is usually reluctant to state assessments that cannot be backed up with data. As noted in the previous section, *ex post* evidence cannot be up to date regarding recent developments in germplasm or technology, and *ex post* evidence cannot address such issues as the nature of the problems and bottlenecks to their solutions. It is thus difficult from *ex post* evidence to assess the current state of germplasm associated with a research problem.

Scientists working on specific problems are in the best position to know about recent germplasmic events and to judge their likely contribution to solving problems. They are also in a position to observe rates of progress toward the achievement of objectives. It is these informal judgements that we hope to capture in SPEs.

Basically there are four procedures for obtaining SPEs. They vary in sophistication and in quality. All are improved by clear problem statement and definition. These procedures in increasing order of sophistication (and quality) are:

1. Ranking techniques.
2. Scoring or scaling techniques.
3. Single point probability estimates.
4. Subjective probability distribution (SPD) estimates.

Ranking techniques are the simplest. Respondents are asked to provide ordinal rankings of alternative projects or RPAs by different criteria. Ranking can be applied to achievement objectives, to time to achievement estimates, research quality or consequences. When properly used the procedure can be helpful. In one way or another, all decisions entail ranking.

Misuse of ranking methods occurs when the wrong criteria are used or are in conflict. It is easy to apply ranking to virtually any goal. For example, if projects or RPAs are ranked according to their 'contribution to national needs' or some such general goal, one may obtain poor-quality information. But, if the objectives of the research activities are clearly stated (e.g. to achieve host plant resistance to a particular disease) and ranking is for this objective, more meaningful results are obtainable.

Ranking is ordinal; only the relative order of projects is stated. Scoring or scaling techniques are cardinal techniques; scores have an origin and a scale. Thus, if scores are say from 1 to 5, this implies that 4 is twice as high as 2. The difficulty with scoring techniques is that respondents translate their own subjective scale onto the general scale. One person may give the first RPA a score of two and the second RPA a score of four. Another person might give scores of one and five even though they agreed on the difference between the projects.

The natural solution to this problem is to use direct quantitative measures instead of scores. For example, the respondent may be asked to estimate the likely yield increase that might be achieved from RPA_1 and RPA_2. This is a single point probability estimate (see below). Respondents, however, are often more comfortable reporting a range of estimates. Scores associated with ranges are one means to achieve this (see Chapters 21 and 22 for applications). It is also possible to normalize scores to correct for the subjective scaling problem.

Single point probability estimates of time to achieve, $(b/u)^*$ or U^* are, as noted, somewhat inconsistent with the inherent uncertainty of subjective estimates. For some respondents this leads to 'forced' estimates, i.e. they provide an estimate because one is required and little careful thought is given to the estimate.

Two point probability estimates are SPD estimates. They have the desired property that they allow the respondent to express the variance or range of his or her uncertainty. This usually produces a more thoughtful, less forced response. Fishel (1970) was a pioneer in this work. In his original work he was motivated by the idea that estimates of the variance of expected pay-off (say a B/C ratio) would themselves be of great interest to research managers. He was later to note that he was disappointed in this regard as research managers did not seemingly place a high value on variance measures (Fishel, 1981).

Fishel's experience with priority setting in the US Department of Agriculture and the State Agricultural Experiment Stations should be sobering to priority setters. Obtaining good SPEs is not easy. It requires care and experience and, in the end, some RPAs simply have a high degree of uncertainty as to research impact. No techniques or reliance on experts will reduce that uncertainty.

The major reason for pressing for two point SPD estimates over single point estimates is that the respondent will give more thoughtful answers. In Chapter 21 of this volume, two point estimates of time to achievement for RPAs using biotechnology techniques are utilized in a priority-setting exercise. Chapter 22 uses both single point but scaled estimates for $(b/u)^*$ estimates and two point SPD estimates for time to achievement estimates.

VII. Issues/Adjusting

Many priority-setting efforts have made adjustments to the basic scores or indexes to reflect special considerations. These include adjustments for:

1. Sustainability.
2. Unfavourable ecosystem.
3. Orphan RPAs.
4. Equity.
5. Weak national agriculture research (NAR) programs (compensation).
6. Environmental effects.

In some cases the adjustment should be a regular component of priority setting. In others they entail a real trade-off between the program pay-off and a non-research objective. It will be argued below that the sustainability issue and the environmental issues should be an integral part of priority setting. As for the others, real trade-offs are involved. For these, the priority setter must determine whether it really makes sense to pursue a low pay-off program to meet the objective in question. Existing practice in making adjustments is somewhat *ad hoc*.

A. *Sustainability*

The term sustainability is currently in fashion. To the casual observer the term suggests that policy makers (priority setters) have an option between projects that yield benefits streams that are permanent and projects that yield benefits streams that are not permanent, i.e. that decline after some time. The simple recommendation of sustainability is to support the project with the permanent stream and to avoid projects with non-permanent effects (and this is particularly the case if the non-permanent effects have negative environmental effects).

There are three important points to be made here. The first is that the options are not so abundant that one can simply rule out the non-permanent stream options (including those with some probability of being non-permanent). The second is that it is not at all evident (in advance) that certain innovations will be sustainable while others will not be. The third is that project evaluation and priority-setting methods are designed to properly compare both options and allow for rational choice between them. The present value of permanent and non-permanent streams can be rationally compared. This comparison is sensitive to the discount rate. As the discount rate is raised, the PVB with the permanent stream will rise relative to the PVB of the non-permanent stream.

But it certainly does not make sense to rule out projects with non-permanent benefit streams. It makes sense to evaluate the reasons for non-permanence and to seek solutions, but not to rule them out. For research projects in particular, it is not always possible to predict events that create non-permanence. Past history has shown that when these events occur they usually have correction options, and often these correction options can themselves create larger benefits streams.

Consider the problem of developing host plant resistance (HPR) to diseases and insect pests. Past experience suggests that HPR is not permanent. Changes in the biology of pests and pathogens reduce the effectiveness of HPR. Past experience also suggests that new types of HPR can be developed in response to these events. Researchers understand this process well and they appreciate the value of durable HPR. But it would be foolish to rule out HPR research on the grounds that it is not sustainable.

This point applies to all research projects. There is a large element of empiricism – of trial and error in research. In fact, researchers learn from errors and they cannot eliminate the error component; it is part and parcel of doing research. Researchers, can, of course, attempt to avoid errors and they can monitor projects for problems and be ready with corrective action.

Even for more general economic development problems this same principle holds. Trial and error play a large role in economic development. As with research, one learns from errors. Responsiveness to non-permanent impacts is often the chief mechanism by which progress is made. Good project analysis and good priority setting takes this into account. Sustainability is properly handled with these methods.

B. Unfavourable ecosystem

A real trade-off may exist (and probably does) between an RPA with a high germplasm component (e.g. irrigated ecosystems) and an RPA with a low germplasm component (e.g. upland ecosystems). The same expenditure may yield a PVB of $US10 million in the high germplasm RPA and $US1 million in the second. Are there rational reasons to fund the low pay-off project instead of the high pay-off project?

The principle argument for doing so is a distributive argument. Consider Fig. 6.1 as depicting producers in two regions or ecosystems (types 1 and 2). Suppose a high pay-off research program is implemented for region 2 (type 2) but not for region 1 (type 1). This will produce a rightward shift in total supply and a decrease in prices. S_1 is the supply curve for producers in ecosystem 1. S_2 is the supply curve for producers in ecosystem 2. Thus $S_1 + S_2$ is the aggregate supply. The initial price is P_0 and production is Q_1 and Q_2. Now suppose the high pay-off program is chosen and implemented in ecosystem 2. Aggregate supply shifts to $S_1 + S_2$ and price falls to P_1. The supply curve from ecosystem 1 does not shift because the low pay-off program was not pursued (note, that it would have had a small impact had it been pursued).

It is clear that producers in the unfavourable ecosystem have been harmed. Prices have declined, but costs have not. On closer examination one can see that if variable factors, especially labour, are mobile they can move from region 1 to region 2 and escape losses. As they do this, wages will rise in region 1 leaving the full burden of losses to be borne by immobile factors, chiefly land – although it may be family labour as well.

Should one pursue the low pay-off option to help these immobile factors? The IRRI's study of advantaged and disadvantaged regions showed that most labour did more to avoid losses and that land owners bear the losses depicted above. Should the IRRI sacrifice the high pay-off project for landowners? Probably not. In fact, it seldom makes sense to sacrifice the high pay-off project.

This example serves to raise a related consideration: if the problem with the low pay-off RPA is germplasm, it makes economic sense to undertake more germplasmic research either jointly with or preceding more applied research.

C. 'Orphan' research priority areas

It is widely argued that 'orphan' problems require special funding. Orphan RPAs are RPAs where the units, U, are too small to justify much research. The critical issue in this is the potential for growth in U. This should be taken into account in any priority-setting exercise. It needs attention – but there is no escaping the fact that if U is small, the PVB of the RPA is also small.

D. Equity

The ecosystem example discussed above involved an equity issue; the incomes of one group versus another. The more general distributional effects of research include the effects on consumers as well as producers. When research causes a supply increase and a price decline, the benefits are distributed to all rice consumers. And since rice-consumption patterns are such that the share of rice in the expenditure of the household is highest for poor households, it follows that rice research provides benefits disproportionately to the poor.

Are there further 'pro-poor' RPAs that deserve consideration? We have already discussed the ecosystem case, but is not clear that there are more poor farmers in the advantaged ecosystem than in the disadvantaged ecosystem. Regional issues are important, but they are not straightforward equity issues. There are also some rice-quality issues that are related to poverty. If the poor consume a different quality of rice we may wish to emphasize RPAs for the poor. We also know that prices determine the use of technology and hence wage levels are related to research RPAs. In low-wage economies, transplanting – a labour intensive activity – is economic, similarly hand harvesting of rice is economic. In high-wage economies these activities are not economic because machines are available for these tasks. RPAs for improved efficiency of activities that have an advantage in low-wage economies have a pro-poor element.

E. Weak national agricultural research

The example of the ecosystem noted above was discussed as a national problem but with international trade it is an international problem. Countries with little national research capacity are harmed by advances in other countries in the same way that farmers in disadvantaged ecosystems are harmed. Should the IRRI choose RPAs to attempt to compensate for weak NAR?

The IRRI, as noted in Chapter 4, has a comparative advantage as a germplasm supplier. So if there is a responsive NAR in place, the IRRI's germplasm assists and stimulates the NAR programs. The IRRI has historically played a role in stimulating and assisting weak NAR. It has a number of options to do this. Many of these activities complement its larger comparative advantage as a germplasm supplier and gatekeeper to the world of higher science. However, for some there is a real cost to the compensatory applied research that the IRRI does because it reduces its germplasmic research and this has to be factored into priority setting.

F. Environmental effects

As with sustainability, environmental concerns and politics are in fashion and they will probably remain so; and, as with sustainability, good project evaluation and priority setting internalize effects on the environment. If there are damages to the environment, they should be measured and included in the calculation.

It is often overlooked that the research at the IRRI and in NAR has contributed (along with irrigation fertilizer and other inputs) to increasing yields per hectare and thus has contributed enormously to improved environments, by enabling rice production increases without crop land expansion.

G. Summary

To sum up, sustainability and the environment should be integral parts of priority setting. Distributional issues are important – but if it comes down to choosing between significantly different high and low pay-off projects it usually does not make sense to choose the low pay-off project. Rice researchers, in the end, are specialists, in a narrow range of activities. Countries have numerous institutions and organizations to deal with equity, mobility and related issues.

References

Fishel, W.L. (1970) Uncertainty in public research administration and scientists' subjective probability estimates about changing the state of knowledge. PhD dissertation, North Carolina State University, Raleigh, North Carolina.

Fishel, W.L. (1981) Changes in the need for research and extension evaluation information. In: *Evaluation of Agricultural Research*, Miscellaneous Publication 8-1981. Minnesota Agricultural Experiment Station, University of Minnesota, Minneapolis.

II Country Studies

7 Prioritizing the Rice Research Agenda for Eastern India

D.A. Widawsky[1] and J.C. O'Toole[2]
[1] Food Research Institute, Stanford University, Stanford, CA 94305, USA;
[2] Rockefeller Foundation, BB Building, Suite 1412, 54 Sukhumvit Soi 21 (Asoke), Bangkok 10110, Thailand

I. Introduction

This chapter reports the results of an in-depth study of technical constraints to rice production in eastern India, including the states or regions of Bihar, eastern Madyha Pradesh, eastern Uttar Pradesh, Orissa and West Bengal.* The investigation involved identifying the technical constraints to rice production in eastern India, ranking their importance in terms of crop losses and assessing their suitability for being 'solved' through varietal improvement.

The study was motivated by several factors. Primarily, the aim was to identify constraints which might be ameliorated through genetic improvement of rice. Genetic improvement can provide an effective way to transfer appropriate technological advances in agronomy, entomology and plant pathology to farmers since the basis of the transfer is, simply, the seed. Rice yields in eastern India are limited by technical constraints such as pests, soils and climate which have the potential to be 'solved' through genetic improvement of rice, as opposed to socioeconomic constraints like price policy or fertilizer availability which are solved through other means. Our results show that some constraints limit rice yields more than others. Research priorities are derived from an analysis of those constraints, when minimized, which will lead to the largest increases in rice production.

Understanding technical constraints to rice production in eastern India requires a knowledge of the geographic and seasonal rice-cropping patterns in the region. The next section provides a short introduction to the geographic

---

* Assam was not included for the following reasons: it has less than half the cropped area of any of the other states in eastern India; the system of agricultural research makes it difficult for an 'outsider' to conduct research; the area is politically unstable; and travelling to rice-growing areas is discouraged.

patterns of rice production in eastern India, the agroecological environments in which rice is grown and the technical constraints endemic to each rice-growing environment. A description of the methodology of the study then illustrates how crop losses in eastern India are estimated and how loss estimates inform the research-prioritization process. The resulting estimates of yield gaps and crop losses are presented with respect to rice-growing environments and individual constraints. Finally, the crop-loss estimates are used to provide the basis for rice research prioritization in eastern India.

II. Synopsis of Rice Production in Eastern India

A. Production and yields in eastern India

Increases in agricultural productivity in eastern India since the 1950s have lagged behind the nation as a whole. Rice yields are low (1.2 tons ha^{-1}) compared to other regions (1.9 tons ha^{-1} in the north and 2.1 tons ha^{-1} in the south) (Government of India, 1985; Fertiliser Association of India, 1988). For example, eastern India accounted for 60% of India's cropped rice area in the 1986/87 season yet produced only 50% of India's rice (Muralidharan et al., 1988). In West Bengal and Orissa, total production and yield fluctuate more severely than in Bihar and eastern Uttar Pradesh (Government of India, 1979, 1986). The latter two states have experienced, with an exceptional year or two, a steady rise in productivity.

The low and fluctuating yields in these four states are partly attributable to the high proportion of rainfed rice area in eastern India (Huke, 1982). Rainfed rice fields are subject to water-control problems, both drought and flooding, and a host of associated biotic and abiotic constraints (described in detail below). Yields in irrigated areas are more stable and higher than in rainfed areas. Seasonal yield differences are substantial, where yields in the kharif (mostly non-irrigated monsoon) crop season are much lower than yields in the rabi (irrigated dry) crop season (Table 7.1). There are continuing efforts to bring greater areas under rabi season production through a variety of means such as water harvesting, soil-water conservation and shorter duration varieties to take advantage of residual soil moisture after the monsoon season.* However, though rabi season yields are considerably higher, the area under rabi production is quite small relative to the total cropped area and the kharif season still accounts for the vast majority of annual rice production in eastern India. The kharif season, therefore, commands the greatest amount of research effort in varietal improvement.

Rainfed rice production also limits management opportunities and the choice of rice varieties, which contributes to lower yields. High-yielding modern varieties require greater management than traditionally grown varieties

---

* Note that nearly all fields in eastern India with irrigated rice in the rabi season had a rice crop in the preceding kharif season.

Table 7.1. Rice area, production, and productivity in eastern India, 1984–85.

State	Area (million ha) Kharif	Area (million ha) Rabi	Production (million tons) Kharif	Production (million tons) Rabi	Yield (tons ha^{-1}) Kharif	Yield (tons ha^{-1}) Rabi
Bihar	5.11	0.056	7.89	0.09	1.54	1.56
Orissa	4.13	0.24	5.78	1.02	1.40	4.25
West Bengal	4.73	0.47	10.23	1.91	2.16	4.05
Eastern Uttar Pradesh	2.84	–	4.70	–	1.65	–

Source: Government of India, November 1985, pp. 747–749.

in most aspects of production including water, insects, diseases, weeds and soil fertility (Maurya et al., 1988). However, farmers in eastern India tend to be some of the poorest in the country and often lack the resources required to ameliorate some of these potential problems (Muralidharan et al., 1988). They instead rely on traditional varieties which are not as productive but have more stable yields than modern varieties under conditions of uncertain water availability and low input use (Barker and Pal, 1979).

B. Rice environments and production zones

For ease of analysis, eastern India was partitioned by political boundaries (by state), physiographic region (within each state) and rice environment.* The state and physiographic divisions segregate eastern India into unique elements. The rice environments delineate these elements into standard hydrological classes.

For the purpose of this study, the hydrologically based environment descriptors used to discriminate rice cultural systems in eastern India include two subdivisions of irrigated rice (kharif, rabi) and three subdivisions of rainfed, non-irrigated rice (deepwater, upland, lowland). The definitions given in Table 7.2 were adapted from those used by the IRRI.[†] There are great similarities within an environmental classification across all of eastern India, and distinct differences among the classes. Therefore, it is important, in determining the relative importance of constraints, to look at variations among environments in terms of area, production and productivity.

---

* Rice environment has also been called 'rice ecosystem' and 'rice agroecological zone'. Agroecological zone is an effective way to describe the combination of hydrology, climate and soil of rice environments. However, the Government of India Planning Commission uses the term for other purposes, so rice environment is used here to avoid confusion. In addition, the IRRI uses the term rice environment in a handbook on terminology (IRRI, 1984).
† Since the definitions cannot account for every idiosyncrasy of a location, scientists involved in the study were invited to adjust definitions, if they felt the need, to fit their locale.

Table 7.2. Environmental definitions (adapted from IRRI and Indian publications).

Irrigated wet season (kharif)	Fields are bunded and puddled. Rice is transplanted. Water is added to the fields from canals, river diversion, pumps, tanks, etc. to supplement the rains
Irrigated dry season (rabi)	Similar to the wet season, but water must be supplied from storage reservoirs or from pumps. Solar energy flux is normally much higher in the dry season
Rainfed lowland	Maximum water depth from tillering to flowering ranges from 30–100 cm. Fields may be bunded and puddled. Rice is often transplanted
Rainfed upland	Maximum water depth from tillering to flowering ranges from 0–30 m, although there may be no standing water at any time. Fields are often unbunded or levelled. Broadcast seeding is usually done
Deepwater	The maximum water depth from tillering to flowering exceeds 1 m and may run as high as 6 m. Rice seeds are normally broadcast in dry, unbunded fields before the onset of the rains

Irrigated ricelands

Irrigated rice in eastern India accounts for about 30% of the cropped rice area (including both kharif and rabi systems). Yields are highest and range from less than 2 tons ha^{-1} to over 5 tons ha^{-1}. Modern varieties are extensively planted and relatively high fertilizer use results in not only higher yields but lush foliage which attracts an array of insects and diseases.

The wide range of yields in eastern India is largely due to variations in local definitions of 'irrigated' and the idiosyncrasies of irrigation system infrastructure and management. Some land classified as irrigated during the kharif season may have water available at all stages of growth (i.e. water is not limiting). In contrast, other irrigated kharif lands may experience periodic water shortages, receiving supplementary irrigation only when reservoirs, tanks or river diversion schemes are full from monsoon rains. Land classified as irrigated rabi (post-monsoon or dry season) usually has water available during the crucial seedling and flowering growth stages due to proximity to a large reservoir or tube well. Within the irrigated kharif and rabi land classifications, there are frequently locational differences in available soil moisture, depending on the distance to a canal, canal headwall or lateral channel. Thus, even in some areas classified as irrigated, drought may be a limiting constraint to rice growth and yield.

Rainfed ricelands

Rainfed rice accounts for 70% of the cropped rice area in eastern India. Beyond the fact that rainfed uplands, lowlands and deepwater areas are all unirrigated and have much lower average yields than irrigated areas, there are few similarities among the three types of rainfed rice environments.

Rainfed lowlands make up the greatest proportion of rice land in eastern India, accounting for 40–50% of the cropped area. Productivity varies widely for a number of reasons, but averages 1.2–1.5 tons ha^{-1}, which is nominally higher than upland or deepwater rainfed areas. Pests and diseases are also more problematic than in the other two rainfed zones. While droughts occur with less frequency than in the uplands, they can cause chronic damage, particularly at the seedling and flowering stages. Rainfed lowlands are also subject to periodic flooding necessitating rice varieties which are both drought- and submergence-tolerant.

Rainfed upland rice constitutes 22–26% of the cropped rice area in eastern India, yet 90% of India's total upland rice area. Therefore, improvements in the productivity of rainfed upland rice will be concentrated in the target region of eastern India. Yields in rainfed upland areas average 1.1 tons ha^{-1}, and vary with soil type, fertilizer use, rainfall and agronomic practices. Due to drier climatic conditions, diseases and insects in upland rice tend to be less of a problem than in rice environments where there is more standing water. On the other hand, poor soil fertility can severely limit yields.

Upland soils in Orissa, Madhya Pradesh and southern Bihar suffer from low fertility and high porosity leading to problems with nutrient availability and water-holding capacity, both critical to rice production. In northern Bihar and eastern Uttar Pradesh, upland soils are more favourable and respond well to irrigation or adequate rainfall. No division is made in this study between favourable and unfavourable rainfed upland environments.

Estimates of the extent of deepwater rice area vary from 6 to 9% of the rice-growing area in eastern India, varying dramatically among states. In eastern Uttar Pradesh, the area of deepwater rice is about 12% of the total rice area, while in eastern Madhya Pradesh it is negligible. With an average productivity of about 1 ton ha^{-1}, the annual production from the deepwater area amounts to only 1.96 million tons and represents approximately 8% of the rice production for this region of India.

C. Technical constraints

Constraints to rice production for various areas within eastern India have been studied extensively. Technical constraints encompass biotic and abiotic factors that limit rice yields and are distinct from socioeconomic constraints in that many are potentially solvable through genetic improvement of rice.* Technical constraints in eastern India can be characterized on the basis of:

---

* It is not intended that yield-limiting constraints and their effects are definitively bifurcated by the terms 'technical' and 'socioeconomic', since interactions among them are important in understanding the rice system. But, the distinction is crucial in aiding yield-loss estimates and prioritizing constraints.

1. Adverse climate/weather.
2. Adverse soils.
3. Insects.
4. Diseases.
5. Other (weeds, lodging, birds, etc.).

Adverse climate

The first category describes unfavourable weather factors, the most important being droughts and floods. In predominantly rainfed eastern India, these climatic constraints result from the unpredictable nature of the monsoon. The two components of seasonal monsoons which most influence the success of a given growing season are the total amount and distribution of precipitation.

In nearly every year, sufficient total precipitation is received during the growing season to support the physiological needs of the rice crop. However, this precipitation may be distributed over a limited number of rainfall events with heavy precipitation during certain periods leading to floods, and dry periods in between leading to drought conditions. Some parts of eastern India will experience at least one 15-day period of drought every other year (D.K. Paul, personal communication, 1989).

This bi-annual drought can depress yields if it occurs during vegetative growth and can cause catastrophic loss to the crop if it occurs during flowering. Access to irrigation greatly reduces the effects of these periodic droughts but irrigation facilities are unavailable to most farmers in the region. Therefore, droughts and floods are chronic causes of yield instability. Other weather-related constraints are cold at seedling and anthesis, but this is a minor problem in eastern India.

Adverse soils

The complex mosaic of soils in eastern India is characterized by a wide continuum of fertility from those soils favourable to rice cultivation, such as alluvium in lowland environments, to soils strongly inhospitable to rice cultivation, such as red laterite soils with a profile less than 15 cm thick. Poor soil fertility is common in rainfed uplands where yields are constrained by lateritic soils with high iron and low nitrogen content, and a pH occasionally below five. Soil problems in other areas include salinity, alkalinity and/or zinc, sulphur or iron deficiency. While some soil problems may be efficiently solved with affordable soil amendments, other constraints such as alkalinity and salinity cause great yield losses which might be partially averted through tolerant high-yielding varieties.

Insects

Insects are one of three categories of biotic constraints. The most severe insect losses usually occur in areas where the rice foliage is densest and the productivity is highest; namely irrigated and rainfed lowland fields. Nearly all of eastern India's major insect pests are found at some level of infestation in lowland areas, and the profitability of rice farming induces farmers to rely on pesticides as a common form of pest control. Upland and deepwater environments also experience significant losses from insects, but the range of pests is narrower and two to three species are usually responsible for most of the damage. Since there exist insect pests in all regions and rice environments, and there is limited natural resistance in locally cultivated varieties of rice, varietal improvement through biotechnology offers critical alternatives to insecticide use.

Diseases

Diseases are the second category of biotic constraints. Like insects, diseases are most prevalent in irrigated and rainfed lowland areas where rice productivity is highest. Unlike insects, however, there are few effective simple prophylactic measures analogous to insecticides. Thus, the need for host plant resistance to diseases is acute. Different diseases are severe under different conditions of soil moisture status and plant health, so diseases endemic to drought-prone uplands may not be seen in flood-prone areas and vice versa. As breeders develop rice varieties with resistance to specific diseases, diseases for which there are no sources of genetic resistance account for a higher proportion of disease losses (e.g. sheath blight and sheath rot). Thus, in assessing disease constraints, it is important to distinguish between diseases which are gaining in importance as a result of limited host plant resistance and those which are on the decline due to widespread availability of host plant resistance.

Other constraints

Biotic constraints that do not fit into the insect and disease categories constitute the class of other constraints. The most severe of these is weed competition. Weeds cause substantial crop losses in upland rice fields and are significant in all other ecosystems as well. Broadcast seeding and seasonal labour scarcity are two factors leading to heavy weed infestation and the attendant yield losses.

Birds, crabs and rodents are also included in this category and may cause heavy crop losses, albeit in very localized areas.

Lodging is another common problem in many of the medium to tall varieties traditionally grown in India. Because rice straw has value in rural areas in many applications, farmers often resist growing semidwarf varieties to avoid losing the valuable straw by-product. Therefore, there may be a need for lodging-resistant, medium height (90–130 cm) varieties.

III. Survey Background and Methodology

A. *Some basic assumptions*

There are several fundamental assumptions in setting and prioritizing the elements of a research agenda for technical constraints. Briefly stated, these assumptions are as follows:

1. *There is a quantitative difference between actual rice production and potential production.* This difference is called the *yield gap*.

2. *The yield gap is not identical for all environments and varietal interactions, and a geographic area of study must be separated into similar units to assess the yield gap.*
3. *The yield gap is composed of a number of different constraints.* These constraints are either technical or socioeconomic (including management differences). Together the losses account for all of the yield gap.
4. *The specific contribution of the technical constraints can be measured and cumulatively account for part, but not all, of the yield gap.* These measurements take the form of yield-loss assessments.
5. *Based on values obtained from yield-loss assessments, technical constraints can be ranked in importance.* Importance is conferred by severity of yield loss associated with the constraint, with yield loss summed across environmental and/or geopolitical units.
6. *Based on these rankings, a research agenda can be formulated.* Transformations may be used to weight or adjust the rankings and loss estimates in order to incorporate factors such as equity, probability of success, costs of research, potential benefits and a host of other adjustments. However they are applied, these are all used to achieve a list of priority research topics.

Based on these assumptions, a short description of methodology begins with a brief description of yield gaps in eastern India.

B. Yield gaps in eastern India

A major international effort was initiated at the IRRI in the 1970s to characterize yield gaps in rice in a number of different countries (IRRI, 1977, 1979; De Datta el al., 1978). The underlying hypothesis stated that the yield gap could be divided into two parts. Yield gap I is the difference between an experiment station's maximum yield and an on-farm experiment's maximum yield. This yield gap arises from differences in the environment which cannot be managed or eliminated in the farmers' fields. Yield gap II was a primary concern because it is the difference between actual farmer yields and yields attained in on-farm experiments. This gap reflects biological constraints, soil and water constraints and socioeconomic constraints.

In this study, the central idea of a yield gap is crucial to understanding the benefits of varietal improvement. Such technology can help bridge part of yield gap II by lessening the impact of technical constraints. Measuring the technical components of yield gap II is the basis for prioritizing a rice biotechnology research agenda.

Rice scientists at 12 agricultural institutes distributed throughout all regions of the study area participated in the survey. They were asked to estimate, based on experiments or their expert knowledge, the following (for the environments in their area):

1. The maximum yield from experiment stations for both modern and traditional varieties.
2. The maximum yield from on-farm experiments for both modern and traditional varieties.

Table 7.3. Estimated yields and yield gaps in eastern India.

Average yields (kg ha^{-1})	Irrigated kharif	Irrigated rabi	Rainfed lowland	Upland	Deepwater
Experiment station					
Maximum Yield (MV)	5573	4619	4430	2941	1698
Maximum Yield (TV)	3342	2369	2800	1873	1327
On-farm experiment					
Maximum Yield (MV)	5020	3836	4038	2622	2150
Maximum Yield (TV)	3197	1885	2505	1639	1097
Actual estimated yield	2759	2969	2151	1316	617
Yield gap I (MV)	553	783	391	319	(451)
Yield gap II (MV)	2262	866	1887	1306	1533
Yield gaps I and II (MV)	2814	1650	2279	1626	1081
Yield gap (%)	102	56	106	124	175

MV, modern variety; TV, traditional variety.
Yield gaps are based on yield potentials with modern varieties.

3. Actual average yields for farmers in each of four different cultivation practices (combinations of inputs and varieties).
4. The distribution of these cultivation practices in each ecosystem in their area.

The results of these estimates (Table 7.3) illustrate a considerable yield gap and its variability across environments, as stated in the assumptions. The yield gap is smallest (in percentage terms) in favourable areas where one expects to find modern varieties, high inputs and assured water control or management; namely, the irrigated rabi crop. In absolute terms, the largest gap is found in irrigated kharif, lowland and upland environments, those areas where experiments show that constraints can be overcome through management. The highest yield gap in percentage terms occurs in deepwater environments, areas with the lowest adoption of high-yielding varieties and least responsive to high inputs. The existence of a considerable yield gap suggests there are barriers to improved management.

Total production gaps illustrate the impact of yield gaps in eastern India based on the extent of different rice environments in the region (Table 7.4). Estimates of total production in each rice environment were calculated by multiplying estimated yields by the extent of the environment in each state. Total production figures include both potential and actual production. Results were summed across states to derive values for eastern India. Actual production for a rice environment in a state or physiographic region was adjusted by the yields and distribution of the various farming practices.*

---

*Farming practices were separated into four types with respect to input use and variety: (i) modern varieties and high inputs; (ii) traditional varieties and high inputs; (iii) modern varieties and low inputs; and (iv) traditional varieties and low inputs.

Table 7.4. Estimated production and production gaps in eastern India.

Total production (million tons)	Irrigated kharif	Irrigated rabi	Rainfed lowland	Upland	Deepwater
Experiment station					
Maximum yield (MV)	24.56	4.64	26.64	13.12	2.45
Maximum yield (TV)	14.53	2.49	17.21	8.19	2.09
On-farm experiment					
Maximum yield (MV)	21.94	4.36	24.23	11.82	2.29
Maximum yield (TV)	13.39	2.31	15.43	7.32	1.70
Actual estimated yield	11.96	3.07	11.87	5.48	1.15
Yield gap I (MV)	2.63	0.30	2.42	1.29	0.16
Yield gap II (MV)	9.98	1.29	12.36	6.35	1.13
Yield gaps I and II (MV)	12.61	1.59	14.78	7.64	1.29
Yield gap (%)	105	52	125	140	112

MV, modern variety; TV, traditional variety.
Yield gaps are based on yield potentials with modern varieties.

The magnitudes of total yield loss show that the percentage yield gap or per hectare yield gaps must be considered jointly with total area affected to assess the total impact of production constraints in a region. Even with some margin of error, the potential for increasing regional yields is highest through improvement in irrigated kharif, lowland and upland yields.

These results confirm the first two assumptions of the study: a yield gap exists and it varies across rice environments. The next step was to identify the composition of the gap in terms of technical constraints and measure the contribution of these constraints. Estimates of crop loss by individual constraint provide a quantitative comparison of constraint severity across geopolitical and environmental zones and provide the rationale for research investment.

IV. Estimates of Crop Losses

A questionnaire-based survey was carried out in four of the states.* The major physiographic regions in all the states were identified and an attempt made to visit research stations or institutions in each. Twelve stations participated in the survey (Widawsky and O'Toole, 1990). In the course of the survey, more than 100 rice scientists participated in the estimation of crop loss. The information

---

* Orissa, Bihar, eastern Madhya Pradesh and eastern Uttar Pradesh were surveyed intensively. An attempt was made to enlist the cooperation of scientists in West Bengal, but difficulty in obtaining government permission and logistics made this impossible. As stated earlier, ecosystems are similar across eastern India, and the distribution of rice crops area is similar in West Bengal to the rest of the region.

they gave was based on experimental results and their extensive experience. Scientists were sought who had a long history of field experience. They were asked to make loss estimates on all technical constraints only in rice environments where they had personal knowledge and experience regarding the constraints and severity regarding crop yield.

The questionnaire was designed to elicit two crucial aspects of yield loss for each type of biotic and abiotic constraint: the percentage of an environment affected by a constraint and the average yield loss associated with the presence of the constraint. The product of these estimates represents the average yield loss of a constraint in a particular rice environment. The extent of an environment that was described by an individual scientist depended on the command area of his or her institute. An example of a yield loss calculation is given in Table 7.5. Estimates of participating scientists were averaged for each environment and constraint in a locality. Total yield loss was calculated from these averages.

Total loss can be summed by environment, locality or both. Rice research is often organized on the basis of environment and research priorities are based on potential productivity improvements within a rice environment. Therefore, yield loss is presented for the five rice environments in eastern India by constraint (Table 7.6) and by environment (Table 7.7). Percentage yield loss by constraint category is given in Table 7.8.

A. Irrigated kharif losses

In irrigated kharif environments, total losses from technical constraints were 854 kg ha^{-1} (see Table 7.6) and account for a large proportion of the estimated yield gap. The total loss from technical constraints accounts for 30.9% of the yield gap between experiment stations and actual yields (sum of yield gap I and II). It accounts for 39.1% of the yield gap between on-farm experiments and

Table 7.5. Calculating yield loss from survey data.

Example: Bacterial leaf blight in northern Bihar in an irrigated kharif environment

Data:
Area affected: 9.5%
Loss per hectare: 200 kg ha^{-1}
Irrigated kharif area in northern Bihar: 634,200 ha

Calculation:

(Area affected) × (loss) = average loss per hectare

0.095 × 200 kg ha^{-1} = 19.0 kg ha^{-1}

(Average loss) × (land area) = regional ecosystem loss

19.0 kg ha^{-1} × 634,200 ha = 12,049,000 ha

Result:
Yield loss from bacterial leaf blight in irrigated kharif environments in northern Bihar is estimated to be 12,049 tons year^{-1}

Table 7.6. Estimated yield losses due to technical constraints (kg ha^{-1}) in eastern India.

	Rainfed ecosystems			Irrigated ecosystems		
				Wet season	Dry season	Eastern
Constraints	Upland	Lowland	Deepwater	(kharif)	(rabi)	India
Weeds	183	86	70	80	134	111
Anthesis drought	143	109	–	45	–	88
Vegetative drought	75	26	28	42	27	43
Yellow stemborer	6	35	82	38	190	39
Seedling drought	63	42	35	10	14	38
Acid soils	70	8	–	44	84	37
Bacterial leaf blight	3	26	42	82	25	36
Lodging	25	41	37	32	5	33
Blast	54	37	21	10	9	32
Submergence	8	45	63	20	–	28
Zinc deficiency	1	33	–	29	22	21
Armyworm	26	9	–	39	13	21
Green leafhopper	19	15	–	38	4	21
Rodents	13	15	20	32	9	19
Alkaline soils	15	15	78	11	4	17
Brown spot	32	8	12	16	3	17
Cold at anthesis	3	28	–	29	1	16
Iron deficiency	34	8	–	11	18	16
Birds	33	7	21	9	3	15
Gall midge	–	8	–	39	27	14
Sheath rot	2	15	10	28	4	14
Gundhi bug	24	3	–	23	–	13
Short duration	35	–	–	5	–	11
Saline soils	17	4	–	10	30	10
Sheath blight	8	7	8	14	13	9
White-backed planthopper	3	5	–	15	58	9
Brown planthopper	4	7	–	16	29	9
Iron toxicity	5	14	–	7	–	8
Caseworm	2	13	–	10	4	8
Leaf folder	2	9	4	14	6	8
Total (top 30 constraints)	908	678	531	798	736	761
Total (all 44 constraints)	933	696	568	854	774	795

actual farmer fields (yield gap II). The total estimated yield loss in irrigated kharif is 3.902 million tons.

These results are important for two reasons. First, a large portion of the yield gap was attributable to technical constraints, demonstrating that solution of these constraints is not trivial. Secondly, loss estimates, when aggregated, are not larger than the yield gap. The latter is a hazard in summing loss estimates for individual constraints, but was not encountered in this study.

Table 7.7. Comparison between estimated yield loss and estimated yield gap by rice ecosystems in eastern India.

Ecosystem	Estimated yield loss (kg ha^{-1})	Rice area under four states* (million ha)	Total production loss (thousand tons)	Production loss as percent of Yield gap II	Production loss as percent of Yield gap I
Irrigated					
Wet season (kharif)	854	4.57	3,902	39.1	30.9
Dry season (rabi)	774	0.77	595	46.2	37.2
Rainfed					
Lowland	696	6.62	4,610	37.3	31.1
Upland	933	4.57	4,265	67.2	55.8
Deepwater	568	1.05	596	52.6	46.0
Eastern India	795	17.58	13,968	44.9	36.8

*The States included are eastern Uttar Pradesh, Bihar, Orissa and West Bengal.

Table 7.8. Distribution of yield loss among constraint categories in eastern India.

Category	Estimated loss (kg ha^{-1})	Percent of total losses	Percent of yield gap II	Percent yield gaps I and II	Percent of average farm yield
Insects	157	19.7	8.8	7.2	8.7
Diseases	120	15.1	6.8	5.5	6.7
Water regime	200	25.1	11.2	9.2	11.1
Soils	110	13.8	6.2	5.1	6.1
Other (weeds, lodging, etc.)	209	26.3	11.8	9.7	11.6
Total	795	100	44.8	36.7	44.2

The ten top constraints account for 56% of the losses due to technical constraints in the irrigated kharif environment. Bacterial leaf blight causes the greatest damage (82 kg ha^{-1} or 382,000 tons). However, two of the top five constraints are drought-related (vegetative and anthesis drought) and when they are summed, drought is the most damaging constraint (98 kg ha^{-1}). Why is there drought in an alleged irrigated rice environment? The reason is that irrigated kharif land in eastern India does not always have assured irrigation. Irrigation is usually supplementary and drought years lead to *in situ* droughts in areas nominally designated as irrigated. The other water regime constraint, submergence, was ranked number 16 with overall losses of 90,000 tons.

The next most damaging constraint is weeds (80 kg ha^{-1} or 365,000 tons). In spite of the fact that irrigated kharif is some of the most productive rice land in eastern India, weeding practices are insufficient to prevent large losses.

After weeds, acid soils were estimated to cause the most damage (44 kg ha^{-1}). It is difficult to characterize acid soils as a single constraint because they are more appropriately viewed as a set of problems ranging from iron toxicity to drought-exacerbating soil textures. In terms of solving the problem of acid soils, biotechnology offers limited solutions.

Insects represent the next four constraint rankings. Gall midge, armyworm, green leafhopper and yellow stemborer all cause similar levels of yield loss. Canopy cover and standing water in irrigated kharif environments promote insect infestations.

The tenth most damaging constraint is rodents. Rice research appears to provide few immediate answers to the rodent problem. There is more potential for the eleventh constraint, lodging, which accounts for 32 kg ha^{-1} yield loss. Varietal improvement has focused in the past on lodging resistance, and the prospects for incorporating lodging resistance in medium-statured plants are favourable.

The relative magnitudes of losses in the irrigated kharif environment illustrate the common pattern of technical constraints in eastern India. The top three constraints (drought, bacterial leaf blight, and weeds) all cause losses greater than 350,000 tons year^{-1}. They can be considered the top tier of constraints because the next most damaging constraint causes about half as much loss as any of these (acid soils with 201,000 tons year^{-1}). These are top priorities for a research agenda.

After acid soils, losses gradually decrease for the next four constraints (gall midge, armyworm, green leafhopper and yellow stemborer). These five constraints are the second tier and represent important areas of potential research.

B. Irrigated rabi losses

Total losses in irrigated rabi environments were much less than in irrigated kharif environments (more than six times less); understandable given the smaller land area devoted to irrigated rabi production (see Table 7.7). Total losses for eastern India from technical constraints in irrigated rabi environments were estimated to be 595,000 tons year^{-1}. As in irrigated kharif, there is a wide range of losses among constraints.

Losses from technical constraints represented a sizeable proportion of the total yield gap. More so, in fact, than in irrigated kharif environments. The losses estimated for technical constraints accounted for 37% of the total amount of yield gaps I and II in irrigated rabi environments, and 46% of production yield gap II alone.

It is not surprising that a higher proportion of the yield gap is attributable to technical constraints in irrigated rabi environments. These are the most favourable rice-cropping lands and adoption of modern varieties and practices is widespread. Socioeconomic and infrastructural constraints are not as limiting, given the high investment in inputs. Therefore, technical constraints assume a greater role in limiting yield and the contribution of the top ten constraints to total yield loss is much greater than in irrigated kharif environments. The top ten factors accounted for 81% of total yield losses from technical constraints.

In this environment, losses are dominated by insect, weed and soil problems. The insect pests in the top ten constraints account for almost 40% of total losses from technical constraints. Yellow stemborer is responsible for the greatest loss (190 kg ha^{-1}). White-backed planthopper causes considerable damage (58 kg ha^{-1}), as do the brown planthopper (29 kg ha^{-1}) and gall midge (27 kg ha^{-1}).

Weeds are the second most damaging factor (134 kg ha^{-1}). This shows that even in the most favourable rice-growing environment in eastern India, where investment in production is highest, weed control is still insufficient to prevent substantial losses.

Four soil conditions cause high-yield losses with acid soils accounting for the most loss (84 kg ha^{-1}). The other soil-related constraints are salinity (30 kg ha^{-1}), zinc deficiency (22 kg ha^{-1}), and iron deficiency (18 kg ha^{-1}). Losses due to soil-related constraints are highest for this ecosystem compared to others. This suggests that as farmers increase yields with intensification of rice production, soil-related constraints become more important.

Losses from vegetative drought are substantial enough to rank in the top ten, anomalous since it is usually assumed that irrigated rabi environments have assured irrigation. In fact, this is not the case for all land described as irrigated rabi. As described earlier, tail-end users of irrigation systems often receive insufficient water. Moreover, water shortages sometimes occur due to things such as high seasonal demand or low levels of storage, ground or river water.

The only disease included in the ten most damaging factors was bacterial leaf blight (25 kg ha^{-1}). The dry weather during rabi season inhibits the propagation and spread of rice diseases. The list of yield losses (see Table 7.6) shows that there are losses from a number of diseases, but they are only a small fraction of the losses from insects and other constraints.

Crop-loss assessment in irrigated rabi environments requires looking at the absolute as well as the relative magnitude of losses. It would seem that the top five constraints (stemborer, weeds, acid soils, white-backed planthopper and salinity) are top priorities of research. But loss in irrigated rabi environments from the most damaging constraint is only equal to that of the tenth most damaging constraint in irrigated kharif environments and is clearly a function of land area devoted to rabi rice production. This is not to say that the priority constraints in rabi environments are not research priorities for eastern India, but that they must be assessed in the context of all the rice environments.

C. Rainfed lowland losses

Total losses due to technical constraints were higher in rainfed lowlands than in any other environment. The total annual loss of an estimated 4.6 million tons partly reflects the fact that the lowlands encompass more land than any other rice environment in eastern India. It also reflects the great potential for increasing yield in rainfed lowlands.

Yield losses from technical constraints were a large proportion of the estimated yield gap in rainfed lowlands. They accounted for 37% of yield gap II, that between actual and potential yields as measured in on-farm experiments. They accounted for 31% of yield gaps I and II, between actual and potential yields as measured in experiment station trials (see Table 7.7). Losses from technical constraints in rainfed lowlands are a smaller proportion of yield gaps than in irrigated rabi environments, showing that much of the lowland yield gap can be solved through both socioeconomic schemes and higher adoption of modern varieties.

As in irrigated rabi environments, the ten key constraints account for 68.8% of the total losses, with drought clearly the dominant constraint. Three of the top ten constraints were some form of drought. In order of importance they were: anthesis drought (109 kg ha^{-1}), seedling drought (42 kg ha^{-1}), and vegetative drought (26 kg ha^{-1}). Added together, they account for 25.5% of total losses from technical constraints in the rainfed lowlands of eastern India.

Losses from the other top seven constraints are much lower than drought-related losses. Weed losses in rainfed lowlands (86 kg ha^{-1}) are about half as much as drought losses. Submergence is a major problem in the rainfed lowlands (45 kg ha^{-1}) as is lodging, which causes considerable damage and is characteristic of the medium to tall varieties which are preferred.

Three diseases account for most of the disease loss in rainfed lowlands. Blast is the most damaging followed by bacterial leaf blight. These two are included in the top ten constraints. The third disease, sheath rot, causes a substantial loss and is ranked as the 13th most serious technical constraint in rainfed lowlands.

Yellow stemborer is the only insect that is a primary constraint, causing damage in many environments. Other insects cause significant damage too, but were *not* among the top ten constraints. These include the green leafhopper, caseworm, armyworm and leaf folder.

The dominance by the top two constraints clearly stratify the research priorities. With drought more than twice as damaging as the next constraint (weeds), and weeds almost twice as damaging as any of the next tier of factors, these two offer the greatest potential for yield improvement. They have also proven to be two of the most difficult yield-limiting constraints to solve.

D. Rainfed upland losses

Total production loss from technical constraints in rainfed upland environments (4.26 million tons year^{-1}) are higher than in any other environment except rainfed lowlands. Since rainfed uplands cover less area than rainfed lowlands, yield losses are therefore more acute (933 kg ha^{-1}). Of the total difference in yield between experiment station potential and actual farmer yield (yield gaps I and II), technical constraints comprise 55.8%. They constitute an astonishing 67.2% of the estimated yield gap II, indicating that of all possible constraints to rice production, technical constraints described here are the most critical in terms of yield foregone in eastern India.

The top ten constraints account for 77% of the total losses from technical constraints in this environment, showing that losses are concentrated. Three forms of drought were among the five most damaging constraints in the rainfed uplands. Together they account for 30.1% of the total losses from technical constraints. With drought a real possibility at all phases of growth, water status is the single overriding concern in upland rice areas.

Weeds are the second damaging constraint in rainfed uplands, as they are in lowland environments. However, the magnitude of weed losses in upland environments (183 kg ha^{-1}) is greater than in lowland environments, and adequate weeding is not done for a number of socioeconomic and technological reasons.

Acid soils cause the greatest loss in the next tier of constraints (70 kg ha^{-1}) with iron deficiency causing about half as much. These soil problems are concentrated in two different locations. Acid soils are a major problem in Orissa, southern Bihar and Madhya Pradesh. Iron deficiency is typical of upland areas in Northern Bihar and Uttar Pradesh.

Blast and brown spot are the major diseases of rainfed upland environments. Unlike certain soil constraints, these two diseases seem to be a problem in every upland location in eastern India. Blast is the most damaging (246,000 tons year^{-1}) and brown spot slightly less so (146,000 tons year^{-1}).

Two other factors were included among the ten key constraints of this ecosystem. Birds cause significant losses (151,000 tons year^{-1}). The other factor was called 'short duration' in the questionnaire. The losses attributed to this constraint (159,000 tons year^{-1}) arise from the fact that farmers grow short-duration varieties to escape drought conditions at anthesis. Since yield declines with a shorter growing season, all other factors being the same, the constraint captures this aspect of drought management.

As in the last section, both relative losses among constraints are as informative as absolute magnitudes in rainfed upland environments. Drought and weeds are three to four times as damaging as the next most severe constraint. All other constraints pale in comparison. The other top ten key constraints are much larger than the less damaging constraints, with losses dropping off rather quickly. Hence the 77% of total losses accounted for by the top ten constraints.

E. Deepwater losses

Total losses from technical constraints in deepwater environments are only a fraction of those in rainfed upland rice environments. Total yield loss was estimated to be 596,000 tons year^{-1}, and losses from individual constraints range from 86,000 tons (yellow stemborer) to 108 tons (hispa). In addition, several constraints which cause losses in other environments pose no problem to rice production in deepwater environments (see Table 7.6).

Losses due to technical constraints account for about half of yields foregone, as measured by yield gaps. Technical constraints make up 52.6% of yield gap II and 46% of yield gaps I and II. Thus, about half of the potential yield that remains unrealized is due to losses from technical constraints.

The top ten constraints, in terms of yield loss, account for 83.9% of total losses from technical constraints. Unlike in the other rice environments, no single constraint dominates and yield losses are spread among insects, diseases, water regime and soils.

The stemborer claims the greatest yield loss among the technical constraints (82 kg ha^{-1}). Alkaline soils are close behind (78 kg ha^{-1}), followed by weeds and submergence. These top four constraints account for over half of total yield losses in deepwater areas.

Water regime problems include both submergence and drought. Deepwater areas are normally typified by submerged plant conditions, but submergence loss refers to losses when the depth of water is greater than the plant's ability to elongate or survive the submerged conditions. Rice in the deepwater environment grows under dry conditions for 8–10 weeks before the floods arrive. Drought at the seedling stage causes up to 37,000 tons of annual loss in deepwater environments. Vegetative drought is another of the ten key constraints and occurs in instances when the monsoon is late in arriving.

Major diseases in deepwater environments include bacterial leaf blight and blast. Birds consume the same amount of rice in deepwater environments as is lost to blast.

The relative magnitude among technical constraints in deepwater environments do not show a clear stratification in terms of severity. Moreover, absolute levels of yield losses are low enough so that it is unclear how these constraints fit into the research priorities. Aggregated yield loss data, over all rice environments, can provide some clarification.

F. Yield losses for all of eastern India

Losses from technical constraints were aggregated across rice environments to characterize yield losses in the whole of the region. As expected from yield gap theory, total losses from technical constraints account for part of, but not all, the yield gap. Total losses from technical constraints in eastern India were estimated to be about 795 kg ha^{-1} amounting to 13.9 million tons year^{-1} (US$2.1 billion at constant prices). This makes up 45% of yield gap II and 37% of yield gaps I and II (see Table 7.7). Clearly, elimination of some of the major technical constraints could substantially increase rice production and narrow the yield gap.

The aggregate data for total losses in all environments show which classes of constraints are critical with respect to total losses. There is a vast difference among constraints (see Table 7.6). The top ten most damaging constraints are droughts, weeds, yellow stemborer, acid soils, bacterial leaf blight, lodging, blast, submergence, zinc deficiency and armyworm, and these account for two-thirds of the total losses due to technical constraints in eastern India. They can be divided into three strata, based on the magnitude of yield loss.

Drought and weeds constitute the first stratum of technical constraints. Weeds are nominally the most severe factor but are second to total drought factors, as in irrigated kharif and upland environments. Three forms of drought losses are among the top five constraints in eastern India. Anthesis, vegetative and seedling drought account for about 3 million tons of annual loss in the four states which comprise the survey area, constituting 21.5% of all losses from technical constraints. An enormous amount of rice, 1.95 million tons year^{-1}, is lost to weeds in the four states we surveyed. Solving, or partially mitigating, these two constraints would considerably increase rice production in eastern India.

The results have shown that the estimated yield losses from technical constraints in eastern India are substantial and account for a large proportion of yield foregone (via the yield gaps). Some constraints cause losses in every rice environment and some only in particular environments. The loss estimates also show great differences in the magnitude of losses among constraints in most rice environments and certainly for the whole region of eastern India. Based on these patterns of yield loss, it is possible to formulate a list or agenda of research priorities. These research priorities are delineated in Chapter 8 as the final exercise in this phase of the prioritization procedure.

V. Research Priorities, Current Progress and Future Needs

Based on the magnitudes of crop losses due to technical constraints, it is possible to identify a partial research agenda for a rice biotechnology program.* The constraints or factors to which research should be directed for a program in eastern India are presented in Table 7.9.

The major technical constraints to rice production cover all five of the categories studied. Mechanisms of tolerance to drought at all three growth stages (seedling, vegetative, anthesis) are crucial to increased productivity in the region. Genetic resistance to insects has a twofold advantage; it serves to increase yields and reduces the dependency of farmers on insecticides, reducing economic and health problems associated with insecticides. Significant losses from the stemborer, green leafhopper, armyworm, gall midge, gundhi bug and leaf folder demonstrate the acute need for genetic resistance. Varietal resistance to disease is needed since yield losses are high and alternative control methods are unavailable in many instances. The priority diseases in eastern India are bacterial leaf blight, blast, sheath rot, sheath blight and brown spot. Submergence tolerance is crucial since many areas experience periodic floods.

---

* We use the term 'partial' because this agenda applies only to those research topics related to field-oriented problems. There is an obvious need for research in basic sciences to develop and perfect techniques before work is possible on those field-oriented problems.

Table 7.9. Research priorities for rice biotechnology in eastern India.

Drought tolerance	Seedling, vegetative, anthesis
Insect resistance	Stemborer, green leafhopper, gundhi bug, gall midge, leaf folder, armyworm
Disease resistance	Bacterial leaf blight, blast, sheath rot, sheath blight, brown spot
Submergence tolerance	Seedling, vegetative
Weed control	
Adverse soil tolerance	Low fertility, acidity, salinity, alkalinity

Elongating varieties tend to lodge after flood waters recede, so the ability to withstand short-term inundation could prove very beneficial to large areas of eastern India.

Some mechanisms to reduce losses due to weeds are clearly necessary. Although many conventional methods are available, they are not successful enough to prevent huge yield loss. Early seedling vigour would help rice compete with weeds for light and nutrients. Selective herbicide resistance would allow rice to survive while non-rice plants are eliminated with herbicides.

Soils are another constraint for which there are some conventional solutions, but these methods of soil reclamation can be very costly and take many years. Biotechnology embraces a range of technical possibilities, the future potential of which is still being hypothesized. It is currently unknown how biotechnology may help ameliorate yield losses associated with adverse soil complexes such as the acid soils syndrome. However, basic research is progressing at a rapid rate and the transfer of genes for tolerance to alkaline and/or saline soils into rice is a distinct possibility.

Elimination or partial solution of these priority constraints would have a major impact on rice production in eastern India. Since the top ten constraints alone account for 61% of losses due to technical constraints, and technical constraints account for 44% of the average farm level yield, many millions of tons could be produced that are now lost.

An important aspect of these results is their wider applicability. These research priorities for biotechnology are based on estimates of yield loss from technical constraints. Other research programs intent on increasing rice production may find these estimates useful in directing their own research or conducting similar exercises in prioritizing a research agenda. It is possible to perform economic analyses on these data by adjusting the loss values for such things as: potential for solution via biotechnology, anticipated years to solution, cost and benefit streams over time and equity weights, which are attempted in section IV of this chapter. These manipulations might help in estimating returns to research in rice biotechnology, but have some severe limitations because of the amount of speculation involved. In addition, such analyses severely limit the accessibility of these findings to the wide range of scientists in rice research. For these reasons, the data were analysed at the level of yield loss. It is hoped that the enumeration of yield losses from technical constraints will also serve to increase the awareness, among rice scientists and administrators, of their relative importance in rice production.

References

Baker, R. and Pal, T.K. (1979) *Barriers to Increased Rice Production in Eastern India.* IRRI Research Paper Series No. 25. International Rice Research Institute, Los Baños, Philippines.

De Datta, S.K., Gomez, K.A., Herdt, R.W. and Barker, R. (1978) *A Handbook on the Methodology for and Integrated Experiment–Survey on Rice Yield Constraints.* International Rice Research Institute, Los Baños, Philippines.

Fertiliser Association of India (1988) *Fertiliser Statistics 1970–71 to 1987–88.* Fertiliser Association of India, New Delhi.

Government of India, Directorate of Economics and Statistics (DES) (1979, 1986) *Area and Production of Principal Crops in India.* DES, Government of India. New Delhi.

Government of India, Directorate of Economics and Statistics (DES) (1985) *Agricultural Situation in India.* DES, Government of India, New Delhi.

Huke, R.E. (1982) *Rice Area by Type and Culture in South, Southeast, East Asia.* International Rice Research Institute, Los Baños, Philippines.

International Rice Research Institute (IRRI) (1977) *Constraints to High Yields on Asian Rice Farms: an Interim Report.* IRRI, Los Baños, Philippines.

International Rice Research Institute (IRRI) (1979) *Farm Level Constraints to High Rice Yields in Asia: 1974–77.* IRRI, Los Baños, Philippines.

International Rice Research Institute (IRRI) (1984) *Terminology for Rice Growing Environments.* IRRI, Los Baños, Philippines.

Maurya, D.M., Bottrall, A. and Farrington, J. (1988) Improved likelihoods, genetic diversity, and farmer participation: a strategy for rice breeding in rainfed areas of India. *Experimental Agriculture* 24, 311–320.

Muralidharan, K. *et al.* (1988) Rice in eastern India – a reality. *Oryza* 25, 213–245.

Widawsky, D. and O'Toole, J.C. (1990) *Prioritizing the Rice Biotechnology Research Agenda for Eastern India.* The Rockefeller Foundation, New York.

8

Yield Gaps, Production Losses and Priority Research Problem Areas in West Bengal, India

N.K. Saha[1], S.K. Bardhan Roy[2], S.K. Ghosh[2] and M. Hossain[3]
[1] Directorate of Agriculture, Government of West Bengal, 17 S.P. Mukherjee Road, Calcutta 70025, India; [2] Directorate of Agriculture, Government of West Bengal, Calcutta, India; [3] Social Sciences Division, International Rice Research Institute, PO Box 933, 1099 Manila, Philippines

I. Introduction

West Bengal is the largest rice-producing state in India, and if it were a country, would rank sixth in the world with respect to land allocated to rice production – after China, Indonesia, Bangladesh, Thailand and Vietnam. Rice is the staple food of the people, it accounts for 95% of the foodgrains produced in the state, compared to 54% for India as a whole. The growth in rice production was relatively sluggish until 1975. Since then, an impressive growth has been achieved due mainly to large-scale private investment in shallow tubewells, strengthening of local government institutions and tenancy reforms. Despite this recent progress, rice yield remains low at 3.1 tons ha^{-1}, compared to 5.0 tons in Punjab and 4.7 tons in Tamil Nadu, the most agriculturally progressive states in India.

West Bengal has yet to achieve self-sufficiency in rice production despite the per capita availability of foodgrains from domestic production being about 250 kg year^{-1}. Rice is still imported from the surplus of the Indian states of Punjab and Andhra Pradesh, the amount depending on the rice harvest which fluctuates from year to year due to irregular monsoons. The state's population is still growing at 2.2% per year and is projected to reach 108 million by year 2025, from the 68.1 million recorded in the 1991 population census. The demand for rice is expected to increase by about 68% over the next 30 years, 58% due to population growth and 10% due to income growth-induced increase in per capita rice consumption. From the present level of 18.2 million tons, rice production must increase to 30.5 million tons by 2025 in order to avoid a further increase in imports and/or an increase in the price of rice relative to other foodgrains. As the land frontier has already been exhausted, the rice yield must increase to 5.2 tons ha^{-1} to meet that challenge.

This chapter complements that by Widawsky and O'Toole (Chapter 7). It aims to study constraints to increasing the productivity of rice land in order to identify priority areas for rice research. The approach is to identify the gaps between the potential and the actual farm yield in different rice-growing environments and agroclimatic zones and to assess the contribution of different biotic and abiotic factors that explain the gap. The study is based on yield data obtained from experiments in zonal adaptive research stations and the official statistics on average farm-level yields for different agroclimatic zones. Production losses due to various biotic and abiotic constraints have been estimated from a 1992 household survey for a study of changes in land-use pattern and productivity of rice conducted by the Department of Agriculture in collaboration with the IRRI. The sample consists of 586 parcels of land operated by 220 randomly selected households from 11 sites representative of different agroecological zones.

This chapter is organized as follows. Section II describes the rice research infrastructure in West Bengal and the focus of the research program. Section III describes the rice-growing environments in the State. Section IV estimates the yield gaps for different rice-growing environments and agroecological zones and assesses production losses due to various biotic and abiotic stresses to identify the priority research problem areas.

II. Research Infrastructure and Programs

The Rice Research Station at Chinsura operated by the State Department of Agriculture is the backbone of the rice research program in West Bengal. The Indian Council of Agricultural Research (ICAR) substation located at Canning is engaged in basic and strategic research of national importance for coastal wetlands and saline affected areas. The State Agricultural University at Kalyani is engaged in basic and strategic research on rice under various programs of the national agricultural research system for different agroclimatic zones of West Bengal. Basic research on rice biotechnology is conducted at the Bose Institute, University of Calcutta and the Vivekananda Institute of Biotechnology under Ram Krishna Ashram in the District of South 24-Parangas. Research on dissemination and adoption of rice technology, constraints to adoption, cost–benefit analysis and overall socioeconomic impact at the farm level in different agroclimatic subzones/subregions is currently conducted by the evaluation wing of the State Department of Agriculture.

The Chinsura Rice Research Station is presently engaged in research to evolve location-specific rice varieties for different ecosystems, to conduct adaptive and on-farm varietal trials and to demonstrate the efficient use of fertilizers and integrated pest and water management. The station also produces breeder seeds of released rice varieties for subsequent multiplication and dissemination to the farmers in different agroclimatic environments through the State Seed Corporation. There are six zonal adaptive research stations and

52 subdivisional/adaptive research farms. The focus of research in these stations is to develop short-duration rice varieties that are photoperiod insensitive, insect–disease resistant and tolerant to drought and flood, as well as improved soil and crop management techniques suited to different ecosystems under varied agroclimatic zones. Rice research programs currently in operation in the state are as follows.

Germplasm improvement
1. Genetic enhancement of short-duration high-yielding rice varieties suited to different rice-based cropping systems.
2. Breeding for development of drought-tolerant high-yielding rice varieties for upland ecosystems.
3. Long-duration submergence-tolerant high-yielding rice varieties for rainfed lowlands.
4. Photoperiod-sensitive varieties tolerant to flooding for flood-prone environments.
5. High-yielding rice varieties for wet and dry seasons for irrigated conditions.
6. Cold tolerance at different stages of growth for dry seasons in irrigated conditions.
7. High-yielding rice varieties with genes for resistance to major insects and diseases for all rice-growing conditions.
8. Hybrid rice varieties for irrigated conditions.

Rice physiology and changing plant architecture
1. Evaluation of rice varieties for low light intensity, waterlogged conditions and elongation ability.
2. Screening for cold and salinity tolerance, both at vegetative and reproductive stages.
3. Study on factors for heterosis (vigour) in rice hybrids.
4. Moisture stress in rice cultivation in upland situations.

Nutrient management
1. Micronutrient management for deficient soils.
2. Mineralization of nitrogen applied through organic and inorganic sources in relation to nitrogen-use efficiency.
3. Evaluation of rice varieties for phosphate-deficient soils to assess phosphorous utilization efficiency.

Plant protection
1. Screening for resistance to bacterial blight and sheath blight, leaf and neck blast and tungro virus.
2. Understanding the ecology of the stemborer, brown planthopper, green leafhopper and leaf folder.
3. Evaluating potential biocontrol agents for pests and diseases in irrigated ecosystems.
4. Trials on integrated pest management.

III. Rice-Growing Environments

Rice is grown in different environments from the saline affected coastal areas in the south to the terraced land in the hills in the Himalayan range in the north at an altitude of 1300 m, and from the subhumid drought-prone plateau region in the southwest with an annual precipitation of 1100 mm to the humid alluvial plains in the northeast with an annual rainfall of 3600 mm. The state is divided into two broad agroecological zones and seven subregions on the basis of differences in soil characteristics, topography, rainfall and temperature. The geographical boundaries covered by these zones are listed in Table 8.1.

Rice is grown under diverse ecological situations in three seasons. The premonsoon rice, known as aus, covers April to July in the northern region and May to September in the southern region and accounts for about 10% of total rice area. This is a low-yielding relatively drought-tolerant upland crop with a yield of 1.5–2.0 tons ha^{-1}. With the expansion of irrigation facilities, the area under this crop has been declining. The monsoon rice, known as aman, is grown from July to December. It accounts for 75% of the total rice area, and is grown under rainfed conditions in the medium-deep and deep-flooded land (mostly traditional varieties) and under irrigated conditions in the flood-free shallow lowlands (mostly modern varieties). The remaining 15% of the rice area is covered by the winter or dry season rice known as boro. This is an irrigated rice crop with the entire area cropped with modern varieties. The growing season for this crop overlaps with aus. With the expansion of irrigation facilities, farmers have been releasing land from aus and deepwater aman rice for raising the boro crop.

The distribution of the rice-harvested area by ecosystem is shown in Table 8.2. Rainfed rice is 63% of the area, 43% in the lowlands and 10% each in the uplands and in deep-flooded areas. Farmers practice intensive double-cropped irrigated rice cultivation (boro followed by aman) in 15% of the area, mostly in the new alluvium Gangetic plains and in coastal zones. The double cropping of rice has been encouraged by private investment in shallow tubewells and

Table 8.1. Agroclimatic zones in West Bengal, India.

Agroclimatic zones	Districts covered
Eastern hills and plateau	
Hills	Darjeeling
Terai	Jalpaiguri and Cooch Behar
Plateau	Purulia
Lower Gangetic plains	
Old alluvium	West Dinajpur and Malda
New alluvium	Murshidabad, Nadia, Hooghly, Burdwan and North 24-Parganas
Lateritic	Birkum, Bankura and Midnapore (West)
Coastal saline	South 24-Parganas, Howrah and Midnapore (East)

pumps. Rice grown with supplementary irrigation during the monsoon season accounts for 21% of the area. In the eastern hills, terai and plateau (part of the Chotanagpur Plateau that covers southern Bihar and eastern Orissa) most of the rice is grown under rainfed conditions.

Rice yield varies widely across seasons and across ecosystems and agroclimatic zones (Table 8.3). It is the lowest in the uplands (2.8 tons ha^{-1}) and in the flood-prone ecosystem (1.9 tons ha^{-1}). In the uplands farmers grow mainly direct-seeded aus rice which depends on the vagaries of the monsoon, resulting in frequent droughts and crop failures. For traditional varieties that are resistant to drought farmers get a yield of 1.5–2.0 tons ha^{-1}. Some farmers grow modern varieties by irrigating the crop during the establishment and early vegetative stage and depending on monsoon rains during the latter stages and get 4–5 tons of yield. In the flood-prone environment farmers grow mostly traditional direct-seeded aman varieties that are tall and can elongate with rising flood waters. In normal years it gives a yield of 2.5–3.0 tons ha^{-1}, but crop failures are common due to abnormal flooding and uneven distribution of rainfalls. In the rainfed lowlands rice is mostly transplanted. Farmers choose traditional or modern varieties depending on the elevation of individual parcels, the expected duration of the monsoon season and the availability of supplementary irrigation facilities. Traditional varieties yield 2.5–3.0 tons ha^{-1} and

Table 8.2. Rice area by ecosystem and agroclimatic zones in West Bengal, India (1992–1993).

Agroclimatic zones	Irrigated Wet season	Irrigated Dry season	Rainfed Upland	Rainfed Lowland	Flood-prone	Total
Eastern hills and plateau						
Hills	–	0.001	0.046	–	–	0.047
Terai	–	0.013	0.144	0.413	0.010	0.580
Plateau	0.028	0.001	0.004	0.255	–	0.288
Lower Gangetic plains						
Old alluvium	0.207	0.115	0.059	0.310	0.018	0.709
New alluvium	0.389	0.384	0.199	0.585	0.184	1.741
Lateritic	0.388	0.130	0.106	0.583	0.101	1.308
Coastal saline	0.205	0.218	0.012	0.307	0.281	1.023
West Bengal	1.217 (21.4)	0.862 (15.1)	0.570 (10.0)	2.453 (43.1)	0.594 (10.4)	5.696 (100.0)

Figures within parentheses are percent of total rice area.
Source: Department of Agriculture, Government of West Bengal, 1994.

modern ones 3.5–4.5 tons. A late season drought with early recession of the monsoon rains (in October) frequently reduces the yield of both traditional and modern varieties. In the boro season (irrigated dry ecosystem) all farmers grow modern varieties with full irrigation. The crop benefits from low cloud cover and high sunshine and cool nights, as in temperate regions, and gives from 5.0–6.5 tons ha^{-1}, depending on the quality of the soil and the amount of chemical fertilizers applied. But the yield is sometimes reduced by abnormal low temperatures in January and February, localized hail storms during March–April and early flash floods in May.

The average rice yield achieved in the state is the result of the composition of area in different ecosystems and growing seasons, as well as the rate of adoption of modern varieties and the yield of traditional and modern varieties for specific ecosystems and seasons. The estimates of yield gaps and production losses thus should be made for specific environments and ecosystems.

IV. Yield Gaps and Production Losses

This section first assesses the yield gaps between the potential and actual farm yields for different rice-growing environments and agroclimatic zones, and then estimates the losses due to various biotic and abiotic stresses. The yield gap can be divided into two parts (IRRI, 1979). Yield gap I is defined as the difference between the yield obtained in experimental stations and that obtained in on-

Table 8.3. Rice yield by ecosystem and agroclimatic zones in West Bengal, India (1992–1993).

Agroclimatic zones	Irrigated Wet season	Irrigated Dry season	Rainfed Upland	Rainfed Lowland	Flood-prone	Total
Eastern hills and plateau						
Hills	–	–	2.2	–	–	2.2
Terai	–	–	2.1	2.0	1.8	2.1
Plateau	3.7	5.0	1.5	2.3	–	2.4
Lower Gangetic plains						
Old alluvium	2.6	4.9	2.2	2.5	1.5	2.8
New alluvium	4.0	5.2	3.2	4.0	2.0	4.0
Lateritic	4.2	4.8	3.2	3.1	1.8	3.5
Coastal saline	2.4	4.5	3.6	3.7	2.0	3.1
West Bengal	3.5	4.9	2.8	3.1	1.9	3.3

farm experiments, i.e. the gap due to environmental factors that cannot be managed by farmers. Yield gap II is defined as the gap between the maximum yield in on-farm experiments and the average farm yield. This gap reflects biological constraints, soil and water constraints and socioeconomic constraints. The primary focus of this section is estimation of yield gap II and the contribution of different biotic and abiotic stresses to this yield gap.

Table 8.4 reports the yield gaps for different rice ecosystems and illustrates several points. First, the yield obtained in experimental stations varies greatly across rice ecosystems, from about 4.0 tons ha^{-1} in the uplands and flood-prone ecosystems to 6.9 tons ha^{-1} in the irrigated ecosystem for the dry season. For the monsoon season the difference in the experimental yield between the irrigated and the rainfed crop is relatively low (0.3 ton ha^{-1}), but for the irrigated conditions the difference between the wet season and the dry season crop is about 1.9 tons ha^{-1}. This suggests that environmental factors such as cloud cover, sunshine, variations in day and night temperatures and continuous saturation of land (rather than alternate wet and dry conditions as with irrigation) make a big difference in the yield potential. Second, yield gap I, i.e. the difference in yields between experimental stations and on-farm trials, is relatively low for West Bengal. The average is only 0.37 ton ha^{-1} and it varies from 0.14 ton for the rainfed lowlands to 0.86 ton for the uplands. Third, the gap between the yield in on-farm trials and the average obtained in farmers' yield, i.e. yield gap II, is quite large. It varies from 0.4 ton in the uplands to about 1.6 ton in the rainfed lowlands and the irrigated dry season system. The average yield gap II is estimated at 1.32 tons ha^{-1}. If it were eliminated rice production in the state could increase by 39%. An earlier study done by the Bidhan Chandra Agricultural University (1984) estimated the yield gap for 1982 at 1.82 tons ha^{-1} for the rainfed lowlands and 1.0 tons ha^{-1} for the uplands.

Table 8.4. Estimates of yield gap by ecosystem in West Bengal, India (in kg ha^{-1}).

Rice ecosystem	Experimental station yield	Estimated on-farm trials	Actual average farm yield	Yield gap I	Yield gap II
Irrigated					
Wet season	5055	4355	3569	700	786
Dry season	6936	6652	4954	284	1698
Rainfed					
Upland	4063	3201	2800	862	401
Lowland	4806	4665	3104	141	1561
Flood-prone	3916	3680	1949	236	1731
West Bengal	5036	4670	3350	366	1320

Source: personal communications from rice research stations and official statistics from the Department of Agriculture, Government of West Bengal.

Chapter 7 assessed rice production constraints in eastern India by interviewing rice scientists with a structured questionnaire. For West Bengal they found the major constraints for rainfed lowlands as (i) submergence and droughts at the seedling stage; (ii) the stemborer, brown planthopper, leaf folder and green leafhopper as the major insects; (iii) bacterial leaf blight, brown spot and sheath rot as the major diseases; and (iv) soil salinity and weeds as the other constraints. For the uplands, the major constraints were identified as droughts, the stemborer and green leafhopper, blast, brown spot, acid soils and weeds. Drought was found to be a major constraint even under irrigated conditions because the quality of irrigation is poor and there could be prolonged moisture stress even during rainy seasons due to the lack of rain for a number of consecutive days and the early recession of the monsoons. Weeds are generally controlled by manual labour, but some farmers abandon weeding due to heavy infestation after prolonged rains and the very low marginal productivity of weeding labour. For eastern India, the loss due to drought was estimated at 123 kg ha^{-1} for rainfed lowlands and 300 kg ha^{-1} for the uplands. Weeds were found to be the second major constraint with a loss estimated at 60 kg ha^{-1} for the rainfed lowlands and 195 kg ha^{-1} for the the uplands. The most important disease was blast which caused a production loss of 25 kg ha^{-1} for the lowlands and 58 kg ha^{-1} for the uplands.

We estimated the yield losses from various biotic and abiotic stresses from a farm household survey conducted by the Department of Agriculture in 1992. The estimates for the two major seasonal varieties, the wet season aman rice and the dry season boro rice are reported in Table 8.5. For the aman rice the actual farm yield was estimated at 2.56 tons ha^{-1} while the production losses due to

Table 8.5. Yield losses from biotic and abiotic stresses, estimates from farm household survey, 1992.

Variables	Dry season irrigated rice	Wet season rainfed rice
Number of samples	274	312
Average size of rice farm (ha)	0.73	0.79
Average yield (kg ha^{-1})	5176	2555
Estimates of yield loss (kg ha^{-1})	760	1014
Normal yield (kg ha^{-1})	5936	3569
Percent of yield loss due to		
Insects	36.7	5.4
Diseases	24.2	21.1
Weeds	0.0	2.3
Drought	36.8	0.0
Submergence	0.1	71.2
Others	2.8	0.0

Source: Farm household survey, Department of Agriculture, Government of West Bengal, 1992.

various biotic and abiotic stresses were 1.0 ton ha^{-1}. Thus, if the losses are controlled, aman production could be increased by another 40%. For the boro season the sample farmers reported an average yield of 5.2 tons ha^{-1}, and a loss of 0.76 ton ha^{-1} mostly due to insects and diseases. If these losses are controlled, the irrigated rice production could be increased by another 15%. Thus, controlling the yield losses in the rainfed aman crop deserves higher priority, because of the importance of aman rice and the higher yield losses.

The severity of various biotic and abiotic stresses, the probability of their occurrence and the loss of production when they occur are reported in Tables 8.6 and 8.7, respectively, for the aman and boro crop. The problems that commonly occur during the aman season are reported as weeds, submergence, the brown planthopper, leaf roller and stemborer, bacterial blight, sheath blight and blast. The highest intensity of damage in the affected land, if unprotected, are caused by submergence, tungro virus, sheath blight and the brown planthopper. But the probability of the occurrence of tungro virus is reported as very low. For the dry season the most common problems affecting the rice crop were reported as the gundhi bug, rice hispa, leaf roller, caterpillar and drought, but the intensity of loss in the affected land are the highest for the brown planthopper, blast, caterpillar, drought and bacterial leaf blight.

Table 8.6. Estimates of crop losses in rainfed lowland (aman) rice, 1992.

Problem areas	Percent of farms reporting losses from problem in 1992	Probability of occurrence of problem	Production loss in affected area (kg ha^{-1})	Production loss per unit of area under crop (kg ha^{-1})
Insects				
Brown planthopper	3.2	0.38	2026	25
Caterpillar	2.6	0.14	376	15
Leaf roller	1.9	0.32	330	1
Stemborer	1.3	0.30	768	1
Hispa	1.0	0.20	162	1
Diseases				
Tungro virus	4.5	0.05	2580	113
Bacterial leaf blight	4.2	0.30	620	63
Sheath blight	4.5	0.30	645	28
Blast	1.9	0.30	2087	9
Weeds	1.8	1.00	1566	28
Abiotic stresses				
Submergence	28.0	0.50	2589	722

Source: Farm household survey, Department of Agriculture, West Bengal, 1992.

Table 8.7. Estimates of crop losses in irrigated dry season (boro) rice, 1992.

Problem areas	Percent of farms reporting losses from problem in 1992	Probability of occurrence of propblem	Production loss in affected area (kg ha^{-1})	Production loss per unit of area under crop (kg ha^{-1})
Insects				
Brown planthopper	12.5	0.14	2906	231
Caterpillar	7.3	0.46	17836	38
Leaf roller	5.1	0.40	833	6
Stemborer	4.0	0.93	848	2
Hispa	0.7	0.65	612	2
Gundhi bug	0.7	0.70	450	–
Diseases				
Blast	21.2	0.27	2436	168
Bacterial leaf blight	3.6	0.28	1687	16
Weeds				
Abiotic stresses				
Drought	24.8	0.30	1836	200
Heat	14.6	0.26	360	80
Cold	4.7	0.12	617	16
Flash floods	0.7	0.04	331	1

Source: Farm household survey, Department of Agriculture, West Bengal, 1992.

The ranking of major problem areas for production loss in the rainfed lowland and irrigated dry season rice has been made on the basis of the estimated loss per unit of area under aman rice (Table 8.8). Submergence tops the list causing maximum yield loss. Stagnation of rainwater in poorly drained areas and frequent flood are responsible for this constraint. Salinity is also widely prevalent in the waterlogged coastal zone. A breakthrough in these problems can be achieved by evolving suitable varieties tolerant to submergence, flooding and salinity. The brown planthopper also causes severe yield losses; as a research problem it ranks first for irrigated dry season rice and fifth for rainfed lowland rice. Other insects causing crop losses are the caterpillar, stemborer, gundhi bug, leaf roller and rice hispa. Rice blast is endemic in areas where humidity is high. It ranks third as a problem area for the irrigated dry season rice and eighth for the rainfed lowland rice. Tungro virus is severe in lowland areas causing substantial yield loss and it ranks second as a problem area for the rainfed lowland rice. The other diseases causing yield loss in the lowlands are bacterial leaf blight and sheath blight which rank third and fourth, respectively. Weeds are another disturbing aspect in lowland rice

Table 8.8. Ranking of rice yield constraints by ecosystem in West Bengal, India.

Rank	Production constraints in Rainfed lowland rice	Production constraints in Irrigated dry season rice
I	Submergence	Brown planthopper
II	Tungro virus	Drought
III	Bacterial leaf blight	Blast
IV	Sheath blight	Heat
V	Brown planthopper	Caterpillar
VI	Weeds	Bacterial leaf blight
VII	Caterpillar	Cold
VIII	Blast	Stemborer
IX	Leaf roller	Gundhi bug
X	Stemborer	Leaf roller
XI	Rice hispa	Flash floods

Source: Farm household survey, Department of Agriculture, West Bengal, 1992.

production; they are very acute in water stress areas. Weeds rank sixth in order of priority because farmers' usually control weeds with manual labour and hence the actual production loss is much lower.

Drought at the anthesis stage due to water stress causes severe yield loss in irrigated dry season rice. The other abiotic stresses in the irrigated dry season rice, causing significant yield loss, are heat and cold which rank fourth and seventh, respectively. There is an urgent need for research on the major problem areas to combat crop damage and yield loss due to major insects, diseases and abiotic stresses. The other problem areas like soil acidity in the terai and lateritic zones, imbalance of fertilizer use, water management in the dry season, late planting and the use of aged seedlings in the lowlands, scarcity of labour, power shortages and lack of credit facilities to resource-poor farms also require research and policy interventions to boost rice productivity.

The survey conducted by the Department of Agriculture also asked farmers the reasons for choosing varieties grown in different parcels of land in the wet and the dry season. The responses are reported in Table 8.9. During the dry season farmers grew modern varieties in all parcels, while in the aman season they were grown in 60% of the parcels. In the dry season the criteria for choosing cultivars were reported as high yield (38%), good eating quality (24%), short duration (21%) and disease and insect resistance (11%). But for the wet season the most important criterion for selection was submergence tolerance, (66%), followed by eating quality (12%), high yield (9%) and elongation ability (7%). Thus, high yield and resistance to insects and diseases, which rice breeders consider as the most important criteria for selecting and releasing improved varieties, are not the same criteria that farmers use, particularly for the wet season.

Table 8.9. Farmers' criteria for selected rice cultivators in wet and dry seasons, 1992.

	Percent of responses	
Criteria	Wet season ($n=1613$)	Dry season ($n=959$)
High yield	8.7	37.5
Disease and insect resistance	4.5	11.0
Submergence tolerance	66.3	3.7
Elongation ability	7.2	0.0
Short duration	0.3	21.1
Taste, milling recovery, quality and grain expansion during cooking	11.5	23.8
Others	1.5	2.9

Farmers reported multiple responses regarding the criteria for choice. We considered the first three responses in order of importance. The 'other' category includes grain size, aromatic flavour and drought resistance at the anthesis stage.
Source: Farm household survey, Department of Agriculture, Government of West Bengal, 1992.

V. Summary and Conclusions

In West Bengal the demand for rice is expected to increase by about 68% over the next 30 years. As the land frontier has already been exhausted, rice yields must increase to about 5.2 tons ha^{-1} from the present level of 3.1 tons ha^{-1}, in order to meet the growing demand for staple grains. A large part of the demand could be met with existing technology if further investment in irrigation and drainage helped to transform the rainfed environment (now about 63% of rice area) into irrigated land. Farmers have already achieved an average yield of 5.2 tons ha^{-1} in the best irrigated environment (in new alluvium lower Gangetic plains) while the average yield is 3.1 tons ha^{-1} for the rainfed lowlands and 1.5–2.0 tons ha^{-1} in the flood-prone and upland ecosystems.

A substantial increase in production could also be achieved if rice research helps reduce production losses due to various biotic and abiotic stresses in specific rice ecosystems. The gap between the maximum yield obtained in on-farm experiments and the average that farmers get is about 50% of present yield for the rainfed lowland ecosystem and 34% for the dry season irrigated rice ecosystem. A large part of the gap is from production losses due to biotic and abiotic stresses; about 40% of the present yield for the wet season rainfed crop and 15% for the dry season irrigated crop. The priority research problem areas are submergence, tungro virus, bacterial blight, weeds and the brown planthopper for the wet season; and the brown planthopper, drought, blast and heat for the dry season.

Farmers consider a large number of factors in choosing rice cultivars besides high yield, short duration and resistance to insects and diseases – the ones that breeders consider to be the most important in selecting and releasing improved varieties. Other factors that farmers give more weight are milling (recovery of head rice), eating quality, expansion during cooking, resistance against temporary submergence and resistance to prolonged droughts. Breeding strategies must consider these factors for a wider acceptance of modern varieties among farmers.

References

Bidhan Chandra Agricultural University and Indian Statistical Institute (1984) *Constraints to Technology Progress in Rice Cultivation: an Experiment Survey Research in West Bengal (mimeo)*. Bidhan Chandra University, Kalyani, West Bengal, India.

Government of West Bengal (1994) *Estimates of Area and Production of Principal Crops in West Bengal, 1992-93*. Socioeconomic and Evaluation Branch, Department of Agriculture, Calcutta.

International Rice Research Institute (IRRI) (1979) *Farm Level Constraints to High Rice Yields in Asia: 1974–77*. IRRI, Manila, Philippines.

9. Constraints to Higher Rice Yields in Different Rice Production Environments and Prioritization of Rice Research in Southern India

C. Ramasamy, T.R. Shanmugam and D. Suresh
Department of Agricultural Economics, Tamil Nadu Agricultural University, Coimbatore 641 003, India

I. Introduction

Rice is the staple food for most of the Indian population and particularly for southern and eastern Indians. There has been a giant leap in rice productivity and production in the 1960s and 1970s after the introduction of high-yielding, short-duration and photo-insensitive varieties. The increase in area of rice cultivation, generation of improved rice production technologies and the enthusiasm of rice farmers in adopting improved production technologies were major factors behind increases in rice production which matched the demand from a growing population.

However, the trend in rice production in southern and northern India over recent years indicates this that scenario is changing. Rice production in southern India has developed slowly in recent years due to many constraints. The prospects for higher productivity in rice appear gloomy in view of the fact that farmers' yields on irrigated land are about to reach the potential of existing technologies in this environment. A breakthrough is possible only by the revolution of another kind in rice production by way of increasing biological efficiency through hybridization of rice varieties, and by unfolding new frontiers of production technologies with biotechnology and wide hybridization.

It is important that we identify the constraints which operate to keep rice yields significantly below their potential maximum and to channel efforts to increasing yields by solving the constraints, particularly in the 'low-yield' regions and in farms with low yields within given regions. The approach should be to identify the nature of the yield gap between the potential and what farmers actually get. Once the magnitude and nature of the yield gap is established, the second aspect is to identify biological, physical and socio-economic factors that explain the gap.

This chapter seeks to identify production constraints which contribute to major yield losses in rice, to prioritize research areas and to explore the possibilities of introducing biotechnological initiatives to solve such problems in southern India. While prioritizing research areas, the study also recommends research investments to solve each constraint based on cost–benefit analyses. The study also suggests research alternatives which could develop cost savings and ecologically sustainable technology, particularly biotechnological methods to resolve the constraints in rice production. The southern states of India (i.e. Andhra Pradesh, Karnataka, Kerala, Tamil Nadu and Pondicherry (Union Territory)) are important rice-producing states in India. The study will help to prioritize rice research areas in southern India and guide the allocation of research funds granted by state, national and international development agencies.

II. Methodology

The present study broadly adopted the procedure suggested by Herdt and Riely (1987). The steps as implemented in the present study are detailed below.

A. Defining the challenge and constraints

An exhaustive list containing all the possible constraints to higher yields of rice in different production environments, with special reference to southern India, was prepared on the basis of the knowledge and experience of the scientists involved in rice research projects. A short list of leading problem areas was prepared after consulting the scientists and managers of agricultural extension activities. Thus, 43 constraints were identified, on which information was collected. The constraints include biotic and abiotic factors that limit rice yields. They are distinct from socioeconomic constraints in that many are potentially solvable through biotechnology or enhanced genetic improvement of rice varieties. We categorized the technical constraints as:

1. Insects and pests.
2. Diseases.
3. Adverse soils/agronomy.
4. Genetic/physiological.
5. Adverse climatic and environmental factors.

B. Estimation of intensity/severity of constraints

The severity of each constraint was estimated through the quantification of yield loss. The effects of constraints account for the gap between potential farm yield and actual farm yield. For each agroclimatic region – rice production environment – the absolute quantity of yield loss attributed to each constraint was estimated by knowledgeable scientists. These scientists were also required to estimate the proportion of area affected by each constraint to the total rice area of the region. While identifying constraints in the region, scientists were reminded to be specific to sub-production environments within the region. To be precise and authentic in the estimation of yield losses, scientists were sought who could estimate yield loss due to constraints which were relevant to their area of specialization. For instance, an entomologist was asked to estimate yield loss due to insects and pests above the economic threshold level. The total area affected by each constraint in each region was arrived at by summing up the areas affected in different sub-production environments within a region. The scientists were asked to indicate the average yield loss due to a particular constraint over the last 5 years. In order to cross-check yield-loss estimates due to production constraints, views of extension personnel who had long field experience in rice production in that region were obtained. The same procedure was used for the estimation of loss by the extension personnel.

The average value was considered as the loss due to the constraint. Sums of losses across the regions gave the total production foregone to each constraint in each state. Production foregone was then multiplied by the prevailing market price of rice in the state to obtain the value of foregone production due to each constraint.

Prices as recommended by the Commission of Agricultural Cost and Prices (CACP) for rice were used to forecast prices for the next 5 years. Forecasted prices were used to derive the future stream of benefits which will accrue by solving the constraint.

C. Estimating costs

Each constraint is associated with a cost of research and extension required to solve the problem. The stream of annual expenditures required for research and extension to solve the constraint, assuming a given probability of success in a given time involving several years, were elicited from senior scientists and extension specialists who have a fair idea of the costs incurred in research and extension activities. If a given problem could be addressed by alternative research approaches, the alternative which has the lowest externality was considered in the decision process. The conventional cost–benefit approach was used to evaluate each problem. Scientists were asked to add 10% of total research cost as environmental cost wherever it is was applicable to the cost of solving each technical constraint so as to reflect economic costs or returns to the society.

D. Estimation of net present value (NPV) and benefit–cost ratio (BCR)

The present worth of the net benefits of a research project is obtained by deducting the present worth of research costs from the present worth of research benefits. It may be interpreted as the present worth of the net benefit stream generated by the research investment in solving the constraints to higher rice yield. Mathematically,

$$NPV = \sum_{t=1}^{n} \frac{B_t}{(1+i)^t} - \sum_{t=i}^{n} \frac{C_t}{(1+i)^t} \qquad (9.1)$$

where B_t = benefit obtained by solving each constraint in years t. This is calculated by multiplying the annual production loss due to each constraint in the zone by price of rice in years t. It is a proxy measure for the productivity gain. C_t is the research and extension cost allocated for undertaking the research project to solve each constraint and the extension cost allocated for the dissemination of technology which emerges out of research efforts in year t. n is the number of years to achievement and the number of years in which benefit would flow due to solving constraints. Research costs are incurred for the first 5–6 years. During that period, benefits are assumed to be zero. Benefits may accrue for an infinite number of years after the research accomplishment, but it is limited to 5–6 years here, as this is considered to be a reasonable period of time. After that time, the period value of benefits becomes insignificant because of the high rate of discounts. During the period of benefit flow extension costs are incurred for transmitting technologies. The discount rate i is the opportunity cost of capital. The discount rate chosen for this economic analysis is 15%.

The research problem areas which have positive NPVs can be identified and prioritized. A given production constraint can be addressed by many research alternatives with different NPVs. Also, the same research alternatives may have differential impacts across rice production ecologies. In situations where resources are scarce and projects are mutually exclusive, the NPV is a more appropriate measure than the BCR.

The BCR is the ratio obtained when the present worth of a benefit stream of a project is divided by the present value of the research cost stream:

$$BCR = \frac{\sum_{t=1}^{n} \frac{B_i}{(1+i)^t}}{\sum_{i=1}^{n} \frac{C_i}{(1+i)^t}} \qquad (9.2)$$

B_t, C_i, n and i are as explained above.

If BCR worked out to be greater than one, then the present worth of research would have exceeded present worth research costs. The constraints which have BCR greater than one could be identified and prioritized.

E. Addressing the challenges through biotechnological initiatives

The potentials of biotechnology to solve rice production constraints is still under test. Nevertheless, its cost effectiveness at the farm point, after establishing its potential through on-farm trials, is beyond doubt. This is the fundamental reason we seek to explore the possibilities of introducing some biotechnological initiatives to solve rice production constraints. Judgements of knowledgeable scientists were sought to determine the rate of success of technological alternatives to solve a given constraint. Based on this, the BCR was worked out for solving each of the constraints through biotechnological alternatives and other approaches for biotic, abiotic and socioeconomic constraints. (See Chapter 21 for a further evaluation of biotechnology research.)

III. Rice Research in India: Institutions and Projects

A. Institutions and projects

India has one of the largest agricultural research systems in the world with a large number of research scientists. In 1989 the Indian Council of Agricultural Research (ICAR) Review Committee, headed by Dr Rao, estimated that India spent 0.17% of its gross agricultural domestic product on agricultural research. The National Agricultural Research System (NARS) in India has two distinct components – ICAR at the national level and the State Agricultural Universities (SAUs) at state level. ICAR institutions are engaged in basic and strategic research of national importance, while the SAUs are responsible for the generation and diffusion of farm technologies relevant for the agricultural development of respective states. The ICAR institutions engaged in rice research are the Central Rice Research Institute, Cuttack and Directorate of Rice Research, Hyderabad. All the states have regional research stations to conduct research on location-specific problems. These regional research stations had been strengthened during the 1980s under the National Agricultural Research Program (NARP) with funding from World Bank. The NARP covered 120 agroclimatic zones of the country. The major research emphasis was on applied research, mainly aiming at increased food production. Rice research was given top priority among foodgrains research.

Emphasis in rice research was on the development of high-yielding varieties, evolving pest and disease resistance in rice plants and designing improved soil and crop management techniques. The following is a synopsis of rice research projects currently in operation in India with relevance to major areas.

B. Rice varietal improvement

- Tissue culture techniques to hasten the Basmati rice breeding program by using somaclonal variation for identifying productive plant types and resistance to biotic stresses have been taken up at Hissar (Haryana), the Indian Agricultural Research Institute (New Delhi), Raipur (Madhya Pradesh) and the Directorate of Rice Research (Hyderabad). The genetic enhancement of yield and quality rices for export are being investigated at Hissar (Haryana), Pantnagar (Uttar Pradesh), the Directorate of Rice Research (Hyderabad) and Kapurthala (Punjab). The systematic screening of released varieties of proven merit is being done at the Indian Agricultural Research Institute (New Delhi), Kapurthala (Punjab), the Central Rice

Research Institute (Cuttack) and the Directorate of Rice Research (Hyderabad).
- There are breeding programs for the development of high-yielding export-quality non-basmati rices incorporating pest and disease resistance at Nellore (Andhra Pradesh), Faizabad and Varanasi (Uttar Pradesh), Raipur (Madhya Pradesh), Banswara (Rajasthan) and Siruguppa (Karnataka).
- Studies on the influence of the environment (ageing, storage and shelf life) on quality indices and standardization of post-harvest handling of non-basmati rices for high milling turnout are being undertaken at Kapurthala (Punjab), Hissar (Haryana) and the Directorate of Rice Research (Hyderabad).

C. Rice crop management

- There is research into zinc management in irrigated rice-based cropping system being done at Moncompu (Kerala), Navsari (Gujarat), Raipur (Madhya Pradesh) and Ramachandrapuram (Andhra Pradesh).
- The integrated management of alkaline soils for rice cultivation is being researched at Mandya (Karnataka), Karnal (Haryana) and Kanpur (Uttar Pradesh).
- Studies on the mineralization of nitrogen applied through organic and inorganic sources in relation to nitrogen use efficiency are being undertaken at Mandya (Karnataka), Faizabad and Kanpur (Uttar Pradesh).
- Maruteru (Andhra Pradesh), Mandya (Karnataka) and Aduthurai (Tamil Nadu) are evaluating rice genotypes for phosphorous utilization efficiency.
- Long-term soil fertility in rice-based cropping systems is being evaluated at Maruteru (Andhra Pradesh), Mandya (Karnataka) and Faizabad (Uttar Pradesh).
- The relative efficiency of partially acidulated phosphate rock in wetland rice soils is being explored at Maruteru (Andhra Pradesh), Faizabad and Kanpur (Uttar Pradesh), Raipur (Madhya Pradesh) and Moncompu and Pattambi (Kerala).
- The management of iron toxic soils for improving rice yields is being investigated at Moncompu and Pattambi (Kerala).

D. Rice physiology

The following research projects are being conducted by the Central Rice Research Institute (Cuttack), the Directorate of Rice Research (Hyderabad) and the SAUs:

- Evaluation for moisture stress in rice cultivation.
- Evaluation for low light intensity.
- Evaluation for water-logged conditions.
- Analysis of factors for heterotic vigour in rice hybrids.
- Evaluation of synchronization of flowering among promising parental lines.

E. Rice plant protection

The experiments on plant protection conducted by state and central institutes are given below.

- Screening for resistance to blast, brown spot, sheath blight, sheath rot, bacterial blight, rice tungro virus and false smut.
- Screening for resistance to the planthopper, gall midge, stemborer, leaf folder, ear head bug, leafhoppers and nematodes.
- Evaluation of potential biocontrol agents against pests and diseases in rice ecosystems.
- Integrated pest management trials.

The major constraints presented above are addressed by both biotechnology and non-biotechnology measures.

F. Rice biotechnology

In recent years, biotechnology alternatives have been realized to solve rice production constraints taking into account environment and equity considerations. The rice biotechnology research program was initiated in India in 1988 at the Directorate of Rice Research, Hyderabad. It addressed several major constraints relevant to India, particularly in the rice-intensive regions. With a growing shortage of research resources, private sectors are also helping rice biotechnology research; they concentrate mostly on hybrid seed production and the formulation of botanical pesticides.

These laboratories have begun basic research towards tissue and anther culture, wide hybridization and restriction fragment length polymorphism (RFLP) markers for numerous biotic and abiotic constraints. They have already set objectives which will ultimately bring better adapted and more productive rice varieties to the farmers of India.

IV. Estimates of Yield Gaps

The yield gap is not identical for all environments. It is explained by a number of constraints – biological, physical and socioeconomic. All these constraints together account for all of the yield gap. It can be divided into two parts: yield gap I and yield gap II. Yield gap I is the difference between an experiment station's maximum yield and an on-farm experiment's maximum yield. This yield gap arises from differences in the environment which can not be managed in the farmer's field. Yield gap II, which is the primary concern of the present study, is the difference between actual farm yield and the yield attained in on-farm experiments. This gap reflects biological, soil and water, physiological, genetic and socioeconomic constraints. The purpose of the present study is to quantify the contribution of each major constraint to yield gap II or production loss, and estimate the benefits that might accrue by overcoming the constraints.

The estimates of rice yield obtained from experimental stations and farmers' fields for different rice ecosystems for the states in southern India are reported in Table 9.1. The maximum rice yield obtained in experimental stations under irrigated conditions varied from 6.0 tons ha^{-1} in Kerala to about 8.6 tons ha^{-1} in Tamil Nadu and Andhra Pradesh. The average yield obtained by farmers is less than half of that amount. It varies from 3.3 tons ha^{-1} in Kerala to 5.2 tons in Tamil Nadu. The yield gap is 39% of yield in Tamil Nadu, 45% in Kerala, 51% in Karnataka and 55% in Andhra Pradesh. Most of the gap is, however, due to environmental factors, which cannot be managed in farmers' fields. Yield gaps due to biotic, abiotic and soil-related constraints are relatively small. They vary from 0.16 tons ha^{-1} in Kerala to 0.66 tons ha^{-1} in Andhra Pradesh.

The yield achieved in experimental stations is substantially lower under rainfed conditions than under irrigated conditions (see Table 9.1). For example, in Tamil Nadu, the experimental yield for rainfed situations is only 4.5 tons ha^{-1} compared to 8.5 tons ha^{-1} for irrigated conditions, i.e. lower by nearly 47%. The case in the other three states is similar. Yield gap I, the gap due to environmental factors, is relatively low for the rainfed ecosystem compared to the irrigated; it varies from 1.4 tons ha^{-1} in Tamil Nadu to 0.8 tons ha^{-1} in Karnataka. The gap due to biotic and abiotic stresses are, however, higher under rainfed conditions than under irrigated; it varies from 1.3 tons ha^{-1} in Andhra Pradesh to only 0.5 tons ha^{-1} in Tamil Nadu.

V. Estimates of Yield Losses

The respondents, 120 scientists engaged in rice research and an equal number of extension personnel, provided estimates of yield losses due to each constraint

Table 9.1. Estimates of yield gaps in southern India for irrigated and rainfed ecosystems (kg ha^{-1}).

Particulars	Andhra Pradesh	Karnataka	Kerala	Tamil Nadu	Southern India
Irrigated (kharif)					
Experiment station yield	8600	8200	6000	8500	8348
Average estimated potential yield	4550	4590	3482	5682	4781
Average actual farm yield	3892	4012	3320	5159	4201
Yield gap I	4050	3610	2518	2818	3564
Yield gap II	658	578	162	523	580
Rainfed					
Experiment station yield	4400	4450	3000	4500	4175
Average estimated potential yield	3452	3637	2061	3084	3250
Average actual farm yield	2170	2490	1470	2540	2191
Yield gap I	948	813	939	1416	925
Yield gap II	1282	1147	591	544	1059

for their respective agroclimatic zones. However, this does not mean that all these constraints are major and occur simultaneously. These may occur in the most severe form in any one of the regions in one of the seasons. The yield loss indicated by them is when the constraint is occurring beyond the economic threshold level and is at least moderately severe. Only the constraints causing major production losses are discussed below for the four states surveyed (Tables 9.2–9.4).

Andhra Pradesh

In Andhra Pradesh six agroclimatic regions were selected for the purpose of constraint analysis. The total production of rice in the state was 10.56 million tonnes in 1989/90. The total losses due to all constraints worked out to about 2.06 million tonnes or about 539 kg ha^{-1}, which is about 20% of the total production. There therefore exists a 20% potential to increase the productivity of rice, through elimination of the losses.

The leaf folder tops the list of damages by causing a maximum yield loss accounting for about 7.7% of total losses. The other insects, in order, are the ear head bug, yellow stemborer, brown planthopper and gall midge. Among diseases, blast is the major problem causing about 5.1% of total losses.

Table 9.2. Estimates of yield losses from insects and diseases by states in southern India (kg ha^{-1}).

Constraint	Andhra Pradesh	Tamil Nadu	Karnataka	Kerala	Southern India
Insects					
Leaf folder	42	64	27	36	44
Ear head bug	34	50	25	30	35
Stemborer	31	34	29	31	32
Gall midge	22	34	22	30	25
Brown planthopper	23	22	22	24	23
Green leafhopper	19	18	17	23	19
Thrips	13	18	18	21	16
Rodents	13	18	18	21	16
Others	4	0	11	4	3
Total insects	203	264	189	222	215
Diseases					
Blast	28	37	30	30	30
Bacterial leaf blight	18	21	17	23	19
Sheath blight	18	18	16	29	19
Brown spot	16	23	25	23	19
Sheath rot	16	18	17	27	18
Tungro	15	20	18	24	17
Grain discolour	11	18	19	23	15
Total diseases	122	154	137	180	137

Table 9.3. Yield losses from abiotic stresses, adverse soils and management constraints by states in southern India (kg ha^{-1}).

Constraint	Andhra Pradesh	Tamil Nadu	Karnataka	Kerala	Southern India
Abiotic stresses					
Scarcity of irrigation water	23	37	24	28	26
Drought	18	23	18	0	18
Cold at anthesis	0	6	14	0	4
Lodging	28	28	17	28	26
Low light intensity	0	3	11	0	3
Total abiotic stresses	23	37	24	28	26
Adverse soils					
Soil salinity	23	22	22	27	23
Low fertility	17	29	18	18	20
Zinc deficiency	15	25	23	0	18
Acid soils	0	9	10	27	6
Alkalinity	0	12	0	0	3
Iron toxicity	0	6	0	0	2
Total adverse soils	55	103	72	88	72
Management Practices					
Weeds	25	30	25	10	25
Imbalanced use of fertilizers	19	41	26	0	24
Aged seedlings	7	7	0	0	5
Varietal problems	0	0	26	28	7
Total management practices	51	78	77	38	61
Socioeconomic constraints	39	64	111	142	66

Salinity is an important adverse soil constraint contributing about 4.3% to the total losses. Of the agronomic constraints, lodging, weeds and water management are the major constraints causing yield losses of about 5.2, 4.6 and 4.3%, respectively.

Karnataka

The total rice production of Karnataka in 1989/90 was about 2.41 million tonnes. Of this, the production loss due to all constraints worked out to about 0.91 million tonnes or 669 kg ha^{-1}, accounting for about 37% of total production. Insects, diseases, adverse soils, agronomic and socioeconomic constraints contributed 28.1, 20.4, 10.8, 24.1 and 16.6%, respectively. Among insects, the yellow stemborer, leaf folder, ear head bug, brown planthopper and gall midge are the major pests. Rice blast is the major disease causing maximum economic damage in Karnataka. It ranks first and accounts

Table 9.4. Rice production loss by ecosystem in southern India.

Ecosystem	Area (thousand ha)	Actual farm yield (ton ha^{-1})	Actual production (thousand tons)	Production loss (kg ha^{-1})	Potential production (thousand tons)
Irrigated					
Andhra Pradesh	3,531	3.76	13,264	496	15,014
Karnataka	1,066	3.69	3,910	629	4,580
Kerala	425	2.61	1,111	693	1,405
Tamil Nadu	1,742	4.51	7,782	747	9,083
Rainfed upland					
Andhra Pradesh	216	2.17	468	1,282	746
Karnataka	106	2.49	263	1,152	386
Kerala	11	1.47	15	590	22
Tamil Nadu	50	2.54	126	547	154
Rainfed lowland and deepwater					
Andhra Pradesh	75	1.20	90	1,200	121
Karnataka	192	1.87	360	636	482
Kerala	128	1.70	218	845	326

for about 4.4% of total loss. Salinity and zinc deficiency are common adverse soil problems. Among the agronomic constraints, an imbalanced use of fertilizers, weeds and water management are the important problems contributing to yield loss by 3.9, 3.7 and 3.6%, respectively.

Kerala

The yield losses were estimated for four agroclimatic zones. Total losses in Kerala were about 1.01 million tonnes or 725 kg ha^{-1}. The yield loss accounts for about 39% of total production. The relative share of losses to production is highest in this state. Insects, diseases, adverse soils, agronomic and socio-economic constraints account for about 30.6, 24.8, 12.1, 12.9 and 19.5%, respectively. Among pests, the leaf folder tops the list followed by the stemborer, ear head bug and gall midge. The losses due to diseases in Kerala are the highest among the four states because of its humid climate. Rice blast is the major disease and causes economic losses of about 4.1%. Next to blast, sheath blight is an endemic disease in deepwater production environments and accounts for about 4.0% of losses. Salinity and acid soils are common adverse soil problems in Kerala because of coastal and hilly tracts, and their shares of total losses are about 3.7 and 3.7%, respectively. Among agronomic and other constraints, varietal degeneration, lodging and water management problems contributed about 3.9, 3.9 and 3.9% of losses, respectively.

Tamil Nadu

In Tamil Nadu six zones were considered for the present study. The total losses in Tamil Nadu were 1.35 million tonnes or about 758 kg ha^{-1}; this amounts to about 24% of production. Pests, diseases, adverse soils, agronomic and socio-economic constraints contributed to the production loss by 34.5, 20.4, 13.6, 25.4 and 6.1%, respectively. Among pests, the leaf folder is the most serious followed by the ear head bug, stemborer and gall midge. Rice blast is a endemic disease in Tamil Nadu where relative humidity is high, causing about 4.9% of losses. Among adverse soil problems, low fertility is a common phenomenon in Tamil Nadu; it contributes to about 3.8% of losses. Losses from soil-related problems in Tamil Nadu are the highest among the four states in southern India. This suggests that with the increase in rice yield and intensive use of chemical fertilizers soil-related problems are intensified. Imbalance in the use of fertilizers, water management, weeds and lodging are important agronomic constraints sharing about 5.5, 4.6, 4.0 and 3.7% of production losses. Scarcity and an unreliable supply of irrigation water are also important problems in Tamil Nadu.

The estimates of total production losses per unit of land for different rice ecosystems and states are reported in Table 9.4. The losses are higher for the rainfed lowlands and uplands than for the irrigated ecosystems. The highest loss is estimated for the uplands in Andhra Pradesh and Karnataka, at about 1.2 tons ha^{-1}, where the actual farm level yield is only 2–2.5 tons ha^{-1}. Thus, the yield would increase by 50–60% from this low level if production constraints could be eliminated. The yield losses are also high in the flood-prone and coastal areas of Andhra Pradesh and Kerala – about 50–100% of the actual farm yields. For the irrigated ecosystems, which account for over four-fifths of the total rice land in southern India, the production losses are relatively low; the average loss varies from 0.5 ton ha^{-1} in Andhra Pradesh to 0.75 tons ha^{-1} in Tamil Nadu.

VI. Priority Research Problems

The ranking of the top 15 research problem areas on the basis of the estimated loss of production can be seen from Table 9.5. Since the ranking is ordinal, it is difficult to compare the severity of constraints when the relative ranking changes across states.

The leaf folder tops the list of damages by causing maximum yield losses in Andhra Pradesh, Kerala and Tamil Nadu, indicating the urgency for research intervention to contain the damage. It is also one of the major constraints in Karnataka, ranking third. Other insects causing major losses are the ear head bug, yellow stemborer, brown planthopper and gall midge. Rice blast is particularly endemic in regions where humidity is high and it ranks first in Karnataka, fourth in Tamil Nadu and Kerala and fifth in Andhra Pradesh. The other disease causing large yield losses in the deepwater regions of Kerala is sheath blight. Lodging is a common constraint in the coastal regions of southern

Table 9.5. Ranking of rice yield constraints by state in southern India.

Rank	Andhra Pradesh	Karnataka	Kerala	Tamil Nadu
I	Leaf folder	Rice blast	Leaf folder	Leaf folder
II	Ear head bug	Yellow stemborer	Yellow stemborer	Ear head bug
III	Yellow stemborer	Leaf folder	Ear head bug	Fertilizer imbalance
IV	Lodging	Fertilizer imbalance	Rice blast	Rice blast
V	Rice blast	Varietal problem	Gall midge	Water management
VI	Weeds	Ear head bug	Sheath blight	Yellow stemborer
VII	Brown planthopper	Weeds	Varietal problem	Gall midge
VIII	Salinity	Water management	Lodging	Weeds
IX	Water management	Zinc deficiency	Water management	Low fertility
X	Gall midge	Brown planthopper	Salinity	Lodging
XI	Green leafhopper	Gall midge	Acid soils	Zinc deficiency
XII	Fertilizer imbalance	Salinity	Sheath rot	Thrips
XIII	Bacterial leaf blight	Brown spot	Brown spot	Drought
XIV	Drought	Grain discolour	Brown planthopper	Brown spot
XV	Sheath blight	Thrips	Rice tungro virus	Salinity

India; frequent cyclones and floods are responsible for this constraint. The problem is such that it warrants research into varieties resistant to lodging through wide hybridization. There is another, rather disturbing aspect of rice production – that of weeds. The problem is very acute in tail-end farms where water stress is quite a common phenomenon. It ranks sixth in Andhra Pradesh, seventh in Karnataka and eighth in Tamil Nadu. The adoption of efficient water management and cultural practices has become crucial for the successful control of weeds.

Salinity and water management problems are prevalent in water-logged and coastal zones of Southern India. Both problems are due to poor drainage systems and the occurrence of cyclones, especially in coastal areas. Though drainage facilities were well conceived and implemented several decades ago, they are not properly maintained; large-scale encroachments have made drainage channels inoperative. According to scientists, a breakthrough in controlling these problems can be achieved by evolving varieties tolerant to salinity through wide hybridization and proper water management training of farmers.

A. Alternative research methodologies and costs

Biological and social scientists involved in rice research assessed yield losses due to various constraints in their respective states. Scientists were asked to suggest alternative research methodologies and the costs involved in solving these constraints. It is evident from the results that there are 24 important constraints identified for the whole of southern India which need immediate attention for policymakers and agricultural research managers. Scientists were asked to assess the cost of research for four research alternatives – biotechnology, conventional breeding, wide hybridization and chemical and cultural methods. The costs and marginal productivity gain estimates formed the basis to evaluate each problem by each research alternative by computing the NPV. Since rice research is decentralized at the state level, we considered only the top 15 constraints to prioritize rice research in each state (see Table 9.5).

It is interesting to note that biotechnology emerges as the preferred method for a majority of the constraints. Good results are expected as biotechnology ensures both economic (cost-effective) and environmental benefits.

B. Research priorities

Based on the magnitude of crop losses due to constraints, it is possible to identify a research agenda for a biotechnology program. As seen in the list of top 15 constraints for southern India, the major constraints identified are found in most of the four states studied. Mechanisms of tolerance to pests at all stages of crop growth are crucial to increased productivity. Genetic resistance to insects and diseases has a twofold advantage: it serves to increase yields and reduces the dependency of farmers on pesticides, thereby addressing environmental concerns. Significant losses from the leaf folder, ear head bug, yellow stemborer, thrips, brown planthopper, green leafhopper, rice blast, brown spot, bacterial leaf blight, sheath rot, rice tungro virus and sheath blight demonstrate the urgent need for genetic resistance and biotechnological research intervention. The NPV analysis suggested the need for biotechnological research to control the above pests and disease constraints.

Lodging tolerance is crucial since many areas experience periodic cyclones and floods. Elongating varieties tend to lodge after flood water recedes. So, the ability to withstand short-term inundation through genetic mechanisms could prove very beneficial to large areas of southern India.

Some mechanisms to reduce losses due to weeds are clearly necessary. Although there are many conventional methods available, they are not successful enough to prevent yield losses. Early seedling vigour would help rice to compete with weeds for light and nutrients. Selective herbicide resistance would allow rice to survive while non-rice plants are eliminated with herbicides.

Drought tolerance is crucial since many areas in southern India experience periodic water stress during summer seasons and pre-monsoon periods. Salinity and acid soils are other constraints for which there are many conventional solutions, but some of these are costly and take many years to reclaim soils. Hybridization embraces a wide range of technical possibilities and may help ameliorate yield losses associated with adverse soil complexes. In this regard, transferring genes for tolerance to salinity and acid soils is desirable.

Important constraints for which chemical and cultural methods are the solutions include the imbalanced use of fertilizers, low fertility, water management and zinc deficiency.

VII. Conclusions

Rice is the major crop in southern India and it will remain so as long as it remains the staple food of almost the entire population of this region. There has been a significant growth in rice productivity and production in southern India after the mid-1960s due to factors such as an increased area under high-yielding varieties, an increase in input use (particularly chemical fertilizers and pesticides), enhancement of irrigation infrastructure, improved crop management techniques and an expansion in rice area. However, great imbalances in both rice production and productivity have been observed among the production environments within the regions/states. Despite the adoption of improved cultivation practices, rice yield has been found to be stagnating in almost all rice production environments.

Rice scientists are highly concerned about this trend since there is a great challenge to feed an ever-increasing rice-eating population. A growing population demands more and more rice for its consumption. Apart from this, many dietary surveys conducted in southern India revealed that the people switch over to rice-based diets from the diets based on millet and other coarse grains as their standard of living increase. The only way to match this growing demand for rice is to increase its production. However, the limiting factor is availability of land, particularly with assured irrigation; in fact, due to urban industrial and infrastructural expansion, the availability of land for cultivation has been decreasing. So increasing rice production can only be realized through growth in productivity in the existing areas. It is possible that the existing modern rice varieties with their present yield potential cannot be sustained for too long. Hence, it is increasingly realized that the yield potential of varieties should be increased by improving their photosynthetic efficiency. This could only be possible by way of DNA transfers through genetic engineering and exploitation of hybrid vigour.

With due acknowledgement to high-yielding varieties and modern rice production technologies we can say that the pace of rice production has kept ahead of the growing demand of the population so far. Recent developments indicate that current technologies seem to be inadequate to boost rice production further. It is argued that the wide gap between potential yield (experiment stations) and farm yield is yet to be bridged. It is also pointed out that socioeconomic factors at farm level inhibit the efforts to narrow the yield gap. Most modern rice technologies are resource- or input-intensive and put the small farmers of southern India in a disadvantageous position. Hence, there must be a thrust on designing technologies which are scale neutral and cost-effective.

In this regard, biotechnologies with their inherent low-cost and resource-neutral characteristics could be a viable technology alternative for small farmers to help them realize the yield potentials of modern rice varieties. Apart from cost effectiveness, biotechnologies have another strong factor in their favour – they are best suited for eco-friendly farming systems essential for ensuring sustainable agriculture. The objective of the rice biotechnology program is also to evolve high-potential varieties which are more responsive to bio-inputs.

Recombinant DNA technology and protoplast fusion may help to achieve genetic diversity, essential to obtain plant types which utilize soil nutrients, resist damage by pests and diseases, withstand a wide range of soil toxicities and deficiencies and allocate energy more efficiently. Research on rice biotechnology, therefore, deserves a high priority. Genetic resistance of new varieties could substitute high-cost inputs and synthetic technologies, which would be a great incentive to the resource-poor rice farmers in southern India. In this part of India high doses of chemical inputs (above the national average) are currently being used to obtain maximum rice yield. Widawsky and O'Toole (1990) observed that the contribution of rice biotechnology research to eastern India could very well apply to southern India also. As they said, the returns to research in biotechnology may be realized through higher yields at low costs; costs which would otherwise be incurred through input procurement and distribution and training programs in proper input use. Such innovations are particularly important to resource-poor and environmentally handicapped rice farmers.

References

Herdt, R.W. and Riely, F.Z. (1987) International rice research priorities: implications for biotechnology initiatives. Prepared for the Rockefeller Foundation Workshop on Allocating Resources for Developing Country Agricultural Research, Bellagio, Italy, 6–10 July, 1987.

Widawsky, D.A. and O'Toole, J.C. (1990) *Prioritizing the Rice Biotechnology Research Agenda for Eastern India.* Rockefeller Foundation, New York.

10 Rice Production Constraints in China

Justin Yifu Lin[1] and Minggao Shen[2]
[1]China Center for Economic Research, Peking University, Beijing 100871, People's Republic of China, [2]Food Research Institute, Stanford University, Stanford, CA 94305, USA

I. Introduction

Rice is China's most important grain crop. In 1991, 30% of grain acreage was planted to rice and 42% of total grain output consisted of rice. Because of the importance of rice as a foodgrain, rice research possesses a significant position in China's agricultural research.*

The history of China's organized agricultural research is rather short. Under the Nationalist Government's rule in the 1920s to 1940s, a small decentralized research network was established. This decentralized system continued after the socialist take-over in 1949. In 1957, the Chinese Academy of Agricultural Sciences was founded in Beijing, while each of the 29 provinces established its own academy of agricultural sciences. Each of the national and provincial academies consisted of ten to 30 independent research units. Most prefectures have also founded their own agricultural research institutes. The division of labour within this three-level research system is rather broad and there are considerable overlaps. The research institutes in the Chinese Academy of Agricultural Sciences emphasize basic and applied research with national significance and are responsible for technical supervision and coordination of provincial programs. Institutes in provincial academies stress applied research in accordance with the ecological conditions of the province. Prefecture institutes engage mainly in crop selections and adaptive research. The institutes and their research projects in all three levels are mainly funded by government budgets in the corresponding levels.

---

*Previous studies show that research resource allocation in China was consistent with the pattern predicted by the Schmookler–Griliches hypothesis of market demand-induced innovation, which suggests that the research resource allocation to a crop is a positive function of its size and price (Lin, 1991a, 1992a).

Varietal improvements have been the focus of agricultural research in China from the very beginning.* In the early 1950s, emphasis was given to the selection and promotion of the best local varieties; whereas new varieties of rice, wheat, cotton, corn and other crops were also imported from abroad.† A major breakthrough in rice breeding began providing benefits when China began the full-scale distribution of fertilizer-responsive, lodging-resistant dwarf rice varieties with high-yielding potential. This breakthrough occurred 2 years before the release of IR8, the variety which launched the green revolution in other parts of Asia, by the IRRI in the Philippines. At about the same time, hybrid corn and sorghum, improved cotton varieties and new varieties of other crops were also released and promoted. The high-yielding varieties were accepted rapidly. A second major breakthrough in rice breeding occurred in 1976. In that year, China became the first country to commercialize the production of hybrid rice. The innovation and commercial development of hybrid rice was heralded as the most important achievement in rice breeding in the 1970s (Barker and Herdt, 1985, p. 61). By 1979, high-yielding varieties of rice covered 80% of the area, wheat covered 85%, soybeans 60%, cotton 75%, peanuts 70% and rape 45% (Ministry of Agriculture, 1989, pp. 348–349).

The contribution of seed-improvement research to the increase in rice yield in China has been substantial. In the 1950s, rice yield averaged about 2.5 tons ha‡. After the introduction of semidwarf varieties in the early 1960s, yields increased gradually to a level of 3.5 tons ha^{-1} in the mid-1970s. Partly because of the introduction of hybrid rice in 1976 and partly because of the decollectivization of the farming system in 1979, yields increased rapidly from 3.5 tons ha^{-1} in the mid-1970s to 5.5 tons ha^{-1} in the mid-1980s (Lin, 1991b, 1992b). China's rice yield in 1990 was about 60% higher than the world's average yield and close to the level achieved by the most advanced countries, such as Japan and South Korea. This achievement is especially significant because about half of the rice areas in China grow two crops of rice each year instead of just one crop.

One of the challenges to the rice research community in China is how to sustain the pace of yield increase. The acreage sown to both grain and rice has been declining since the mid-1970s. Such a decline is most likely irreversible because it is a natural adjustment to the process of economic growth of farmers

---

*We will only briefly summarize seed-improvement research in this chapter. For other aspects of agricultural research and technological change in China, see Wiens (1982).
†In the 1950s and 1960s, 3776 new varieties were imported from more than 30 different countries, and imports of new varieties continued to increase. In the 1970s, 43,674 varieties were imported from 85 countries and international organizations (Zhu Rong, 1988, p. 464). During the 1950s two problems were noted: when borrowing from abroad (or the best seeds of a particular locality) and when borrowing without adaptation often resulted in crop failures. Too much attention was given to high-yielding varieties requiring unusually favourable conditions, to the neglect of varieties that perform well under poor conditions (Wiens, 1982, p. 103).
‡The yield in 1958–1962 fell substantially below the level of 2.5 tons ha^{-1}. The sudden decline in yield was attributed to the imposition of forced collectivization in 1958, which distorted the farmers' incentive structure. For a further discussion of this see Lin (1990).

seeking to meet the growing demand for higher value crops. However, the demand for rice, as well as grain, is expected to rise for several more decades with continued population and per capita income growth. Yield improvement is the major measure for meeting the increasing demand.

Several issues related to rice production are important to the research community:

1. What are China's rice production constraints?
2. What should be the priorities of China's rice research?
3. What is the best research strategy according to these priorities?

This chapter reports the rice production constraints obtained from a study carried out in all of the rice-producing regions of China during October 1991 to April 1992. The findings indicate that technical constraints constitute a substantial portion of estimated yield losses, and that these constraints are concentrated on a few factors. Most of the constraints are soil and weather related and are not easily handled by conventional methods of variety improvements, so Chinese researchers look to biotechnology as one prospect for solving these issues.

The chapter is organized as follows: section II presents an overview of the rice production environment in China. Section III briefly discusses the methodology used in the survey. Major findings of the rice production constraints at the national level are summarized in section IV, and some concluding remarks are presented in section V.

II. Rice Environments and Natural Conditions in China

China lies in the northern half of eastern Asia on the west coast of the Pacific Ocean. Its area covers approximately 9.6 million square kilometres, nearly one-fifteenth of the world's land. After the former USSR and Canada, China is the third largest country in the world.

The climate of China is greatly affected by its proximity to the sea on the east and south and by the huge land mass of Eurasia to the north and west. Monsoon winds from the south, cold winds sweeping in from the north and west, and the physical features of the landscape all have an effect on, and are affected by, the climate. From south to north, the weather may be classified into zones: tropical, subtropical, warm temperate, temperate and cold temperate. Frost-free days per year vary from 365 in the tropical zone to less than 80 in parts of the north and west. From the southeast to northwest, four moisture zones are designated: humid, subhumid, semi-arid and arid. In some of the humid southern coastal regions, annual precipitation exceeds 2000 mm, and in parts of the arid regions of the northwest, it is less than 100 mm.

Since much of China lies within the East Asian monsoon zone, natural conditions such as sunshine, temperature and moisture favour the cultivation of rice. Wherever rainfall is abundant or irrigation water is available there is rice production. It extends from 18° north latitude to 53° north latitude. The main rice production areas, with 94% of the total rice area and production, are south of the Qinling Mountains and the Huaihe River. Rice production areas in China are divided into six ecological zones according to rainfall and temperature conditions and systems of cultivation: southern, central, southwestern, northern, northeastern and northwestern (Table 10.1).

III. Conceptual Framework and Survey Methodology

The basic assumption of setting research priorities is that there exists a yield gap between potential yield and actual yield that research can help to close, or there is a possibility for increasing potential yield through research. One earlier study of rice research priorities involved three issues: (i) identifying production constraints and potential gains in production by overcoming these constraints and identifying other likely sources of yield increase; (ii) estimating the likelihood, possible time length and cost of solving these constraints or attaining

Table 10.1. Rice eological zones in China.

	Criteria			
Rice ecological zones	Accumulated annual temperature (> 10°C)	Precipitation (mm)	Aridity index (E/r)	Cropping system
I. Humid tropical double-cropped rice (southern)	≥ 6500	> 1000	< 1.0	Three maturing double-cropped rice
II. Humid single and double-cropped rice (central)	4500–6500	> 1000	< 1.0	Two and three maturing single- and double-cropped rice
III. Semihumid single-cropped rice (northern)	3500–4500	> 400	1.0–2.0	One or two maturing single-cropped rice
IV. Semihumid early-maturing single-cropped rice (northeastern)	< 3500	> 400	1.0–2.0	One maturing single-cropped rice
V. Arid single-cropped rice (northwestern)	2200–4000	< 400	> 2.0	One maturing single-cropped rice
VI. Humid single-cropped rice (southwestern plateau)	3000–6500	about 1000	< 1.0	One maturing single-cropped rice

Source: Chinese Academy of Agricultural Sciences, 1986, p. 97.

likely sources of yield increase by alternative methods; and (iii) determining equity weights associated with each problem and its potential solution and determining the net present value of equity-weighted expected costs and benefits for all possible problems.* The study reported here focuses mainly on the production constraints.

According to the work done at the IRRI in the mid-1970s (IRRI, 1977, 1979; De Datta et al., 1978) and recent work by Widawsky and O'Toole (1990), the yield gap can be divided into two parts. Yield gap I is the difference between an experiment station's maximum yield and the potential average yield achievable under favourable farm conditions in a region. This yield gap arises from differences in the varieties and production environment which cannot be easily managed or eliminated. Yield gap II is the difference between average farm yields and yields attainable under the most well-managed and favourable conditions for all farm-controlled variables.† Yield gap II is caused by technological constraints and/or sociological constraints which may contribute to the existence of technical constraints. For example, the shortage of herbicides on the market (a sociological constraint) may be one of the causes of damage by weeds (a technical constraint). However, some sociological constraints may also contribute to the yield loss independent of technical constraints, for example, the impurity of seeds and bad water management.

The first step in setting research priorities is to assess yield gaps. In this study we organized two separate surveys, one of research scientists at various rice research institutes and the other of agronomists at agricultural bureaus of local governments. For factors contributing to yield gap I, we relied on the judgement of research scientists because of their knowledge of varietal differences between experimental and field varieties and the environmental constraints that prohibit the use of experimental varieties in field production. For factors contributing to yield gap II, we relied on the judgement of local agronomists responsible for grain production and technological extension because of their knowledge of actual farm practices and local conditions that prevent the realization of existing field varieties in field production.

A. The agricultural bureaus' survey

In China, the government is organized into four hierarchical levels: the central government, the province, the prefecture and the county. The agricultural

---

*For further discussions of the steps for setting research priorities, see Herdt and Riely (1987).
†The definitions of yield gap I and yield gap II here are somewhat different than the definitions used in studies by the IRRI and Widawsky and O'Toole (1990). In China the maximum yield on farms is very close to, and sometimes even higher, than the maximum yield on experiment stations. This is because research institutes are required to do field experiments at farmers' fields and because of the local government's special assistance to a few 'window' farmers to demonstrate the possibility of achieving higher yields in a region.

bureau (ministry) at each level of government is responsible for agricultural production in the region under its jurisdiction. An agricultural bureau has divisions on crop production, crop protection, soil conservation, technology promotion and field experiments. These divisions keep detailed records on information relevant to the issues of rice production constraints. Thus, a cost-effective way of obtaining information about the constraints that contributed to yield gap II is to conduct a survey of experienced agronomists at agricultural bureaus. China is divided into 30 provinces (in addition to Taiwan), 364 prefectures and 2830 counties. Rice is grown in 29 of these 30 provinces. The information at the provincial level is too aggregated for analysis and the number of rice-producing counties too numerous to survey; therefore, this study was conducted at the prefecture level.

In the survey we collected: (i) historical data on sown acreage, total output, average yield and highest yields on demonstration plots and experimental plots; (ii) farm yields, acreage and sources of the three leading varieties in the prefecture; (iii) production losses due to input constraints, pests, disease, weather/climate, and other technical issues; and (iv) the estimated yield potential under favourable conditions for average farms in the prefecture. Information for early season rice, late season rice and single season rice was collected separately.

The survey was administered in two stages. In the first stage, a workshop was organized in each provincial capital to discuss the contents of the questionnaires. A leading agronomist from each prefecture was invited to participate in the workshop. After the workshop, these agronomists were responsible for organizing a team of about five agronomists from major fields in their bureaus to collectively answer the questionnaires. After about a month, the leading agronomists were invited to participate in a second workshop to hand in the questionnaires and discuss the findings with the research team.

With the sanction of the Ministry of Agriculture and the State Science Commission, most prefectural bureaus gave full support to the survey. Summary statistics of the survey are reported in Table 10.2. For the nation as a whole, we obtained valid responses from 98 prefectures producing early season rice, 97 producing late season rice and 152 producing single season rice. The valid responses covered 85, 83 and 80%, respectively, of the cultivated areas of early season, late season and single season rice in the prefectures originally surveyed. The responses were based mainly on official records supplemented by the judgements of agronomists with extensive knowledge of actual rice production in the localities where they work.

B. The research scientists' survey

Nationwide, we surveyed 193 rice research scientists with a full professor or associate professor rank to obtain their judgements about the yield gaps and the factors that contribute to yield gap I. The list of scientists was provided by the Chinese Academy of Agricultural Sciences with the indication that they were actively involved in China's rice research. We received 125 valid responses from

Table 10.2. Summary statistics of the agricultural bureaus' survey.

	Type of rice	South	Central	North	North-east	North-west	South-west	Nation
Number of prefectures surveyed	Early	39	80				6	125
	Late	39	81	1		1	4	125
	Single	17	88	47	31	18	23	224
Valid response	Early	28	66				4	98
	Late	97	28	66			3	97
	Single	11	56	36	21	9	19	152
Surveyed area (thousand ha)	Early	154.50	775.43				2.50	932.43
	Late	169.45	796.94	0.90	0.02	0.83	967.22	
	Single	127.22	665.70	181.28	135.75	27.69	134.35	1271.99
Valid area/ Surveyed area (%)	Early	93.31	83.20				98.59	84.92
	Late	93.21	80.69				97.43	82.89
	Single	95.65	76.28	97.46	65.32	20.08	85.46	

35 research institutes and ten universities. Among the valid responses, 47.3% were from professors and 52.7% were from associate professors. They had been engaged, on average, in rice research for 26.8 years.

The responses were based on both experience and judgement. We used an open-ended questionnaire in which research scientists were asked to provide personal judgements about factors that contribute to yield gap I. They were instructed to consider factors for which they had sufficient knowledge, but did not have to comment on factors for which they had little or no knowledge. In total, 50 factors were mentioned, of these, 18 were mentioned most frequently (Table 10.3), and each scientist commented on at least nine factors.

IV. Rice Production Constraints in China

The purpose of the two surveys was to obtain estimates of yield gaps and the factors which contributed to them. The rest of the chapter discusses these findings.

A. Yield gaps

Respondents to the agricultural bureaus' survey were asked to provide information about maximum yields on experiment stations and on plots of experienced farmers during the past 10 years in their prefectures. The information on

Table 10.3. Summary statistics of the research scientists' survey.

	Frequency of responses (%)		
	Early season rice	Late season rice	Single season rice
Number of valid responses	62	53	56
I. Plant-related constraints			
1. Plant structure	48.4	58.5	44.6
2. Photosynthetic efficiency	50.0	50.9	48.2
3. Growth duration	61.3	56.6	66.1
4. Efficiency of nutrition transformation	37.1	37.7	35.7
5. Structure of spicas	17.7	15.1	12.5
6. Tillering ability	33.9	32.1	30.4
7. Thousand-grain weight	8.1	9.4	10.7
8. Setting percentage	16.2	29.7	19.7
9. Vigour of roots	6.5	0.0	7.1
10. Grain to straw ratio	8.1	5.7	1.8
11. Filling problem	1.6	3.8	1.8
II. Environment-related constraints			
1. Duration of sunshine	43.5	49.1	50.0
2. Aggregated effective temperature	35.5	34.0	25.0
3. Soil condition	32.3	30.2	26.8
4. Diurnal temperature variation	21.0	17.0	19.6
5. Frost-free period	1.6	1.9	1.8
6. Relative humidity	3.2	3.8	3.6
7. Other unfavourable climate	8.1	5.7	8.9

reported maximum yields for each type of rice in farmers' fields and experiment stations and the averages of the maximum yields reported by each agricultural bureau are summarized in Table 10.4.

The 11-year average of reported maximum experiment station yields for early season, late season and single season rice are, respectively, 12,260, 11,220 and 15,790 kg ha^{-1}. The maximum yields for farmers' fields are very close to the maximum yields for the experiment stations, and in some years slightly higher. This is because: (i) varieties grown on experiment stations were still in trial stages and some might not be adaptable to local environments; (ii) farmers with the highest yields often received advice and help from the experiment station, so their yields can be viewed as an on-farm experimentation; and (iii) last, but by no means least, because of the official campaign for 'reaching 15 ton per hectare' (a ton per mu), the local government often gave a few 'window' farmers priorities in the allocation of chemical fertilizers and other inputs in an effort to reach this campaign goal. A correlation analysis indicates that

Rice Production Constraints in China

Table 10.4. The maximum yields reported in the agricultural bureaus' survey.

	Early season rice		Late season rice		Single season rice	
Year	Maximum	Average	Maximum	Average	Maximum	Average
Farmer's field						
1980	11,364	7,483	10,552	6,885	12,885	7,734
1981	12,154	7,566	10,305	6,844	15,441	8,119
1982	11,485	7,767	11,422	7,116	14,158	8,464
1983	11,362	7,893	11,460	7,462	15,214	8,815
1984	11,707	8,136	11,482	7,606	16,423	9,177
1985	12,825	8,070	11,626	7,692	15,150	8,967
1986	11,250	8,166	12,378	7,825	15,750	9,363
1987	11,880	8,349	10,875	7,860	15,343	9,505
1988	11,557	8,499	11,316	8,022	15,450	9,781
1989	11,862	8,619	12,082	8,323	15,885	9,969
1990	12,711	8,832	12,375	8,416	15,345	10,287
Average	11,830	8,125	11,443	7,640	15,185	9,170
Experimental plot						
1980	11,707	7,944	10,855	7,201	14,152	8,476
1981	10,911	7,858	10,012	7,239	14,440	8,736
1982	11,970	8,109	10,507	7,404	14,646	9,057
1983	11,421	8,047	10,575	7,618	16,131	9,244
1984	13,995	8,523	12,187	7,828	15,685	9,619
1985	11,310	8,346	11,841	8,023	15,585	9,709
1986	12,463	8,410	10,377	8,080	15,892	9,834
1987	14,700	8,593	12,307	8,280	17,011	10,080
1988	12,645	8,595	11,356	8,269	17,577	10,182
1989	11,760	8,853	11,535	8,515	15,747	10,405
1990	11,985	8,976	11,910	8,808	16,821	10,852
Average	12,260	8,386	11,220	7,933	15,790	9,654

maximum yields on experiment stations and maximum yields on farms in a prefecture is highly correlated at 0.74. Therefore, it is clear maximum yields on farms in China have the same property as maximum yields on experiment stations in other countries.

In the research scientists' survey, we also asked each scientist to provide us with the highest ever experimental yield that they have knowledge of. The highest experimental yields obtained from this survey are very close to the highest yields reported in the agricultural bureaus' survey. Table 10.5 lists the maximum yield for each type of rice reported in these two surveys and the average of the highest yields given by the agricultural bureaus and research scientists. These two sets of figures, especially the averages, are very close. For the purpose of calculating yield gap I, we use the average highest experimental station yields reported in the agricultural bureaus survey, assuming they

Table 10.5. Highest yields estimated by agronomists and scientists (kg ha^{-1}).

	Highest yield		Average highest yield	
	Agronomists	Scientists	Agronomists	Scientists
Early season rice	14,700	13,695	9,678	9,705
Late season rice	12,378	16,815	9,246	9,773
Single season rice	17,577	18,075	11,363	11,888

represented the biological potential that can be realized under existing scientific knowledge – 12,260 kg ha^{-1} for early season rice, 11,220 kg ha^{-1} for late season rice and 15,790 kg ha^{-1} for single season rice.

The second step in estimating yield gap I and yield gap II is to obtain an estimate of potential yield under favourable conditions for an average farm in a prefecture. This information is available from the agricultural bureaus' survey. The difference between maximum and potential yield for an average farm represents yield gap I and the difference between potential and actual yield is yield gap II.

Table 10.6 reports the information on yield gap I and yield gap II. For the nation as a whole, yield gap I for early season rice is 4584 kg ha^{-1}, for late season rice is 3628 kg ha^{-1} and for single season rice is 6448 kg ha^{-1}, while yield gap II is 2564 kg ha^{-1} for early season, 2981 kg ha^{-1} for late season and 3391 kg ha^{-1} for single season rice. From the comparison of yield gap I and yield gap II, we find that the non-transferable yield gap is substantially larger than the transferable yield gap II for each type of rice. Yield gap I indicates varietal and environmental differences between single-plot-observed maximum yields and farmers' fields. This difference cannot be easily eliminated or managed by an average farmer; however, research may discover ways of enabling farmers to obtain these higher yields.

Table 10.6. Yield gaps I and II.

	Early season rice	Late season rice	Single season rice
Maximum yield*	12,260	11,280	15,790
Average estimated potential farm yield	7,676	7,592	9,342
Average actual farm yield	5,112	4,611	5,951
Yield gap I	4,584	3,628	6,448
Yield gap II	2,564	2,981	3,391

*Maximum yield refers to the highest observation of the maximum yield in the research scientists' survey or the highest maximum yield in the agricultural bureaus' survey.

B. Yield gap constraints

Yield gap I constraints

In the research scientists' survey, we provided the scientists with information about the average potential yield for each type of rice obtained from the agricultural bureaus' survey, and asked them to list the factors that contribute to the yield gap between maximum and average potential yields and then to estimate the percentage of this yield gap explained by each of the factors they listed. We assumed that if a certain factor was not mentioned, it was not important in this scientist's view and was assigned a value of zero for that observation. The percentage of yield gap I explained by each factor was then a simple average of the estimated percentages reported by the research scientists. The average percentage and estimated yield gap I were used to infer the yield loss attributable to each individual factor. Tables 10.7 to 10.9 report the results of this exercise.

From these tables, the estimated yield gap I for all three types of rice are accounted for as follows. Eleven plant-related factors explained 35%, seven environment-related factors explained 20%, and 45% was unaccounted for. Of the 11 plant-related factors, the most important are plant structure, photosyn-

Table 10.7. Contributions to yield gap I, early season rice.

	Percentage explained	Loss attributed (kg ha^{-1})
I. Plant-related constraints	37.99	1741
1. Plant structure	5.81	266
2. Photosynthetic efficiency	7.56	347
3. Growth duration	11.26	516
4. Efficiency of nutrition transformation	3.58	164
5. Structure of spicas	3.05	140
6. Tillering ability	2.77	127
7. Thousand-grain weight	0.74	34
8. Setting percentage	1.81	83
9. Vigour of roots	0.44	20
10. Grain to straw ratio	0.65	30
11. Filling problem	0.32	15
II. Environment-related constraints	19.78	907
1. Duration of sunshine	5.95	273
2. Aggregated effective temperature	5.66	259
3. Soil condition	3.68	169
4. Diurnal temperature variation	2.27	104
5. Frost-free period	0.32	15
6. Relative humidity	0.32	15
7. Other unfavourable climate	1.58	72
III. Unexplained	42.23	1936

Table 10.8. Contributions to yield gap I, late-season rice.

	Percentage explained	Loss attributed (kg ha^{-1})
I. Plant-related constraints	37.99	1378
1. Plant structure	7.75	281
2. Photosynthetic efficiency	7.79	283
3. Growth duration	7.32	266
4. Efficiency of nutrition transformation	3.70	111
5. Structure of spicas	2.49	90
6. Tillering ability	2.62	95
7. Thousand-grain weight	1.13	41
8. Setting percentage	2.64	96
9. Vigour of roots	0.45	16
10. Grain to straw ratio	0.36	13
11. Filling problem	0.85	31
II. Environment-related constraints	18.58	674
1. Duration of sunshine	7.26	263
2. Aggregated effective temperature	4.51	164
3. Soil condition	3.02	110
4. Diurnal temperature variation	1.92	70
5. Frost-free period	0.57	21
6. Relative humidity	0.38	14
7. Other unfavourable climate	0.92	33
III. Unexplained	44.32	1608

thetic efficiency and growth duration, and these account for 24% of yield gap I for all three types of rice. Of the seven environment-related factors, the most important are duration of sunshine, aggregated effective temperature and soil condition, and these account for 15% of yield gap I for all three types of rice.

Environmental factors cannot be changed easily, but improvement of a plant's resistance to adverse environment will reduce the impact on yield. Therefore, yield gap I may be reduced through varieties which have desirable plant-related characteristics or a higher resistance to the adverse environments.

About 45% of the yield gap I for all three types of rice is not accounted for by either plant- or environment-related factors. There are three explanations for this result: (i) many scientists attribute the difference between maximum experimental yield and potential farm yield to the variations in crop management; (ii) some scientists believe that the highest experimental yield for each type of rice is over-reported; and (iii) some scientists stress in their letters to us that some synthesized effects cannot be broken down into individual factors.

Yield gap II constraints

Yield gap II is the difference between actual farm yield and potential farm yield under favourable conditions in a prefecture. In analysing factors con-

Table 10.9. Contributions to yield gap I, single season rice.

	Percentage explained	Loss attributed (kg ha^{-1})
I. Plant-related constraints	35.04	2259
1. Plant structure	7.37	475
2. Photosynthetic efficiency	6.05	390
3. Growth duration	9.41	607
4. Efficiency of nutrition transformation	3.11	201
5. Structure of spicas	2.18	141
6. Tillering ability	2.64	170
7. Thousand-grain weight	0.98	63
8. Setting percentage	2.25	145
9. Vigour of roots	0.73	47
10. Grain to straw ratio	0.14	9
11. Filling problem	0.18	12
II. Environment-related constraints	18.78	1211
1. Duration of sunshine	8.80	567
2. Aggregated effective temperature	3.05	197
3. Soil condition	3.16	204
4. Diurnal temperature variation	2.57	166
5. Frost-free period	0.14	9
6. Relative humidity	0.45	29
7. Other unfavourable climate	0.61	39
III. Unexplained	46.18	2978

tributing to yield gap II, varieties are taken as given. Since knowledge about rice production in a specific location is required, the analysis for yield gap II is based on the information obtained from the prefectural agricultural bureaus' survey. Yield gap II can be explained by socioeconomic constraints and/or technological constraints. Our focus in the study was on technological constraints. Following the conventions of Widawsky and O'Toole (1990), technical constraints were classified into five categories: adverse soils, diseases, insects, adverse climate/weather and other. The respondents were instructed to answer the questionnaire based on the historical records in their prefectures. The results of the survey are summarized in Table 10.10.

Average losses in the nation as a whole attributed to these identified constraints are 1152 kg ha^{-1} for early season rice, 1212 kg ha^{-1} for late season rice and 1463 kg ha^{-1} for single season rice, or 43% of yield gap II. The balance of yield gap II is caused by socioeconomic or technical factors not listed in the questionnaire, such as seed impurity, seed regression, bad management, bad extension service and so on.

Among yield losses due to technical constraints, soil- and weather-related factors have the lion's share. The soil-related factors can be subdivided into issues arising from soil types and soil fertility. In turn, they can be broken down further. Similarly, the weather-related constraints can be subdivided into

Table 10.10. Contributions to yield gap II (kg ha^{-1}).

	Early season rice		Late season rice		Single season rice	
	Percentage explained	Loss attributed	Percentage explained	Loss attributed	Percentage explained	Loss attributed
Yield gap II	100.00	2564	100.00	2981	100.00	3391
Technical constraints	44.93	1152	40.66	1212	43.14	1463
1. Soils	22.66	581	20.03	597	18.72	635
2. Diseases	2.69	69	1.78	53	2.60	88
3. Pests	1.52	39	1.64	49	2.71	92
4. Weather	15.72	403	14.96	446	16.43	557
5. Other	2.34	60	2.25	67	2.68	91
Unexplained	55.07	1412	59.34	1769	56.86	1928

submergence, drought, cold, heat, typhoon, hail and snow; and each can further be divided into seedling period, vegetative period, anthesis period and the whole period. Pests, diseases and other can all be broken down into more detailed factors as well. For the purpose of setting research agenda, a detailed break down is required.

Estimated yield loss from individual constraints for each type of rice is calculated by the following method:

$$\text{Estimated yield loss}_i = \frac{\sum_j \text{average yield loss } 81-90_{ij}}{\sum_j \text{average sown acerage } 81-90_{ij}} \quad (10.1)$$

where i indicates a constraint and j indicates a prefecture. The estimation for the average yield loss is primarily based on official records. In cases where official records were not available, the agronomists were asked to make an estimate. The absolute magnitude of estimated yield loss from an individual constraint is not exact, but the relative magnitudes among different constraints provide a basis for assessing the relative contribution of these constraints to yield gap.

Table 10.11 reports the estimated yield losses at the national level from the top 20 individual technical constraints. For early season rice, the top ten constraints contributed 59% of total estimated yield losses; among these, seven are related to soil conditions and the other three are related to weather. Another 24% of estimated yield loss from technical constraints was attributed to constraints 11 through to 20. These ten constraints were spread among weather, diseases, soils and weeds. For late season rice, the top ten constraints contributed 67% of total estimated yield losses. Again, adverse soil conditions and weather-related factors dominate the top ten list. Constraints 11 through to 20 attribute another 18% of estimated yield losses. Sheath blight, weeds and rodents are among the list. Again, the rest are related to weather and soil conditions. For single season rice, the top ten constraints contribute 59% and an additional 22% are caused by constraints 11 through to 20. Adverse soil and weather conditions predominate the list.

From Table 10.11 we find that, overall, the most important technical constraints are deficiencies in nitrogen, phosphorus, potassium, soil organic matter contents and trace elements. Two explanations may contribute to these deficiencies. On the one hand, it reflects the long history of intensive cropping of rice in China, where soil fertility has been depleted and cannot be recovered by natural processes. On the other hand, it reflects the highly fertilizer-responsive nature of current varieties, making it is possible to increase yield simply by increasing the application of fertilizers. Whether farmers have the incentive to do so depends on the relative prices of rice and fertilizers.

Cold, water-logged soil is also a major cause of yield losses. This may arise because rice is grown with irrigation and a substantial amount is grown in hilly locations where irrigation water is naturally cold.

The second most important cause of yield losses is related to weather, including drought, submergence, cold and heat at the seedling, vegetative and anthesis periods. Lodging caused by wind and storm is also important. Among the various rice diseases and pests, only sheath blight and the striped stemborer are among the top 20 constraints. The dramatic contrast between the situation in most other countries is explained as follows. Rice yield in China is among the highest in the world. Most constraints that are predominant in low-yielding situations and that can be controlled by conventional methods have been eliminated in China. However, rice production is still subject to weather risk in China, and variations in weather cause 5% or more variation in average nationwide yields from year to year.

V. Conclusions

This chapter reports the research design, survey procedure and findings from a study entitled 'Rice research priorities in China: implications for biotechnology initiative'. The study used questionnaire surveys to obtain information regarding the highest experimental yields and actual farm yields and to elicit estimates of potential farm yield under favourable conditions from scientists at rice research institutions and experienced agronomists at prefectural agricultural bureaus. The difference between the highest experimental yields and potential farm yields under favourable conditions is referred to as yield gap I, which arises from differences in the environment at experiment stations and farm fields and in experimental and farm varieties. The difference between potential farm yields and actual farm yields is referred to as yield gap II. Individual constraints that contribute to yield gaps I and II were identified.

For the nation, the average maximum experimental yields are 12,260, 11,220 and 15,790 kg ha^{-1}, for early, late and single season rice, respectively. Most of the gap between the highest experimental yields and actual farm yields belong to yield gap I. A major goal of using biotechnology and other means of varietal improvement is to transfer characteristics that may provide greater independence from environmental limitations.

For losses from yield gap II, the predominant factors are related to adverse soil conditions and adverse weather/climate. Diseases and pests do not cause

Table 10.11. The top 20 constraints at the national level and the losses caused by them (kg ha^{-1}).

Early season rice		Late season rice		Single season rice	
Total loss (kg ha^{-1})	1152 (100%)	Total loss (kg ha^{-1})	1212 (100%)	Total loss (kg ha^{-1})	1463 (100%)
Sum of top ten constraints' loss (kg ha^{-1})	683 (59%)	Sum of top ten constraints' loss	813 (67%)	Sum of top ten constraints' loss	860 (59%)
Loss due to top 20 constraints total loss (kg ha^{-1})	967 (84%)	Loss due to top 20 constraints (kg ha^{-1})	1025 (85%)	Loss due to top 20 constraints (kg ha^{-1})	1185 (81%)
1. Potassium deficiency	145	1. Potassium deficiency	181	1. Organic matter deficiency	109
2. Phosphorus deficiency	90	2. Cold at anthesis period	101	2. Cold water-logged soil	105
3. Nitrogen deficiency	72	3. Nitrogen deficiency	86	3. Nitrogen deficiency	104
4. Cold water-logged soil	69	4. Drought at anthesis period	82	4. Phosphorus deficiency	100
5. Organic matter deficiency	68	5. Phosphorus deficiency	77	5. Potassium deficiency	95
6. Cold at seedling period	63	6. Organic matter deficiency	75	6. Flood	90
7. Flood	50	7. Drought at vegetative period	62	7. Drought at anthesis period	82
8. Acidity	47	8. Cold water-logged soil	54	8. Drought at vegetative period	70
9. Trace elements deficiency	42	9. Flood	53	9. Trace elements shortage	54
10. Drought at anthesis period	38	10. Acidity	42	10. Cold at anthesis period	51
11. Sheath blight	37	11. Trace elements deficiency	42	11. Submergence at anthesis period	42
12. Heat at anthesis period	36	12. Sheath blight	27	12. Rain at harvesting period	40
13. Lodging from wind and storm	34	13. Weeds	22	13. Drought at seedling period	39
14. Submergence at anthesis period	32	14. Drought at seedling period	29	14. Sheath blight	39
15. Drought at vegetative period	31	15. Lodging from wind and storm	18	15. Weeds	38
16. Swamp soil	29	16. Submergence at vegetative period	17	16. Cold at seedling period	30
17. Cold at vegetative period	25	17. Swamp oil	16	17. Acidity	28
18. Rain at harvest	23	18. Submergence at seedling period	16	18. Rats	25
19. Rice blast (spica)	19	19. Submergence at anthesis period	15	19. Striped stemborer	23
20. Weed	18	20. Rats	15	20. Lodging from wind and storm	22

major losses in China. This finding is consistent with China having close to the highest yields in the world. Most losses from factors that can be controlled by conventional methods have been managed already. Biotechnology may provide prospects for solving the constraints arising from low soil fertility, adverse soil conditions and adverse weather/climate.

This chapter represents a first step towards setting rice research priorities in China. From the findings, we can conclude that there are opportunities for further increasing rice yield in China with research focused on a few well-defined constraints. Yield gap I research should concentrate on improving plant structure, photosynthetic efficiency, growth duration and the ability to overcome short duration of sunshine, low aggregated effective temperature and poor soil conditions. The main areas for yield gap II research include low soil fertility, cold water-logged and acid soil, drought, submergence, heat and cold at seedling, vegetative and anthesis periods, lodging, weeds, sheath blight and stemborer damage.

References

Barker, R. and Herdt, R.W. (1985) *The Rice Economy of Asia*. Resources for the Future, Washington, DC.

Chinese Academy of Agricultural Sciences (ed.) (1986) *Zhongguo Daozuoxue (Rice Cultivation in China)*. Agricultural Press, Beijing.

De Datta, S.K., Gomez, K.A., Herdt, R.W. and Barker, R. (1978) *A Handbook on the Methodology for and Integrated Experiment – Survey on Rice Yield Constraints*. International Rice Research Institute, Los Baños, Philippines.

Herdt, R.W. and Riely, F.Z. (1987) International rice research priorities: implications for biotechnology initiatives. Prepared for the Rockefeller Foundation Workshop in Allocating Resources for Developing Country Agricultural Research, Belagio, Italy, 6–10 July, 1987.

International Rice Research Institute (IRRI) (1977) *Constraints to High Yields on Asian Rice Farms: an Interim Report*. Los Baños, IRRI, Philippines.

International Rice Research Institute (IRRI) (1979) *Farm Level Constraints to High Rice Yields in Asia: 1974–1977*. IRRI, Los Baños, Philippines.

Lin, J.Y. (1990) Collectivization and China's agricultural crisis in 1959–1961. *Journal of Political Economy* 98, 1228–1252.

Lin, J.Y. (1991a) Public research resource allocation in Chinese agriculture: a test of induced technological innovation hypotheses. *Economic Development and Cultural Change* 40, 55–73.

Lin, J.Y. (1991b) The household responsibility system reform and the adoption of hybrid rice in China. *Journal of Development Studies* 36, 535–372.

Lin, J.Y. (1992a) Hybrid rice innovation: a study of market-demand induced technological innovation in a centrally planned economy. *Review of Economics and Statistics* 74, 14–20.

Lin, J.Y. (1992b) Rural reforms and agricultural growth in China. *American Economic Review* 82, 34–51.

Ministry of Agriculture, Planning Bureau (1989) *Zhongguo nongcun jingji ziliao tongji daquan, 1949–1986 (A Comprehensive Book of China Rural Economic Statistics, 1949–1986)*. Agricultural Press, Beijing.

Widawsky, D.A. and O'Toole, J.C. (1990) *Prioritizing the Rice Biotechnology Research Agenda for Eastern India.* Rockefeller Foundation, New York.

Wiens, T.B. (1982) Technological change. In: Barker R. and Sinha R. (eds) *The Chinese Agricultural Economy.* Westview Press, Boulder.

Zhu, R. (ed.) (1988) *Dangdai zhongguo de nongzuowuye (The Science of Crops in Modern China).* China Social Science Press, Beijing.

11 Rice Production Constraints in Bangladesh: Implications for Further Research Priorities

M.M. Dey[1], M.N.I. Miah[2], B.A.A. Mustsfi[2] and M. Hossain
[1]*International Center for Living Aquatic Resources Management (ICLARM), MC PO Box 2631, 0718 Makati, Metro Manila, Philippines;*
[2]*Bangladesh Rice Research Institute, Gazipur 1701, Bangladesh*

I. Introduction

Rice production is vital to the Bangladesh economy. Rice contributes about 50% of the agricultural value added and 60–70% of the total agricultural labour force is employed in rice production, processing, marketing and distribution. About 75% of the total cropped area and 83% of the total irrigated area are presently used for rice cultivation. Rice is also the major food item in Bangladesh accounting for an estimated 76% of the people's average calorie intake and 66% of protein intake.

Due to the introduction of high-yielding seed and fertilizer irrigation technology, and expansion of the area under irrigated dry season rice, (popularly known as boro), rice production has increased substantially over the last two decades, from 9.93 million tons clean rice in 1972/73 to 19.41 million tons in 1991/92. Though the country has achieved near self-sufficiency in rice production, the area under other crops, especially pulses, oilseeds, vegetables and others, has been reduced. As a result, Bangladesh has to spend a substantial amount of foreign exchange earnings to import pulse, oilseeds, etc. to meet the need of the consumers. A recent exercise carried out by the Bangladesh Agricultural Research Council, the apex agricultural research body, suggests that the rice area should be reduced to 7.6 million ha from the present level of 10.7 million ha to make more area available for other crops. On the other hand, productivity per unit of area, particularly in the irrigated boro area, shows a declining trend. With the population still growing at nearly 2% per year, a 2–3% per year increase in rice supplies has to be attained with less land to allow crop diversification with, at the same time, an increase in the productivity of labour and a maintenance of the quality of natural resources.

To meet the urgent national need for increased yield per hectare, the problems that affect rice yield must be clearly identified. Planners and policy-makers need information on the relative importance of various problems so they can allocate and redistribute the available resources among various researchable issues. This chapter has the following specific objectives: (i) to identify the gap between the potential and actual yield (yield gap); (ii) to estimate the contribution of various technical constraints to the yield gap in different rice ecosystems; and (iii) to prioritize the emerging problems of rice production under different ecosystems.

II. Background

In Bangladesh, rice is grown throughout the year on upland to deeply flooded land in three seasons – aus (March–July), aman (July–December) and boro (January–June) with overlapping or short turnover periods. Aman rice comprises two groups: broadcast aman and transplant aman. Broadcast aman is sown (sometimes mixed with aus) in the pre-monsoon period (March–May) and harvested in November–December. Modern rice varieties were introduced in Bangladesh in 1966/67 and 1968/69 for the boro, and aus and aman seasons, respectively. Boro rice is irrigated, and by the mid-1990s around 87% of the boro area was covered by modern varieties. Modern varieties covered around 45% of the transplanted aman area, mostly under rainfed conditions. All the broadcast aman area is planted to local varieties. In the aus season, modern varieties are grown as a transplanted crop in irrigation conditions except in high rainfall areas. At present around 32% of the aus area is covered by modern varieties. Most of the rest of the aus crop is covered by local broadcast aus varieties. The average yield of paddy in Bangladesh is low, ranging from 1.25 tons ha^{-1} for broadcast aman to 3.91 tons ha^{-1} for boro. Modern varieties cover 53%, broadcast aman 9% and other local varieties 38% of total area during 1992 (BBS, 1993). The various types are distributed among four different growing environments or ecosystems. Broadcast aman and aus grow in deepwater and upland rice ecosystems, respectively. Boro and transplanted aus are grown under irrigated ecosystems, while transplanted aman is grown primarily under rainfed lowland conditions with some in deepwater environments (where floodwaters exceed 50 cm).

A. Agricultural research system in Bangladesh

Agricultural research in Bangladesh has a tradition dating back to the 1930s. During the British rule, the area presently called Bangladesh benefited from the research and development efforts of the Imperial Council of Agricultural Research, which was established in 1929 for the entire subcontinent. After independence in 1947, the agricultural research and university system of Pakistan faced an acute shortage of skilled manpower due to the loss of large numbers of ex-patriate scientists who had provided most of the leadership and professional input for institutions in both East and West Pakistan. The Pakistan Agricultural Research Council (PARC) proved incapable of raising the level and quality of research output, and few truly meaningful investigations were accomplished in East Pakistan during the 1950s and 1960s. Bangladesh became independent on 16 December, 1971.

Today's agricultural research plant and professional capability in Bangladesh was established during the last two decades. It consists of 21 autonomous or semi-autonomous institutions which conduct research under the administrative responsibility of eight ministries, but most of the varietal improvement and related research on major agricultural commodities is carried out by research institutes under the Ministry of Agriculture. Rice, jute, sugarcane, livestock, tea, fisheries, forestry and nuclear agriculture each have their dedicated research institutes. The Bangladesh Agricultural research Institute (BARI) deals with wheat, potato, pulses, oilseeds, maize, vegetables, fruits and other species. In 1973 the Government of Bangladesh established the Bangladesh Agricultural Research Council (BARC) to coordinate all agricultural research in the country. Though BARC is technically under the jurisdiction of the Ministry of Agriculture, the heads of all the agricultural research institutes are members of its governing council.

B. Rice research institutions

Scientific research on rice in Bengal started in 1911 with the appointment of G.P. Hector as economic botanist of the Bengal Department of Agriculture (RCAI, 1928). By the early 1920s, approximately five researchers worked exclusively on rice (DOAB, 1925). After 1935, as the Imperial Council of Agricultural Research began to fund rice research, the Habiganj research station, now a regional station of the Bangladesh Rice Research Institute (BRRI), was set up to conduct research on deepwater rice.

Most research in this area stopped in 1942/43 due to World War II. After the war, research on rice was begun, but was hampered by the 1947 partition of British Bengal into East Pakistan and West Bengal, as many Hindu researchers migrated to India from East Pakistan. In the mid-1950s, rice research started to resume but suffered another setback in the early 1960s when the Pakistan government took over the experimental station's rice fields to build a second capital. "Thousands of germplasm collected during the previous 50 years at a great cost were lost forever" (Zaman, 1975).

Rice research was started again systematically in 1970 with the establishment of the BRRI. During the 1960s, the academy of Rural Development at Comilla helped in disseminating varieties such as IR8 and Pajam. At present the BRRI is the main research institute responsible for conducting research on rice. The Bangladesh Institute of Nuclear Agriculture (BINA) also conducts research on rice, with around 31% of its manpower devoted to rice research. In addition to these two research institutes, various universities are conducting research on rice on a limited scale. One of the objectives of these research institutes is to develop new varieties with high yields and superior quality, and a relatively high proportion of their staff are involved in plant breeding.

So far, the BRRI has developed 25 modern varieties suitable for growing both in irrigated and rainfed environments. While BINA has developed three modern rice varieties (two for the boro season and one for the transplanted aman season) and the Bangladesh Agricultural University (BAU) has developed one variety for the boro season. In the early years, these institutions emphasized the need to develop short-duration high-yielding varieties. From the late 1970s, the emphasis was shifted to intermediate-tall and short-duration varieties. Development of varieties resistant to insect pests and diseases has always been a research objective and many varieties developed so far are at least moderately resistant to the major insects and diseases.

C. Insects pests and diseases in Bangladesh

Insects, diseases, weeds and abiotic stresses like salinity, cold, heat and drought are the major technical constraints to higher yields in Bangladesh. So far, plant breeding at the BRRI, BINA and BAU have emphasized insect and disease resistance; no serious attempts have been made to develop varieties resistant to abiotic stresses. This section presents a brief account of the insect and disease problems of rice in Bangladesh and the level of resistance of modern varieties to major pests.

So far, 175 species of rice insect pests have been recorded in Bangladesh, out of which 20–30 species are important. The seasonal incidence patterns of 15 major insect pests in rainfed and irrigated environments are also known (Table 11.1). A series of crop-loss assessment trials carried out in the field by the entomology division of the BRRI against major insect pests from 1977 to 1979 have shown an average yield loss of 13% in the boro season, 24% in the aus season and 18% in the transplanted aman season (Alam et al., 1981). In outbreak situations, insects like the brown planthopper (20–44%) and rice hispa (14–62%) can cause devastating losses (Table 11.2). At the BRRI over 20,000 varieties and breeding lines had been screened against major insect pests by 1990 to identity resistant donors and materials, with resistance incorporated into modern varieties. Despite these efforts, in the 29 varieties so far released by Bangladesh institutions, two are rated as moderately resistant to the stemborer, two as moderately resistant to thrips and one as moderately resistant to the green leafhopper, but otherwise they have little resistance (BRRI, 1979, 1985a, 1985b; Karim, 1991).

The important rice diseases in Bangladesh are tungro, bacterial leaf blight, bacterial leaf streak, sheath blight, sheath rot, stem rot, blast, leaf scald, brown spot, bakanae, seedling blight, damping off, ufra and root knot. Crop-loss assessment trials carried out by the plant pathology division of BRRI have shown a 6–98% loss by different diseases depending on their infection rates. Devastating yield losses may be encountered from the attack of tungro virus (55%), ufra (nematode) infection (50–90%) or seedling blight disease (98%) (Table 11.3). Varietal resistance is the mainstay of disease control in rice. From

Table 11.1. Major insect pests and their frequency in irrigated and rainfed environments, 1982–1986.

	Number of districts reporting high incidence of insects		
		Rainfed rice	
Insect	Irrigated rice (boro)	Aus	Transplanted aman
Stemborer	8	6	7
Gall midge	–	–	1
Rice hispa	39	20	38
Leaf roller	1	11	4
Caseworm	–	1	12
Swarming caterpillar	–	–	3
Grasshoppers	1	2	–
Long-horned cricket	–	1	3
Brown planthopper	29	–	21
White-backed planthopper	5	–	1
Green leafhopper	1	–	1
Orange-headed leafhopper	1	–	–
Mealy bug	–	9	22
Rice bug	–	3	1
Ear-cutting caterpillar	–	–	9

Source: Karim, 1989.

Table 11.2. Rice yield loss to insect pests in the field.

Insect pests	Season	Yield loss (%)	Reference
Major insect pests	Boro	13	Alam et al., 1981
Major insect pests	Aus	21	Alam et al., 1981
Major insect pests	Transplanted aman	18	Alam et al., 1981
Major insect pests	Transplanted aman	22–26	BRRI, 1985a
Stemborer and other minor pests	Aus	15	BRRI, 1985b
Brown planthopper	Boro	20–44	BRRI, 1985b
Rice hispa	Boro	11–62	BRRI, 1985a; Karim, 1987

1970 to 1985, BRRI pathologists have screened more than 46,000 entries against the major rice diseases. The identified disease-resistance germs have been combined in suitable modern varieties and the majority of modern varieties released by the BRRI, BINA and BAU possess multiple resistance to different diseases (Table 11.4).

Table 11.3. Rice yield losses in susceptible varieties due to important diseases in Bangladesh (%).

Diseases	Boro	Aus	Transplanted aman
Sheath blight	–	6–36	19
Stem rot	54	16–19	15–25
Tungro virus	–	–	55
Bacterial leaf blight	–	7–26	–
Blast	22	–	–
Seedling blight	98	–	–
Damping off	45	–	–
Sheath rot	–	–	19–46
Leaf scald	–	–	39
Ufra	–	50–90	–
Root knot	–	9–22	–
White tip	–	60	–
Bakanae	21–34	–	–
Bacterial leaf streak	–	13	–

Source: Miah, 1988; BRRI, 1979.

III. Conceptual Framework and Survey Methodology

The fundamental assumptions of the study are: (i) there exists a yield gap between potential yield and actual yield or there is a possibility of increasing potential yield; (ii) the yield gap is not identical for all environments and varieties of rices; and (iii) the yield gap is composed of a number of constraints. According to De Datta et al. (1978), Widawsky and O'Toole (1990), and Lin and Shen (1994) the yield gap can be divided into three parts. Yield gap I is the difference between the potential yield of the existing farm varieties under favourable environments and the highest yield at favourable experimental conditions. This yield gap arises from differences in the varietal traits and biophysical environments (micro and macro) which cannot be adopted or managed at the field level. Yield gap II is the difference between actual farm yield and the potential yield of the farm varieties under favourable environments. Yield gap II is caused by biological and socioeconomic constraints. Yield gap III is the difference between the theoretical potential yield and highest experimental yield. It represents the potential increase in biological efficiency which researchers are aiming at. In this study we focus on the biological constraints component of yield gap II.

We interviewed a sample of extension officers and model (best) farmers to get their views about yield gaps and factors contributing to yield gap II. Rice land was classified into ten ecologies and four land types. One hundred and twenty subdistricts (thanas) were selected (out of 480) as proportionally

Table 11.4. Reaction of Bangladesh's modern variety rices to different diseases*.

Variety	RTV	BLB	BLS	BL	SHB	SHR	SR	LSC	UF	BS
BR1 (Chandina)	R	S	S	MR	S	S	S	S	HS	–
BR2 (Mala)	R	MS	MS	MS	S	MS	MR	S	MR	–
BR3 (Biplob)	S	S	MR	R	MS	MR	MR	MR	HS	MR
BR (Brisail)	MR	MR	MR	MR	MR	MR	MR	MR	HS	MR
BR5 (Dulabhag)	MR	MR	MR	MR	MR	MR	MR	MR	HS	–
BR6	S	S	MR	MS	MS	MS	MS	–	HS	–
BR7 (Bribalam)	MS	MS	MS	MR	MS	MS	S	MS	MR	–
BR8 (Asha)	MS	S	MR	MR	MS	S	MS	MS	S	–
BR9 (Shufala)	–	MS	S	MR	MR	MS	MS	MS	HS	–
BR10 (Progoti)	MR	S	MR	MR	MS	MR	S	MS	HS	MR
BR11 (Mukta)	MS	S	MR	MR	S	MR	S	MS	HS	MR
BR12 (Moyna)	MR	MS	MR	R	MS	MS	MR	MR	S	–
BR14 (Gazi)	S	MS	MS	R	S	–	R	–	S	–
BR15 (Mohini)	MS	MS	MR	MS	S	–	MR	–	S	–
BR16 (Shaibalam)	MR	MS	MS	MR	MS	–	MR	–	S	–
BR17 (Hashi)	S	–	–	–	–	–	–	–	–	–
BR18 (Shahjalal)	S	–	–	–	–	–	–	–	–	–
BR19 (Mongol)	S	–	–	–	–	–	–	–	–	–
BR20 (Nizami)	S	MS	R	S	MR	–	MR	–	–	–
BR21 (Niamat)	S	MS	–	S	MR	–	–	–	–	–
BR22 (Kiron)	S	MS	–	MR	MR	–	MR	–	–	–
BR23 (Dishari)	S	MS	–	–	MR	–	MR	MR	–	–
BR24	–	–	–	–	–	–	–	–	–	–
BR25	–	R	–	–	–	–	–	–	–	–
BR26	–	R	–	–	–	–	–	–	–	–
IR5	–	S	S	MS	S	MS	MR	MR	–	–
IR8	S	HS	–	S	S	–	MR	MR	S	–
IR20 (Irisail)	MR	S	–	S	MS	–	MS	–	–	–
IR50	MR	S	S	S	S	MS	MR	S	–	–
Purbachi	S	HS	S	MS	S	S	MS	–	–	–
Iratom-24	MR	S	–	–	–	–	MR	–	–	–
Iratom-38	MR	S	–	–	–	–	MR	–	–	–
BAU-63 (Bharasha)	S	S	S	S	S	S	MR	–	–	–
Pajam	S	S	S	HS	MS	MR	MR	MR	S	–
Nizersail	MR	MS	–	S	MR	MR	S	MR	–	–

R, resistant; MR, moderately resistant; MS, moderately susceptible; S, susceptible; HS, highly susceptible.
*RTV, rice tungro virus; BLB, bacterial leaf blight; BLS, bacterial leaf streak; BL, blast; SHB, sheath blight, SHR, sheath rot; SR, stem rot; LSC, leaf scald; UF, ufra; BS, brown spot.

Table 11.5. Yield gap II in Bangladesh.

Mode of irrigation	Type of rice	Maximum farm yield (kg ha^{-1})	Actual average farm yield (kg ha^{-1})	Yield gap II (kg ha^{-1})
Irrigated	Aus	4917	3499	1418
	Aman	5250	3740	1510
	Boro	6370	4990	1380
Rainfed	Aus	2426	1851	575
	Aman	3352	2590	762
	Boro	2038	1406	680

representing the 40 sub-ecosystems*. Two model farmers knowledgeable about rice production practices and their constraints were selected from each thana. Farmers were interviewed by the crop production specialists of their respective districts in the presence of local extension workers. Thana extension officers and local-level extension workers identified various constraints and symptoms. The model farmers provided information on the frequency of different constraints, the area affected by the constraints and their resulting production losses.

A workshop was organized in each agricultural regional headquarters to train interviewers and discuss the contents of the questionnaires. All participating district crop production specialists and other relevant people were present at these workshops. Follow-up workshops were also held at regional headquarters to review the results. Finally, a workshop was held at the BRRI to review the findings by rice researchers and the extension specialists.

IV. Technical Constraints to Rice Production in Bangladesh

A. Yield gap II

To assess the yield gap, model farmers assisted by local-level extension workers were asked to report the maximum yield obtained from particular plots (parcels) during the last 10 years, and the lowest as well as normal (average) yield obtained from that plot during the same time period. Maximum farm-level yields as reported in different rice crops were found within the range of potential farm-level yields of different modern varieties as expected by the BRRI. So the maximum farm-level yield can be treated as the potential farm-level yield.

---

* Administratively Bangladesh is divided into 64 districts and each district is divided into a number of thanas.

We calculated yield gap II as the difference between the maximum farm-level yield and the average actual farm-level yield (Table 11.5). The gap is higher in irrigated conditions for all types of rice, as is average yield.

Although data were collected for ten different rice ecologies, we reported the national averages for the three most important: rainfed aus, rainfed aman and irrigated boro, which contribute 13, 30 and 23% of total rice area, respectively. These three dominant rice types cover around 70% of the rice area and represent three distinct rice ecosystems (upland, rainfed lowland and irrigated).

Technical constraints were classified into two broad categories: (i) biotic factors, which have further been classified into insects, diseases, weeds and rodents; and (ii) abiotic factors, divided into climate and soil. Table 11.6 reports the contribution of technical constraints. A portion of yield gap II remained unexplained by technical constraints. In a poor developing country like Bangladesh, socioeconomic and institutional factors often constrain farmers from the benefits of available technology. In the irrigated boro areas the biotic factors together accounted for 60% of yield gap II. The same factors accounted for 66% in rainfed aus and 80% in rainfed aman areas. One reason for the lower contribution of technical constraints in irrigated conditions is the greater potential of increasing yields through purchased inputs like fertilizers, pesticides and irrigation water. It is more likely that farmers would be constrained by socioeconomic factors like availability of credit under such conditions. Biotic factors also contributed more to yield gap II in irrigated than in rainfed environments. In rainfed environments, climatic factors alone contribute

Table 11.6. Contributions to yield gap II in Bangladesh.

	Rainfed aus		Rainfed aman		Irrigated boro	
	Loss attributed (kg ha^{-1})	Percent explained	Loss attributed (kg ha^{-1})	Percent explained	Loss attributed (kg ha^{-1})	Percent explained
Yield gap II	575.0	100.0	762.0	100.0	1380.0	100.0
Technical constraints	370.0	65.7	612.0	80.30	824.0	59.7
A. Biotic factors	139.0	24.2	360.0	43.1	521.0	37.8
Insect	67.0	11.7	182.0	23.9	357.0	8.0
Diseases	57.0	9.9	117.0	15.4	156.0	2.4
Weeds	14.0	2.5	10.0	1.4	8.0	0.4
Rodents	1.0	0.2	19.0	2.5	0.0	0.0
B. Abiotic factors	239.0	41.50	252.0	37.2	303.0	21.9
Climate	217.0	37.7	220.0	33.0	238.0	17.2
Soils	22.0	3.8	32.0	4.2	65.0	4.7
Unexplained	197.0	34.3	150.0	19.7	556.0	40.0

about 35% of total technical constraints reflecting the more uncontrolled environment.

B. The top 20 technical constraints

Individual technical constraints were ranked based on the estimated average yield loss in a normal year, calculated as:

$$\text{Estimated yield loss in normal year}_i = \frac{\sum_j [\text{production loss during 10 years}_{ji}]}{\sum_j [\text{sown acreage during last 10 years}_{ji}]} \quad (11.1)$$

where i indicates a constraint and j indicates a parcel in a particular production environment. A constraint normally does not occur every year nor in all the plots in one year. Estimated yearly yield loss in a normal year has two components: (i) yield loss in the affected area; and (ii) average incidence of the particular constraint. Yield loss in affected areas is computed by dividing the

Table 11.7. The top 20 technical constraints contributing to yield gap II in rainfed aus rice, Bangladesh.

Constraints	Area affected, last 10 years (%)	Average yield loss, normal year (kg ha^{-1})
Drought, vegetative stage	6.30	110.28
Submerged, anthesis stage	2.90	46.68
Drought, seedling stage	3.77	37.14
Bacterial leaf blight	3.11	24.89
Rice hispa	3.46	19.95
Rice bug	2.24	18.31
Stem borer	6.28	18.12
Blast	1.70	17.06
Low organic matter	2.00	16.00
Drought, anthesis stage	0.89	8.81
Submerged, seedling stage	0.72	7.63
Leaf steak	1.02	5.84
Echinochloa sp.	1.15	5.80
Submerged, vegetable stage	1.04	5.66
Leaf roller	1.47	5.14
Cyperus difformis	1.52	4.72
Bacterial streak	1.35	4.20
Sulphur deficiency	0.26	3.61
Mealy bug	0.64	3.48
Tungro	0.75	2.97
Total loss		370.00*

*Values in this column do not add up to total due to rounding of figures.

total production loss due to a particular constraint over 10 years by the total affected area in the last 10 years, giving an estimate of the average loss per hectare when the constraint occurred. The percent area affected by a particular constraint was computed as follows:

$$\left(\sum_j{}_{ij}\right)\text{Percent of area affected}_i = \frac{\sum_j [\text{total area affected during last 10 years}_{ij}]}{\sum_j [\text{total sown acreage in last 10 years}_{ij}]}.$$

(11.2)

Tables 11.7, 11.8, 11.9 list the top 20 technical constraints operating in rainfed aus, rainfed aman and irrigated boro rice seasons, respectively. In the case of rainfed aus the most prevalent and important contributing factors to yield loss are drought, submergence at anthesis stage, bacterial leaf blight disease and rice hispa. Drought at the seedling stage affects the quality of the seedlings and drought at the vegetative stage causes a significant reduction in tiller numbers per plant. The area under rainfed aus rice is predominantly cropped with local

Table 11.8. The top 20 technical constraints contributing to yield gap II in rainfed aman rice, Bangladesh.

Constraints	Area affected, last 10 years (%)	Average yield loss, normal year (kg ha^{-1})
Submerged, vegetable stage	7.33	92.23
Submerged, seedling stage	5.67	78.55
Brown plant hopper	8.35	60.93
Drought, anthesis stage	4.83	58.00
Ufra	5.06	40.91
Stem borer	10.41	40.79
Rice hispa	7.69	37.65
Rice bug	3.25	24.97
Bacterial leaf blight	4.04	21.09
Rats	1.88	18.76
Tungro	1.88	15.98
Low organic matter	1.50	15.75
Drought, vegetative stage	2.40	14.87
Sheath blight	2.67	12.01
Blast	1.82	10.70
Drought, seedling stage	0.76	7.51
Leaf roller	2.45	7.37
Leaf streak	0.98	5.89
Echinochloa sp.	2.04	5.12
Zinc deficiency	1.02	4.55
Total loss		612.00*

*Values in this column do not add up to total due to rounding of figures.

Table 11.9. The top 20 technical constraints contributing to yield gap II in irrigated boro rice, Bangladesh.

Constraints	Area affected, last 10 years (%)	Average yield loss, normal year (kg ha^{-1})
Stem borer	15.10	137.67
Brown planthopper	9.08	136.20
Drought, anthesis stage	2.49	82.56
Submerged, anthesis stage	3.58	72.05
Bacterial leaf blight	4.16	68.28
Rice hispa	3.82	41.07
Sheath blight	2.80	39.35
Rice bug	7.93	34.92
Low organic matter	3.00	26.70
Cold, vegetative stage	1.59	19.54
Ufra	1.67	18.36
Drought, vegetative stage	2.00	17.97
Cold, anthesis stage	1.26	16.20
Blast	1.44	14.91
Sulphur deficiency	2.22	13.59
Cold, seedling stage	1.19	9.99
Stem rot	0.33	9.77
Drought, seedling stage	1.30	7.86
Heat, anthesis stage	0.21	6.95
Zinc deficiency	1.19	6.56
Total loss		824.00*

*Values in this column do not add up to total due to rounding of figures.

varieties which are susceptible to bacterial leaf blight disease and rice hispa – an insect that causes serious damage to the leaves of rice plants.

The abiotic factors submergence and drought are the top two constraints in rainfed aman rice. Submergence at the vegetative stage is common in the aman season and it sometimes causes total crop failure. Submergence at the seedling stage causes deterioration in the seedling quality resulting in a poor stand and causes substantial yield loss. Drought is also common during late September to early October when most of the aman varieties approach the reproductive stage. This hampers panicle emergence as well as the development of spikelets. Brown plant hopper, ufra, stem borer and rice hispa are important pests in rainfed aman in Bangladesh.

Boro rice is mostly grown in an irrigated environment and the coverage of modern varieties is about 90% of the total boro cropped area. A negligible acreage is in a rainfed environment where the traditional boro varieties are grown. In the top 20 constraints for irrigated boro rice, stem borer and plant hopper are the most important. The next most important constraints are drought and submergence at the anthesis stage.

References

Alam, S., Karim, A.N.M.R. and Nurullah C.M. (1981) Insecticide use on rice in Bangladesh. *Bangladesh Journal of Agricultural Research* 6, 37–50.

Bangladesh Bureau of Statistics (BBS), Ministry of Planning (1993) *Statistical Yearbook of Bangladesh – 1992.* BBS, Ministry of Planning.

Bangladesh Rice Research Institute (BRRI) (1979) *Annual Report for 1975–1976.* BRRI, Gazipur–1701.

Bangladesh Rice Research Institute (BRRI) (1985a) *Annual Report for 1981.* BRRI, Gazipur–1701.

Bangladesh Rice Research Institute (BRRI) (1985b) *Annual Report for 1982.* BRRI, Gazipur–1701.

De Datta, S.K., Gomez, K.A., Herdt, R.W., and Barker, R. (1978) *A Handbook on the Methodology for an Integrated Experiment Survey on Rice Yield Constraint.* International Rice Research Institute, Los Baños, Philippines.

Department of Agriculture, Bengal (DOAB) (1925) *Annual Report 1924/25.* DOAB, Calcutta.

Karim, A.N.M.R. (1987) The hispa episode. In: *Proceedings of a Workshop on Experiences with Modern Rice Cultivation in Bangladesh.* Bangladesh Rice Research Institute, Gazipur–1701, pp. 125–160.

Karim, A.N.M.R. (1989) Some recent progress in insect pest management research and problems in rainfed rice. In: *Proceedings of a Workshop on Experiences with Modern Rice Cultivation in Bangladesh.* Bangladesh Rice Research Institute, Gazipur–1701, pp. 134–144.

Karim, A.N.M.R. (1991) Insect pests and diseases: the sources of risk in crop production. Paper presented at the National Workshop on Risk Management in Bangladesh Agriculture, held at BARC, Dhaka, 21–27 August, 1991.

Lin, J.Y. and Shen M. (1994) Rice production constraints in China: implication for biotechnology initiative. Paper prepared for the International Workshop on Rice Research Prioritization in Asia, IRRI, 21–22 February, 1994, Los Baños, Philippines.

Miah, S.A. (1988) Cereal crop losses due to disease. Paper presented at the Third Biennial Conference of the Bangladesh Phytopathological Society, 8–9 December, 1988, Bangladesh Agricultural University, Mymensingh.

Royal Commission on Agriculture in India (RCAI), Great Britain (1928) *Bengal.* RCAI, London.

Widawsky, D. and O'Toole, J.C. (1990) *Prioritizing the Rice Biotechnology Research Agenda for Eastern India.* Rockefeller Foundation, New York.

Zaman, S.M.H. (1975) History of agricultural research in Bangladesh. Paper presented in a seminar on integrated rural development, 29 November–3 December, 1975. Institute of Engineers, Dhaka, Bangladesh.

12 Rice Research in Nepal: Current State and Future Priorities

H.K. Upadhyaya
Center for Environmental and Agricultural Policy Research, Extension and Development (CEAPRED), PO Box 5752, Kathmandu, Nepal

I. Introduction

Agriculture employs more than 80% of the labour force, produces nearly 60% of the gross domestic product (GDP) and generates a significant portion of the export earnings of Nepal. Within agriculture, rice is by far the most important crop, accounting for about 50% of the total agricultural area and production of the country. About 75% of the total rice area is located in the flat plains of the tarai. The hills and high hills, or mountains, occupy only about 23 and 2% of the country's rice area, respectively (Table 12.1).

Efforts to improve rice productivity in Nepal have resulted in the introduction of a large number of modern varieties (MVs) with varying yield potential. At present, about 50% of the total rice areas are planted to these MVs. However, the increase in average productivity during the past three decades has been small. Available data indicate changes in the average national rice yields from 1.80 tons ha^{-1} during 1960–69 to 1.89 tons ha^{-1} during 1970–79 to 1.99 tons ha^{-1} during 1980–89 (DFAMS, 1990).

This low productivity growth in Nepal contrasts with changes in other Asian countries as a result of the adoption of short-statured, early-maturing, higher-yielding MVs developed at the IRRI. Apparently, varietal development work that resulted in the introduction of a number of more promising MVs, including those developed at the IRRI, has not had significant impact on rice productivity in Nepal.

The newer MVs, although having higher yield potentials, did not perform well under the production environments facing the Nepali rice farmers. Relatively higher yields have been obtained in limited – more favourable – areas where the IRRI MVs are planted; but considerable gaps still exist between the actual and potential farm-level yields due to various production constraints. Knowledge of these various production constraints will help policymakers and researchers assign meaningful priorities and allocate resources optimally in rice research.

Table 12.1. Average rice area and production by ecological region in Nepal, 1975–1989.

	1975–1979	1980–1984	1985–1989	1975–1989
Rice area (thousand ha)				
Mountains	26.26	26.28	39.91	29.03
	(2.1)	(2.0)	(2.5)	(2.2)
Hills	199.71	250.39	326.98	254.71
	(15.8)	(19.1)	(23.4)	(19.3)
Tarai	1003.71	1032.92	1036.65	1034.27
	(82.1)	(78.9)	(74.1)	(78.5)
Nepal	1259.67	1309.59	1399.54	1317.46
	(100)	(100)	(100)	(100)
Rice production (thousand tons)				
Mountains	54.97	48.21	65.04	55.43
	(2.3)	(2.0)	(2.3)	(2.2)
Hills	473.98	530.75	650.26	544.62
	(20.3)	(21.5)	(22.7)	(21.5)
Tarai	1805.59	1885.72	2145.07	1931.20
	(77.3)	(76.5)	(75.0)	(76.3)
Nepal	2334.53	2464.68	2860.37	2531.25
	(100)	(100)	(100)	(100)

Figures in parentheses are percentages of the total.
Source: DFAMS, 1990.

In Nepal, where financial resources are largely limited and a majority of agricultural research and development activities are carried out by external assistance or loans, it is particularly important that these production constraints be property identified and research priorities set based on their probable pay-offs. Specifically, what research problems to undertake and how many resources to allocate to each of them are two major questions that must be answered to achieve research efficiency.

The overall objective of this chapter is to identify priority areas for future rice research in Nepal. Specifically, this chapter will assess the current state of rice research in Nepal, identify the existing rice yield gaps across production environments and associated production constraints, and generate a list of priority problem areas for rice research based on their corresponding yield-loss estimates and potential research pay-offs.

This chapter is organized as follows. The following section presents the methodology used for the study. A brief description of the major rice production environments, yield gaps and associated production constraints is presented in section III. Section IV presents an overview of the agricultural research institutions and trends in resource allocation in the country. In section V, major crop-loss constraints – diseases, insects and abiotic constraints – are identified and prioritized for research based on their corresponding yield-loss estimates and the portion of yield gap II explained by them. The last section summarizes the main results.

II. Methodology

Analogous to a firm's production rule, efficiency in resource allocation across various research alternatives or problem areas requires that the ratio of marginal value product (MVP) to marginal factor cost be equated across all such alternatives. The MVP of research on a particular research problem can be obtained from data on yield benefits due to research, area affected by yield change, prices and costs associated with obtaining the yield benefit.

Expected yield increases can be approximated by the intensity of the problem or opportunity of each research problem area (RPA). The extent of yield losses caused by various problems at present can be taken as the expected yield increases from successful research on that problem. It has been observed that some problems do not occur every year; if they do, their severity may differ from year to year. To consider this, estimates of subjective probability of occurrence were obtained for each problem (e.g. in the past 5 years if the problem occurred for 3 years, then the probability of occurrence would be 3/5). The product of unit yield loss and this probability will give the expected yield increase from successful research on that particular problem.

The units affected are the number of hectares of rice area affected by the particular problem. The probability of research success is a subjective judgement about the possibility of overcoming the problem through research, given the time frame and scientific resources. As for other country studies, this chapter reports production and yield benefits from each RPA.

A. Survey data

Three sets of data were generated from interviews with rice research scientists, extension specialists and farmers representing different rice production environments of Nepal. A structured questionnaire was used. Substantial data were also generated from personal field visits by the research team, which included an agronomist, an entomologist, a pathologist, a soil scientist and an economist. This diversity greatly helped in understanding and assessing farmers' production problems and their severity at the farm level.

Researchers of different disciplines working in research stations around the country, who had some experience in rice research, were personally approached by the team members. Major questions included in the questionnaire were discussed to avoid any possible confusion. The questionnaires were then given to the researchers and left with them for a few days to allow sufficient time for the respondents to think and write their responses. The information collected from this source included average rice yields, major rice production constraints and the corresponding area affected, yield losses and frequency of their occurrence by production environment in the country, and the probability of research success and the number of years required to solve these problems in Nepal.

The second set of data were collected from extension specialists (agricultural development officers and some specialists) and provided district-level information by production environment on present rice area, yields and major production problems with corresponding estimates of yield losses and frequency of occurrence. The questionnaires were personally handed over, either by research assistants or team members, and collected the next day. In addition, information on the present trend of adoption of new varieties and maximum adoption rates were obtained through informal discussions with extension workers working in the district.

There are 75 districts in Nepal; each district has a district agricultural development office with at least one graduate agricultural development officer (ADO) at the district headquarters and a number of junior agricultural extension workers who are stationed in different parts of the district.

The third set of data were collected from a sample of farmers representing different rice production environments and provided farm-level information on rice yield trends, varieties, use of inputs, and major production constraints with corresponding yield losses and frequency of occurrence. A majority of team members was present at the time of the farm surveys, so that responses regarding particular problems were correctly understood. In many cases, the study team visited the rice farm and assessed the incidence and severity of various diseases, insects and weeds. During discussion with the respondent farmer, the team also collected information on desired characteristics of a new rice variety, perceived causes of present low yields and the situation regarding inputs availability in the area.

B. Sample size

Thirteen agricultural research stations and six central disciplinary divisions conduct research under the umbrella of the Nepal Agricultural Research Council (NARC). Most of the major research stations are located in the tarai. For this study, almost all of the researchers of these stations who are engaged in rice research were surveyed. The survey also included some researchers who were previously engaged in rice research.

We included ADOs from all 20 districts of the tarai and 14 districts of the hills, which together account for 83% of the total rice area of the country.

We surveyed a sample of 46 farmers in the hills and another 46 in the tarai, representing different rice production environments, to explore and understand actual farm-level production problems as perceived by farmers.

III. Production Environments, Yield Gaps and Production Constraints

A. Production environments

Nepal has a vast climatic diversity which determines land-use limits. Five different climatic zones exist: subtropical, warm temperate, cool temperate, alpine and arctic. The upper limit of arable agriculture is nearly 4000 m above mean sea level with an alpine climate and mean annual temperatures ranging from 3 to 10°C. Nearly 80% of rice production occurs in the tarai, below 1000 m, while the rest occurs in the hills below 2000 m.

There is variation in the amount of rainfall from east to west, which may be attributed to the onset and cessation of the southeast monsoon. The monsoon period begins earlier and terminates later in the eastern part of Nepal, thus the eastern part receives a greater amount of rainfall every year than the western part. The country also receives some winter (dry season) rainfall through the dry west monsoon originating from the Mediterranean Sea.

Almost 85% of rice cultivation occurs in soils derived of alluvium; this includes all of the tarai and the river valleys of the hills and mountains. The remaining rice cultivation occurs in soils developed *in situ* by the weathering of parent rocks. These are on sloping lands, which are bench terraced for rice cultivation. The tarai region consists of recent and post-Pleistocene alluvial deposits – it is the recent alluvial plain where rice is extensively grown.

For the purpose of this study, we follow the popular hydrology-based descriptors to classify rice production environments in Nepal: irrigated, rainfed lowland, upland and deepwater environments. There are generally a great many similarities within one production environment in the tarai or in the hills which makes the production environment within a region more or less a 'recommendation domain' for rice research. Due to climatic differences between the hills and the tarai, however, the same production environments of the hills and tarai may have different technological implications.

About 23% of the rice area is irrigated, either fully or partially. About 77% of the rice is grown under rainfed conditions: 69% in rainfed lowland, 5% in uplands and about 3% in swampy or deepwater areas. For the present study, we consider only the irrigated, rainfed lowland and upland production environments.

B. Yield gaps

The concept of yield gaps stems from the fact that there are quantitative differences between actual and potential rice yields. Such yield gaps may not be identical for all environments and varietal interactions. In more favourable – irrigated – areas, where higher-yielding MVs are adopted, farm-level yields are higher than in less favourable areas, but still far lower than the maximum yields obtained at the experimental level. Within a production environment also, there are considerable differences in rice yields across farms and across seasons within a farm. A number of biotic, abiotic, socioeconomic and other factors may be responsible for these yield differences.

Previous work at the IRRI (1979) divided the yield gap into two parts: one showing the difference between the actual farmer yields and yields attained in on-farm experiments, and the other showing the difference between the experiment station's maximum yield and the on-farm experiment's maximum yield. Since the purpose of the present study is to identify major production constraints and estimate benefits that might accrue from eliminating them, we focus on the gap between actual yields and potential farm yields (see yield gap II in Chapter 1).

On-farm experiments are generally managed by researchers. The difference between high on-farm experiment yields and actual farmer yields may be, in part, the result of management differences which cannot be transferred to farmers. Moreover, available yield data from on-farm experiments cannot be compared with the present level of actual farmer yields due to differences in the environment and time frame to which they correspond. Therefore, we hypothesized that the yield gap and the contribution of various biological, environmental and socioeconomic production constraints can be estimated by comparing the highest, lowest and normal yields obtained by a farmer on the same piece of land during the past 5 years. The differences in these yields will be accounted for by the presence and severity of different production constraints in the corresponding years. This will also largely reflect the effect on yields due to differences in production environments.

Summary data reflecting the highest, lowest and normal yields obtained by farmers under different production environments are given in Table 12.2. As expected, yields in irrigated environments are consistently higher than in rainfed lowland or upland production environments for both the hills and tarai regions. Within the irrigated environment, rice yields are higher for the hills than for the tarai. The yields for other production environments, however, tend to be higher in the tarai than in the hills. The characteristics of rainfed lowland and upland areas are qualitatively different between the two regions. In particular, the upland areas in the hills are generally sloping lands where non-rice crops, mainly maize, are grown during the rainy season; and upland paddy is either grown with maize or as a single crop after maize (in the dry season).

The differences between the highest yield, which is assumed to be potential yield, and normal yield are lower in the irrigated environment than in the other

Table 12.2. Highest, lowest and normal farm-level yields of rice by production environment in the hills and tarai of Nepal, 1990/91.

	Highest yield (tons ha^{-1})	Lowest yield (tons ha^{-1})	Normal yield (tons ha^{-1})	Percent gap between highest and Normal yield	Lowest yield
Tarai					
Irrigated	4.3	2.1	3.4	26	105
Rainfed lowlands	4.0	1.7	2.8	54	135
Uplands	2.1	0.9	1.2	75	133
Hills					
Irrigated	4.9	3.0	3.7	32	63
Rainfed lowlands	3.6	1.1	1.8	100	227
Uplands	1.5	0.4	0.9	67	275

two production environments. As irrigation becomes more scarce, the rice crop is subject to greater risk due to weather fluctuations. The differences between the highest and normal yields across production environments in the tarai range from 26% for the irrigated environment to 75% for the uplands. In the hills, these range from 32% for the irrigated environment to 100% for the rainfed lowland environments. The hills uplands do not show as much yield difference as do the tarai uplands.

The difference between the highest and the lowest yields, which represents potential yield loss caused by various production constraints, was smallest for the irrigated production environments, suggesting the prominent role of irrigation-related constraints in explaining the yield instability in the rainfed environments, particularly in the hills. More specifically, the difference between the highest and the lowest yields obtained in the last 5 years ranged from 63% in the irrigated lowlands to 275% in the rainfed uplands in the hills. In the tarai, this ranged from 105% in the irrigated environment to 135% in the rainfed lowland ecosystem.

C. Production constraints

The differences in observed farm yields within and across production environments can be attributed to: (i) technical constraints such as insects and diseases; (ii) physical constraints such as soil, topography and rainfall; (iii) institutional constraints such as extension services, delivery of inputs and production technology; and (iv) socioeconomic constraints such as low inputs use, markets and relative prices. Each is discussed briefly based on a review of past research findings – from published and unpublished documents.

Technical constraints

Insects, diseases and weeds are the major technical constraints to high rice yields in Nepal. The ecological diversity associated with rice cultivation provides very favourable niches for different pathogenic organisms. More than 19 fungi, five nematodes, four bacteria and two viruses have been reported to be infecting rice plants in Nepal. Among these diseases, some yield-loss assessments were carried out for blast, sheath blight, bacterial leaf blight and sheath rot. The extent of yield losses reported by these few loss-assessment studies are presented in Table 12.3.

Rice entomological investigations carried out since 1972 contributed basic information on rice insects in Nepal. Detailed investigation into yield losses revealed the yellow stemborer and striped borer among the borer species, three species of leaf and planthoppers (*Nephotettix apicalis*, *Sogatella furcifera* and *Cieadela spectra*), mealy bug (*Ripersia oryzae*), gall midge (*Pachydiplosis oryzae*), and hispa (*Dicladispa armigera*) to be major rice insects causing significant yield losses. There are some isolated cases of significant yield losses caused by thrips, seedbed beetle and white ants. Table 12.3 also shows the extent of yield loses caused by some of the major rice insects in Nepal.

Table 12.3. Rice yield loss caused by selected insects, diseases and weeds in Nepal.

Problem	Yield loss	Reference
Insects		
Rice borers	2.6%	Joshi, 1977
	up to 100%	Gyawali, 1973, 1974
	1–3%	Ganesh and Kaphale, 1972
	25–45%	Shah, 1990
Plant and leaf hoppers	60%	Joshi, 1974
Mealy bug	20.6–52%	Jyoti and Bhurer, 1989
	30–40%	Adhi Kary *et al.*, 1986
Gall midge	25–100%	Manandhar, 1983, 1984
	Up to 85%	Joshi, 1974
	Very high	Manandhar and Singh, 1976
Rice hispa	20–60%	Manandhar, 1976, 1986
	6% (local variety)	Bimb, 1978
Gundhi bug	9.9% (Mahsuri)	Bimb, 1978
	20–53% (IR varieties)	Bimb, 1978
Armyworm	30%	Gyawali, 1974
All insects	10–30%	Pradhan, 1980
	21%	Jyoti, 1989
Diseases		
Blast	36 kg ha^{-1} with 1% infection	Manandhar *et al.*, 1985
	20–25%	Manandhar *et al.*, 1989
	50.8%	Shah, 1990
Bacterial leaf blight	14–32%	Manandhar, 1986
Sheath blight	27%	Manandhar, 1986
Sheath rot	40–55%	Manandhar, 1986
Weeds	70–80% (upland rice)	NRRP, Nepal
	20–40% (lowland rice)	NRRP, Nepal

The experimental data suggest that weeds are a relatively minor problem in transplanted rice in the tarai. In Nepal, they are removed manually. Small farmers generally do not consider it a problem since they seed their fields themselves. However, large farmers often experience a shortage of labour to weed their fields.

Physical environmental constraints

Lack of assured irrigation indirectly contributes to low productivity through its impact on crop management. Experimental data indicate that the timeliness of various production activities, such as seeding, transplanting and fertilizer application, is crucial for optimum yields. Delayed transplanting of rice seedlings beyond a certain time causes significant yield reduction. Almost all of the rainfed lowland rice areas in Nepal, which covers about 70% of the total rice area, are subject to this risk.

Because of the high dependence of rice cultivation on rainfall pattern – onset and cessation, amount and distribution during crop cycle – the rice crop grown under rainfed conditions often suffers from drought in one stage or another during crop growth. Even under irrigated conditions, incidences of poor crop yields due to drought are not uncommon due to the unstable supply of irrigation.

Adverse topography and soils also appear to be significant constraints to high rice yields. Deficiencies of some important micro-nutrients, such as zinc, have been adversely affecting rice yields. Isolated cases of soil acidity, poor drainage and siltation problems have also been reported; but their severity at the national level is yet to be explored.

Socioeconomic constraints

Low input use associated with unreliable supply and poor purchasing capacity, underdeveloped market infrastructure and low prices are major socioeconomic constraints to high rice yields in Nepal. The potential for improving farm yields by increasing the level of inputs, particularly chemical fertilizers, is high in Nepal. The average farm yields in the Kathmandu Valley, where the application of chemical fertilizers is highest, are about 5 tons ha^{-1}. Available results of fertilizer experiments over time clearly indicate that rice varieties profitably respond to fertilizer applications of up to 90 kg ha^{-1} of nitrogen. The Central Soil Science Division has recommended the optimum fertilizer doses for both improved and local rice varieties, which are more than five times the present level.

Low levels of fertilizer application in Nepal can be attributed to their unavailability at the right time, the low purchasing capacity of farmers, and also the lack of assured irrigation which imposes a greater risk from investment in cash inputs. Fertilizer supplies in Nepal are controlled by the government and their adequate supply, particularly in remote areas, are seldom ensured. Development of an efficient fertilizer market and efficient access to institutional credit may increase incentives for farmers to invest in fertilizers and increase farm yields significantly even with present technologies.

Institutional constraints

Institutional structure has played an important role in determining the current stage of research and development in Nepal. The development and retention of quality research manpower and directing research toward high-priority areas are primarily functions of the institutional structure within which technology is generated. As will be shown in the next section, however, agricultural research institutions in Nepal, despite frequent attempts to introduce reforms, have not been able to retain trained staff. Due to a lack of adequate incentives and administrative commitments to create a favourable research environment, there is general frustration among most research staff. Although researchers understand the need to plan research according to farmer needs and priorities, they are seldom encouraged to know the actual farm-level constraints to production.

A close and continuous collaboration at all levels between research scientists engaged in generating technologies and extension specialists engaged in disseminating these technologies to the farmer is vital for research to be meaningful and problem-oriented. However, the research and extension systems at present are largely disintegrated, and there is no built-in mechanism by which to establish effective ties between these two institutions. The research and extension systems still operate within the rigid bureaucratic administrative framework, which provides much less freedom and flexibility to individual scientists to be innovative and efficient.

IV. Agricultural Research Institutions, Manpower and Research Planning Process

A. Development of agricultural research institutions

Agricultural research institutions in Nepal have undergone a number of structural changes since 1924, when the institutional base for agricultural research was first initiated with the creation of the Department of Agriculture (DOA) and the establishment of two trial-demonstration farms in the Kathmandu Valley. Until 1966, when the DOA was dissolved and five new departments were created, the rice research efforts were largely concentrated on testing new crop technologies, mainly exotic crop cultivars, borrowed from other countries for their suitability to Nepalese conditions. The major emphasis was on selecting and introducing varieties that were relatively higher yielding under tested conditions. Rice research in Nepal entered into the second stage after the formation of disciplinary divisions, such as pathology and entomology, and their testing of exotic crop cultivars.

Although agricultural research was mostly focused on major cereal crops (i.e. rice, maize and wheat), formal commodity orientation to agriculture research was given only after 1971/72, when the existing five departments were merged into one DOA, and commodity-specific research and development programs were started on rice, maize, wheat, citrus and potatoes. Each of these programs is led by a national coordinator, who is assisted by a multidisciplinary team of researchers (e.g. agronomists, entomologists, pathologists, etc.).

Another major change in the institutional structure occurred in 1985 with the creation of a National Agricultural Research and Services Center (NARSC) to strengthen agricultural research by separating it from extension research. NARSC, which became NARC after about 3 years (with service functions omitted), was to plan and coordinate all research activities on crops, horticulture, livestock and fisheries.

Attempts have been made to make NARC an autonomous research institution. Autonomy of the institution was a precondition for a major United States Agency for International Development (USAID) grant on agro-enterprises and technology systems project within NARC, which increased administrative concerns and ultimately resulted in the creation of the Nepal Agricultural Research Council (NARC) in 1990. The recently formed council is governed by a board, chaired by the Minister for Agriculture, and consists of members representing research scientists, industrialists and farmers, among others. NARC is now viewed as the central research planning and policy making institution of Nepal. Like other commodity programs, the National Rice Research Program (NRRP) operates within the NARC administration.

B. Research manpower

Time-series data on trained agricultural manpower in the country are not available for all years. Available data indicate that the number of agricultural manpower with at least a BSc degree in agriculture has more than doubled during the 1980s, from 539 in 1979 to 1239 in 1989 (Table 12.4). The proportionate increase in manpower is relatively higher in the non-crops subsectors, i.e. fisheries and livestock. Within the crops subsector, the agronomy faculty has the maximum number of trained personnel.

The increase in trained manpower in agricultural research, however, has not been as large. The number of agricultural researchers was 303 in 1977, and

Table 12.4. Comparative situation of trained manpower (with BSc or higher degree in agriculture by discipline and agriculture) in Nepal selected years.

	Agricultural sector manpower (number)		Agricultural research manpower (number)	
	1979	1989	1977	1990
Agricultural sector				
Agronomy	95	213		
Soils	27	50		
Plant pathology	36	104		
Entomology	21	50		
Economics and statistics	50	104		
Agricultural extension	129	183		
Horticulture	54	139		
Fisheries	19	74		
Livestock	52	155		
Veterinary	56	166		
Agricultural research				
National commodity programs			125	91
Central disciplinary division			70	131
Regional research farms/stations			108	374

Source: Ministry of Agriculture, 1977; APROSC, 1981; Rajbhandary, 1989; Winrock International, 1991.

it increased to 374 in 1990. Commodity research programs have suffered from an absolute decline in the number of trained research staff during the period, from 125 in 1977 to 91 in 1990.

C. The existing process for setting priorities

The major focus of the NRRP has so far been on adaptive research on varietal development, i.e. the testing of finished rice varieties from the IRRI and other countries for cultivation in Nepal. Development of new varieties has shown a targeted precedence. As shown in Table 12.5, varietal experiments constituted over 80% of the total experiments conducted in the country on rice. Table 12.6 reports the number of MVs released for general cultivation since 1965. Very few of them have actually been adopted by farmers.

Beginning in the late 1980s, the MVs released also included those which were actually developed in Nepal using local genotypes. Nevertheless, the newer varieties have not been able to replace some of the oldest varieties (obtained from foreign countries), such as CH45, Taichung-176 and Masuli (Mahsuri) introduced before the mid-1970s. Apparently, the breeding program has had limited success in identifying and incorporating farmers' needs and preferences in developing new varieties.

One of the major forums for research planning has been the National Summer Crops Workshop organized every year, where past research results are presented before the researchers, extension specialists and research adminis-

Table 12.5. Number of rice research experiments by discipline in Nepal in selected years.

Year	Varietal improvement	Agronomy	Soils	Plant pathology	Entomology	Total
1978	139	14	12	3	2	170
	(82)	(8)	(7)	(2)	(1)	(100)
1979	242	5	10	18	4	279
	(87)	(2)	(4)	(6)	(1)	(100)
1985	188	1	3	7	1	200
	(94)	(1)	(2)	(4)	(1)	(100)
1988	190	4	3	11	5	213
	(89)	(2)	(1)	(6)	(2)	(100)
1990	134	1	3	10	5	153
	(87)	(1)	(2)	(6)	(4)	(100)
1991	133	3	4	4	5	149
	(89)	(2)	(3)	(3)	(3)	(100)

Figures in parentheses indicate the percentages of total experiments conducted each year.

Table 12.6. Number of modern rice varieties (MVs) released by period and source in Nepal, 1960–1991.

Period	Total number of MVs released	Origin of released MVs			Most popular MVs
		IRRI	Other countries	Nepal	
1965–1969	6	1	5	x	CH45, Taichung-176
1970–1974	5	3	2	x	Masuli
1975–1979	6	4	2	x	Janaki, Sabitri*
1980–1984	5	2	3	x	Bimdeswari
1985–1989	9	2	4	3	
1990 to date	7	2	x	5	Khumal-4

*IRRI origin.
Source: NRRP, Nepal (proceedings of various Summer Crops Workshops).

trators from around the country. These workshops remain routine, and concerns for research planning are overshadowed by setting research targets (number of experiments) for each of the farms and stations. The choice of specific RPAs is predetermined by the concerned coordinator and his team.

There were some opportunities for researchers to identify and integrate farmers' problems into their research agenda. The annual monitoring trip organized by the NRRP before the crop harvest was one of these opportunities to identify actual farm-level problems. Such trips also gave researchers an opportunity to interact with farmers and extension workers. The Cropping Systems Program, initiated in 1977, began documenting farmers' problems for possible consideration by researchers in generating appropriate solutions. In 1979, the Ministry of Agriculture made it mandatory for all station researchers to spend 40% of their time in farmer fields. The bi-monthly and 6-monthly meetings, as part of the Training and Visit System, provided a forum to discuss farm problems and suggest solutions.

The Farming Systems and Outreach Research Division of NARC has been organizing *Samuhik Bhraman* (multidisciplinary group visits) since 1985. It is a multidisciplinary group that interacts with farmers, identifies their major problems, and suggests RPAs on a priority basis. Despite all this, serious research planning from the farmers' perspective has continued to depend on an individual scientist's interest.

Recently, with the formation of NARC, there has been a gradual change in the research planning process. All commodity programs and central disciplinary divisions engaged in research are required to submit research proposals for critical review by a panel of scientists designated by the executive director of NARC. After scrutiny by the panel, the recommended research projects are endorsed by the NARC board for implementation by the concerned agencies. This approach has some promise; but it calls for highly proficient panel members who have a real interest and capability to identify the country's pressing research needs.

D. Trends in research resource allocation

Rice, maize and wheat, the most important cereal crops of Nepal, shared a major proportion of the actual expenditure on crops research in the past; but their shares have gradually declined over time (Table 12.7). Among these three crops, rice research has traditionally received a greater share of research resources than maize and wheat, with maize receiving the lowest share. The declining shares of crops research resources allocated to these three major cereals indicate an increasing emphasis on research in other crops over time.

Table 12.7. Real research expenditures on major cereal crops in Nepal, 1969/70–1990/91 (at 1974–1975 prices) (thousand NRs).

	Rice		Maize		Wheat	
Year	Amount	% of crops research	Amount	% of crops research	Amount	% of crops research
1969/70	2694	35.4	755	9.9	2139	28.1
1970/71	2614	33.0	764	9.6	2220	28.1
1971/72	2601	31.1	654	7.8	1679	20.2
1972/73	2725	32.0	643	7.5	1755	20.6
1973/74	2946	32.7	723	8.0	1926	21.4
1974/75	2756	33.2	566	6.8	1797	21.6
1975/76	4800	36.9	738	5.8	2562	19.7
1976/77	3399	31.7	595	5.5	1768	16.5
1977/78	3202	27.0	594	5.0	1696	14.2
1978/79	2921	32.0	482	5.3	1422	15.6
1979/80	2833	29.5	494	5.1	1468	15.3
1980/81	2501	29.1	579	6.7	1742	20.2
1981/82	2815	30.6	608	6.6	1875	20.4
1982/83	3083	31.1	623	6.3	2005	20.2
1983/84	2950	30.1	612	6.2	1946	19.8
1984/85	2672	29.0	652	7.0	1774	19.3
1985/86	2858	27.5	711	6.8	1969	19.6
1986/87	3261	27.6	808	6.8	2339	19.8
1987/88	3648	28.3	918	7.1	2471	19.1
1988/89	3430	27.6	824	6.6	2260	18.2
1989/90	3633	27.7	835	6.3	2290	17.4
1990/91	3604	26.0	806	5.8	2518	18.1

NR, Nepalese rupee.

V. Yield Loss, Expected Research Benefits and Research Priorities

A. Yield loss

A preliminary list of potential RPAs, which included almost all of the diseases, insects, weeds, soil and physiological problems, animals and other production constraints that can affect rice yields, was presented separately to researchers, extension specialists and farmers to help estimate the rice area affected, yield loss and frequency of occurrence. Only those RPAs that appeared to have some significance, based on the respondents' judgement, have been reported here.

Researchers' and extension specialists' assessments of production loss due to an individual constraint was obtained for individual districts; the area and average rice yields by production environment for each district were obtained from extension specialists, and used to estimate annual production losses due to each RPA in each production environment for each district. The estimated district-level production losses due to each constraint were then summed to arrive at annual production losses due to each constraint at the regional and national levels. To estimate farmers' assessment of production losses, farm-level information on area affected and average rice yields by production environment were used.

Table 12.8 presents the list of major RPAs, with the corresponding total annual rice production losses based on researchers', extension specialists' and farmers' assessments. Since the three groups of respondents did not differ significantly in the identification and ordering of the most important production constraints, we use only the results based on the survey of researchers.

Table 12.9 presents these estimates of average yield loss per hectare due to each of the major technical constraints in the two ecological regions and in the country as a whole. As shown in the table, the estimates of average loss caused by the insects range from 12 kg ha^{-1} for hoppers to 112 kg ha^{-1} for the gundhi bug in the tarai, and from 9 kg ha^{-1} for hoppers to 52 kg ha^{-1} for the gundhi bug in the hills. The gundhi bug appears to be the most damaging insect in both regions. The second and third most damaging insects are the stemborer and rice hispa in the terai, and rice hispa and armyworm in the hills. In order of the weighted average yield-loss estimates, the gundhi bug, stemborer and rice hispa are the three most important rice insects in Nepal.

Among the major diseases identified, rice blast is the most serious disease in both the regions, causing an average yield loss of 125 kg ha^{-1} in the hills and 112 kg ha^{-1} in the tarai. The next three diseases in order of importance are: bacterial leaf blight, brown leaf spot and stem rot in both the regions. In general, the extent of crop loss caused by diseases and insects is higher in the tarai than in the hills.

Among the abiotic constraints, drought is the most serious constraint followed by zinc deficiency in the tarai and Nepal as a whole. In fact, drought stress is the most damaging of all the technical constraints in the tarai region and the country as a whole, causing an average yield loss of 146 kg ha^{-1} in the tarai and 134 kg ha^{-1} in Nepal. In the hills, drought loss amounts to 115 kg ha^{-1}, and is the second most important technical constraint in the region. Weeds are

Table 12.8. Rice research problem areas (RPAs) based on annual production loss (thousand tons) attributed to RPAs by respondent groups, in the tarai and hills of Nepal, 1991.

RPAs	Rice researcher Tarai	Hills	Nepal	Extension specialist Tarai	Hills	Nepal	Farmer Tarai	Hills	Nepal
Insects									
Gundhi bug	162.7	30.0	192.7	218.2	56.0	274.2	304.6	111.9	416.5
Stemborer	94.1	16.8	110.9	115.1	39.3	154.4	59.1	0.0	59.1
Hoppers	16.6	5.0	21.6	41.1	6.8	47.9	0.0	28.1	28.1
Thrips	0.0	0.0	0.0	0.0	9.3	9.3	0.0	0.0	0.0
Rice hispa	48.5	28.6	77.1	104.5	20.8	125.3	182.2	54.1	236.3
Armyworm	30.2	19.1	49.3	33.4	101.7	135.1	108.6	32.4	141.0
Grasshopper	0.0	0.0	0.0	47.4	9.8	57.2	0.0	0.0	0.0
Leaf folder	42.0	10.7	52.7	35.9	22.9	58.8	0.0	0.0	0.0
Mealy bug	27.3	6.2	33.5	28.9	0.0	28.9	0.0	0.0	0.0
Seedbed beetle	0.0	0.0	0.0	12.5	0.0	12.5	0.0	0.0	0.0
Gall midge	29.2	6.0	35.2	0.0	0.0	0.0	0.0	0.0	0.0
Rice weevil	50.0	40.9	90.9	62.1	34.5	96.6	225.0	40.0	265.0
Storage moth	40.0	43.0	83.0	70.4	61.6	132.0	222.9	41.5	264.4
Diseases									
Blast	162.8	72.3	235.1	200.2	34.6	234.8	157.7	164.0	321.7
Brown leaf spot	59.5	19.6	79.1	151.8	8.3	160.1	102.9	0.0	102.9
Bacterial leaf blight	140.7	31.9	172.6	80.3	21.7	102.0	72.9	40.2	113.1
Sheath blight	22.6	7.7	30.3	25.6	10.6	36.2	0.0	0.0	0.0
Seedling blight	0.0	0.0	0.0	61.6	11.9	159.1	0.0	0.0	0.0
False smut	19.2	3.1	22.3	147.2	11.9	159.1	0.0	0.0	0.0
Stem rot	49.0	12.7	61.7	0.0	4.9	4.9	0.0	0.0	0.0
Other									
Rodents	57.8	19.6	77.4	38.2	9.4	47.6	113.9	41.4	155.3
Weeds	103.3	32.5	135.8	52.0	33.9	85.9	46.5	19.9	66.4
Zinc deficiency	139.1	22.2	161.3	194.5	15.9	210.4	82.0	4.9	86.9
Lodging	30.5	10.2	40.7	58.1	12.1	70.2	0.0	0.0	0.0
Drought	237.2	67.0	304.2	220.0	72.3	292.3	291.5	64.9	356.4
Flood	12.8	4.0	16.8	12.8	4.1	16.9	0.0	0.0	0.0

estimated to cause crop losses in the range of 56 kg ha^{-1} in the hills to 71 kg ha^{-1} in the tarai. Zinc deficiency is more serious in the tarai than in the hills, the corresponding crop-loss estimates being 96 kg ha^{-1} and 39 kg ha^{-1} respectively. Lodging and flood problems are relatively less important, judging from their respective yield-loss estimates.

The yield losses caused by all technical constraints add up to 1008 kg ha^{-1} or about 56% of the yield gap in the tarai. Almost the same portion of the gap is explained by all technical constraints in the hills, though the absolute magnitude of the total crop loss is relatively lower in this region. In Nepal, all of the major

Table 12.9. Rice crop-loss constraints and estimated yield losses in the tarai and hills of Nepal, 1990/91.

Constraint	Tarai (kg ha^{-1})	Hills (kg ha^{-1})	Nepal (kg ha^{-1})
Insects			
Gundhi bug	112	52	96
Stemborer	65	29	55
Hoppers	12	9	10
Rice hispa	34	50	39
Armyworm	20	30	21
Leaf folder	29	19	28
Mealy bug	19	11	16
Gall midge	20	10	18
Diseases			
Blast	112	125	118
Brown leafspot	41	34	40
Bacterial leaf blight	98	55	88
Sheath blight	16	13	15
False smut	13	5	11
Stem rot	34	22	30
Other			
Rodents	40	34	40
Weeds	71	56	68
Zinc deficiency	96	39	81
Lodging	21	18	20
Drought	146	115	134
Flood	9	7	8

technical constraints are estimated to cause an yield loss of 936 kg ha^{-1}, which explains about 56% of yield gap II. The remaining portion of the gap may be attributed mainly to socioeconomic and environmental constraints within each region.

In terms of the annual production loss in the country, total loss due to all technical constraints amounts to nearly 1.9 million tons or about 46% of the total annual paddy production. The losses due to diseases and insects are about the same and together amount to nearly 1.2 millions tons per year. Drought alone causes an annual production loss of about 0.3 million tons and accounts for nearly 16% of the total annual production loss due to all technical constraints.

B. Expected research benefits and research priorities

Under known conditions, production loss due to an individual problem at present can reflect the potential benefits of successful research on that problem;

Table 12.10. Probability of research success and expected research benefits by research problem area in Nepal.

Research problem areas	Probability of research success	Expected research benefits (million NRs) Tarai	Hills	Nepal
Insects				
Gundhi bug	0.2	195.3	36.0	231.3
Stemborer	0.3	169.4	30.2	199.6
Plant and leaf hoppers	0.3	29.9	9.0	38.9
Rice hispa	0.2	58.2	34.3	92.5
Armyworm	0.2	36.2	22.9	59.2
Grasshoppers	0.1	0.0	0.0	0.0
Leaf folder	0.1	25.2	6.4	31.6
Mealy bug	0.2	32.8	7.4	40.2
Gall midge	0.3	52.6	10.8	63.4
Diseases				
Blast	0.7	683.8	303.7	987.4
Brown leaf spot	0.6	214.2	70.6	284.8
Bacterial leaf blight	0.7	590.9	134.0	724.9
Sheath blight	0.5	67.8	23.1	90.9
False smut	0.6	69.1	11.2	80.3
Stem rot	0.6	176.4	45.7	222.1
Other				
Rodents	0.0	0.0	0.0	0.0
Weeds	0.0	0.0	0.0	0.0
Zinc deficiency	0.0	0.0	0.0	0.0
Lodging	0.7	128.1	42.8	170.9
Drought	0.6	853.9	241.2	1095.1
Flood	0.0	0.0	0.0	0.0

NR, Nepalese rupee.

and hence, the relative magnitude of production loss can indicate the relative importance of the problem for research. However, the results of future research will be obtained under certain conditions. Future research benefits are, therefore, critically determined by the probability of research success to solve the problem.

The probability of research success in a given period of time may be conditional on a variety of factors such as: the type of research project under consideration; the current stock of research knowledge; the record, age, position and status of the research team; the political and institutional set-up; the outcome of related complementary and competitive projects; and the size and availability of the research budget. In addition, availability of technology from outside a country has been an important source of farm yield gains historically in Asia; and this is expected to be the case for Nepal too.

Table 12.10 presents the estimates of probabilities of research success in solving the identified problems in the next 10 years based on the researchers' subjective judgements. These estimates refer to the probabilities of success in developing, through the conventional breeding technique, rice varieties that are resistant or tolerant to these problems. The estimates of probabilities of research success range from 0 to 0.7. Those problems which cannot be solved by breeding research are rodents, weeds, zinc deficiency and flood; and those associated with the highest probability of research success, of 70%, are blast, bacterial leaf blight and lodging. The remaining disease problems can be solved with a 50–60% probability. The highest probability of research success in solving the insect problem is 30% in the case of the stemborer, plant and leafhoppers and gall midge. Research on the rest of the insects has a probability of success of between 10 and 20%.

These estimates were used to assess the profitability of research on the top ten RPAs, as shown in Table 12.11. The profitability of research on all these ten RPAs is quite high under certain assumptions of research benefits and costs. The net present values of research on these problems range from 820 million rupees for drought to 61 million rupees for sheath blight. The corresponding internal rates of return range from 77 to 41%. The highest break-even investment of 131 million rupees per year is associated with the drought problem. This is the amount of annual research investment that equates the costs and benefits of research on the particular problem. The lowest break-even investment is 10 million rupees per year for rice hispa and sheath blight problems.

The above measures of profitability suggest very high research pay-offs, given the probabilities of success assessed by the researchers. The research costs assumed in the present analysis consider the level of salaries, which, the researchers believe, will increase their incentives to remain in research and,

Table 12.11. Returns to research on the top ten rice research problem areas in Nepal.

Research problem area	Net present value (million NRs)	Internal rate of return (%)	Break-even investment (million NRs year^{-1})
Drought	820	77	131
Blast	739	75	118
Bacterial leaf blight	540	70	86
Brown leaf spot	207	56	33
Gundhi bug	167	53	26
Stem rot	160	53	25
Stemborer	143	51	23
Lodging	116	48	18
Rice hispa	62	41	10
Sheath blight	61	41	10

NR, Nepalese rupee.

hence, will lead to greater probabilities of research success than are considered here. These high research pay-offs clearly call for an increased attention of research planners to assign high researcher priorities to the above problem areas in Nepal.

The ranking of major RPAs based on the expected returns from research investment for different ecological regions and production environments are reported in Table 12.12. The top five problem areas that will give high returns to research investment are drought, blast, bacterial leaf blight, brown spot and stem rot for the rainfed lowlands, which is the most important rice ecosystem in Nepal. The irrigated ecosystem is growing in importance due to the government's attempts to increase rice supplies from limited land resources. For these ecosystems, the top five areas with high returns on research investment are blast, bacterial leaf blight, brown spot, lodging and gundhi bug.

VI. Conclusions and Policy Implications

Almost two decades of Nepalese rice research does not seem to have made much significant impact on rice productivity and the countrywide dominance of low-yield varieties. The number of new varieties tested and introduced for the rainfed production environment, which constitutes more than three-fourths of the total rice-growing areas, is small compared to the number of varieties introduced for the irrigated ecosystem. There is also growing realization that the potential of existing technologies has not yet been fully exploited, leading to significant gaps between the potential and actual farm yields.

The past history of development of agricultural research institutions in Nepal reflects a tendency to change the institutional structure at the centre too frequently, without much appreciable impact on the direction and focus of rice research at the program and station level. There are now considerable opportunities for researchers to identify and integrate farmers problems into their research agendas.

In general, the severity of most of the technical constraints is more pronounced in the tarai than in the hills. In terms of total annual production loss, drought, blast and the gundhi bug are the three most important RPAs for the country. Depending on the respondents, the amount of production loss due to all of the identified technical constraints ranged from 56 to 66% of the country's total rice production at present; 15–19% due to diseases, 14–21% due to field insects, 6–13% due to storage insects, 7–8% due to drought and 0–4% due to other constraints, such as lodging, flood, rodents, etc. Hence, technical constraints seem to be responsible for only part of the gap between potential and actual yields at present.

Table 12.12. Ranking of major research priority areas based on expected pay-off by region and by production environment in Nepal.

Rank	Ecological region			Production environment		
	Tarai	Hills	Nepal	Irrigated	Rainfed lowlands	Uplands
1	Drought	Blast	Drought	Blast	Drought	Blast
2	Blast	Drought	Blast	BLB	Blast	Drought
3	BLB	BLB	BLB	Brown leaf spot	BLB	BLB
4	Brown leaf spot	Brown leaf spot	Brown leaf spot	Lodging	Brown leaf spot	Gundhi bug
5	Gundhi bug	Stemrot	Gundhi bug	Gundhi bug	Stem rot	Brownleaf spot
6	Stem rot	Lodging	Stem rot	Stem rot	Stemborer	Stem rot
7	Stemborer	Gundhi bug	Stemborer	Stemborer	Gundhi bug	Sheath blight
8	Lodging	Rice hispa	Lodging	Drought	Sheath blight	Stemborer
9	False smut	Stemborer	Rice hispa	False smut	Lodging	Rice hispa
10	Sheath blight	Sheath blight	Sheath blight	Rice hispa	Rice hispa	Armyworm
11	Rice hispa	Armyworm	False smut	Sheath blight	Armyworm	Mealy bug
12	Gall midge	False smut	Gall midge	Gall midge	False smut	–
13	Armyworm	Gall midge	Armyworm	Armyworm	Gall midge	–
14	Mealy bug	Plant and leaf hopper	Mealy bug	Plant and leaf hopper	Mealy bug	–
15	Plant and leaf hopper	Mealy bug	Plant and leaf hopper	Mealy bug	Leaf folder	–
16	Leaf folder	Leaf folder	Leaf folder	Leaf folder	Plant and leaf hopper	–

BLB, bacterial leaf blight.
A dash (–) indicates non-existence of the problem.

References

Adhi Kary, R.R., Shrestha, G.L. and Thakur, P. (1986) Review of rice entomology research in Nepal. In: *Proceedings of the 13th Summer Crops Workshop*. pp. 241–278.

APROSC (1981) *Nepal Trained Manpower for the Agricultural Sector*, Vol. I.

Bimb, H.P. (1978) *Biological Study of Rice Bug (Leptocoris vericornis) under Bhairahwa Condition*. Entomology Division, Khumaltar, Nepal.

DFAMS, Nepal (1990) *Agricultural Statistics of Nepal*. Kathmandu.

Gamesh, K.C. and Kaphale, G.P. (1972) *The progress Report on Paddy*. Entomology Division, Khumaltar, Nepal.

Gyawali, B.K. (1973) Progress report of entomological studies. In: *Proceedings of the Xth Summer Crops Workshop*. Department of Agriculture, Harihar Bhawan, Nepal.

Gyawali, B.K. (1974) *Report of Entomology Division, 2031–32*. Entomology Division, Department of Agriculture, Harihar Bhawan, Nepal.

International Rice Research Institute (IRRI) (1979) *Farm Level Constraints to High Rice Yields in Asia: 1974–77*. Los Baños, Philippines.

Joshi, S.L. (1974) *Bharaman Pratibedan (Tour Report)* Entomology Division, Khumaltar, Nepal.

Joshi, S.L. (1977) An assessment of yield loss due to infestation of borer complex on the paddy crop. *Nepalese Journal of Agriculture*, 12, 1–21.

Jyoti, J.L. and Bhurer, K.P. (1989) Relationship between mealy bug and grain yield of rice under field conditions. In: *Proceedings of 14th Summer Crops Workshop and Research Program Planning*. Nepal Agricultural Research Council, Kathmandu. pp. 160–162.

Ganesh, K.C. and Kaphle, G.P. (1972). *The Progress Report on Paddy*. Entomology Division, Khumaltar, Nepal.

Manandhar, H.K. (1986) *Yield Loss Due to Bacterial Leaf Blight, Sheath Rot and Sheath Blight of Rice*. Central Plant Pathology Division, Kathmandu (mimeographed).

Manandhar, H.K., Thapa, B.J. and Amatya, P. (1985) Efficacy of various fungicides on control of rice blast disease. *Journal of Institutional Agriculture and Animal Science* 6, 21–29.

Manandhar, H.K., Thapa, B.J. and Chaudhary, D.N. (1989) Screening of various chemicals against rice blast. Paper presented at the 14th Summer Crops Workshop, Parwanipur, Nepal.

Manandher, D.N. (1976) Hispa problem in Nepal. Seminar presented at The National Rice Research Program, 9 April.

Ministry of Agriculture (1977) *A Proposal for Organizing Agricultural Research in Nepal, Mid Term Preliminary Draft*. Ministry of Agriculture.

Pardey, P.G., Sandra Kang, M. and Elliot, H. (1989) Structure of public support for national agricultural systems: A political economy perspective, *Agricultural Economics* 3: 261–278.

Pradhan, R.B. (1989) Review of entomological work in paddy in Nepal. *Entomological Society Nepal* 1: 91–102.

Rajbhandary, H.B. (1989) An assessment of manpower requirement in the Ministry of Agriculture for the basic need program in agriculture. Paper presented at the workshop on 'Middle Level Manpower for Agricultural Development', 26–28 December, Rampur (Chitwan), Nepal.

Shah, D.N. (1990) Effect of management practices and seed treatment on leaf blast development in two susceptible cultivars. In:*Proceedings of the 4th Summer Crop Workshop and Research Program Planning, Parwanipur, Nepal.*

Winrock International (1991). *Nepal Agricultural Research Study, Phase I, Main Report.* Nepal Agricultural Research Council, Khumaltar, Nepal.

13 Rice Research Priorities in Thailand

S. Setboonsarng
ASEAN Secretariat, Jakarta, Indonesia

I. Introduction

Rice plays at least three major roles in the Thai economy: (i) it engages over 50% of the total cultivable area and labour force; (ii) it is the main staple food crop; and (iii) it is one of the major foreign-exchange earning sectors. Rice research has helped the efficient performance of these roles. Changes in the role of rice in the economy in the future will also change the demand for rice research.

The Thai rice economy faces the following challenges: (i) increase in competition in the international market; (ii) growing competition with other economic activities that increases the cost of production, especially the labour cost; and (iii) degradation of ecological conditions. Rice research has to address these challenges.

In spite of the increasing need for research, there is a trend toward reduction in rice research investment. This has been induced by the decline in the world price of rice and a shift in government interest away from spending revenue on long-term research investments towards activities with short term return. These pressures make it even more important to efficiently allocate scarce research resources. In this chapter information from farmers and scientists, as well as secondary statistical data (and agricultural extension workers), are analysed to contribute to improved rice resource allocation.

II. The Rice Economy in Thailand

The change in the Thai rice market is greatly influenced by its export orientation. While the international market price determines the domestic price, export demand determines the domestic production in terms of quantity and quality. Competition in the international market forces the Thai rice market to be sensitive to consumer needs, both in terms of quality and price.

A. Rice output and consumption

Output of rice in Thailand increased from 12 million tons in 1967–1969 to 20 million tons in 1984–1986, or an average of about 0.4 million tons of rice (about 0.6 million tons of paddy) per year (Table 13.1). The growth rate of production was about 3% per annum during this period. On average, about 4 million tons or about 30% of the total output were exported and about 10 million tons were consumed domestically per year.

The main source of growth of rice output during the 1960s and 1970s came from the expansion of cultivated areas. This source of growth declined in importance during the later part of the 1980s and early 1990s because of the exhaustion of the land frontier. At the same time, the world price of rice has declined and the cost of rice production has increased over time.

Thai rice exports

During 1981–1992, the export of rice from Thailand increased from 3 million tons to about 4.9 million tons (Table 13.2). However, export earnings declined from US$1.2 billion in 1981 to US$0.77 billion in 1986, caused by a sharp fall of the world price of rice. The unit export value of rice dropped from US$400 ton^{-1} in 1981 to US$171 ton^{-1} in 1986. As a result, the share of rice in export revenue declined from 17% in 1981 to 8.8% in 1986.

The world price picked up after 1986 and the quantity of export also expanded. However, the share of export revenue continued to decrease. By 1992, the share of rice export revenue was only 4% of total export revenue of the country. This was mainly caused by the rapid expansion of total exports from the country. At the same time, there was an increase in the amount of rice going to the international market; the export of Thai rice has been facing severe competition from Vietnam, especially in the low-quality rice market.

Table 13.1. Growth of rice output in Thailand.

Year	Output Amount (thousand tons)	Output Growth (%)	Area Amount (thousand ha)	Area Growth (%)	Yield Amount (ton ha^{-1})	Yield Growth (%)
1961–1963	9,501.4		6,260.8		1.52	
		4.40		2.18		2.18
1967–1969	12,375.1		7,135.3		1.73	
		2.03		2.31		− 0.27
1976–1979	15,012.9		8,883.4		1.69	
		3.76		1.64		2.14
1984–1986	19,905.9		10,045.2		1.98	
		− 0.54		− 1.56		1.02
1991–1993	19,171.9		9,006.8		2.13	

Table 13.2. Thai rice exports and their share of total export value and total output, 1981–1991.

		Export of rice				
Year	Total national export value (million US$)	Value (million US$)	Quantity (thousand tons)	Unit price (US$ ton^{-1})	National export value (%)	Rice output exported (%)
1981	7,044	1,214	3,032	400	17.23	30.25
1982	6,975	983	3,784	260	14.09	30.37
1983	6,396	880	3,476	253	13.76	28.87
1984	7,444	1,102	4,616	239	14.80	33.88
1985	7,146	832	4,062	205	11.65	29.18
1986	8,835	775	4,524	171	8.78	31.37
1987	11,628	886	4,443	199	7.62	33.36
1988	15,846	1,376	5,089	270	8.69	42.27
1989	19,917	1,776	6,140	289	8.92	42.01
1990	22,874	1,089	4,017	271	4.76	37.07
1991	28,550	1,201	4,168	288	4.21	33.79
1992	32,522	1,431	4,908	291	4.40	35.40

Source: Bank of Thailand, *Monthly Statistical Bulletin*.

Thailand has been able to retain its position in the international export market and survive the price decline during the 1980s because of the relatively low cost of its rice production, below US$120 ton^{-1} (Office of Agricultural Economics, 1990). There was a downward trend in the cost of production during 1984–1988; this was in response to the general decline in the market price of rice. The share of labour, which accounts for over 60% of the cost of rice production in the rainfed area, showed a slight decline, while the labour share rose in irrigated rice areas. This change was caused by a rapid increase in the wage rate in the irrigated areas which are located closer to urban areas. This increase in cost has become a major issue regarding the competitiveness of rice farming during the 1990s, because of the rapid increase in the wage rate stimulated by the increase in economic growth.

B. Rice production environments in Thailand

Rice production environments in Thailand can be characterized into four types based on water regime: irrigated, rainfed dryland, rainfed lowland and deep-water. The cultivated areas, output and yield of these four production environments in 1992 are shown in Table 13.3.

Prior to the 1970s, most of the irrigated areas were in the valleys in the northern region of the country. The construction of dams and canals in the 1970s and 1980s created most of the irrigated areas in the central plain. There were about 719,000 ha of irrigated area in Thailand in 1992, which accounted for about 7.5% of the total rice area. However, the irrigated area contributed

Table 13.3. Area, output and yield of rice (by production environments) in Thailand in 1992.

	Area		Output		
Environment	Thousand ha	Percent	Tons	Percent	Yield (tons ha^{-1})
Irrigated	719	7.53	2,881,528	14.12	4.007
Rainfed dryland	6,711	70.29	12,243,685	60.01	1.824
Rainfed lowland	1,922	20.13	4,851,100	23.78	2.524
Deepwater	195	2.05	426,107	2.09	2.181
Total	9,547	100.00	20,402,420	100.00	2.137

Source: Agricultural statistics, Office of Agricultural Economics (1993).

only 14% of the total rice output. The average yield is about 4 tons ha^{-1}, which is the highest among the production environments.

The rainfed dryland rice area refers to cultivated areas in the north and northeast plateau. Rainfed dryland rice covers about 6.7 million ha and accounts for about 70% of the total cultivated area of rice in Thailand. It spreads from the northern region to the northeast and some provinces in the west and upper south regions. It accounts for 60% of the total output of the country although the average rice yield of 1.8 tons/ha^{-1} is below the national average. The rainfed dryland area faces a long dry period and a monsoon rain which is short and sometimes intense. Farmers use mostly traditional varieties which are photoperiod-sensitive. Cultivation methods and varieties vary depending on how soon the rain comes. This leads to variations in the yield of rice across regions and over time.

Rainfed lowland rice is found in the rice cultivation areas in the Chao Phya River delta. The rich deltaic soil is the main asset of rice cultivation in this production environment. Water from the Chao Phya River system determines the supply of water for rice cultivation in this production environment. When the river is dammed up for hydroelectric generation purposes, the supply of water in the delta is determined by the release of water from the dam; timing may not be consistent with the water requirement for crop cultivation. This environment covered about 1.9 million ha or about 20% of the total rice area in 1992. Rice output from this production environment was about 2.5 tons ha^{-1}, this has increased in recent years due to the adoption of modern varieties and the increased use of fertilizers and chemicals.

Deepwater rice, or floating rice, is grown in flooding conditions in the central and eastern plain areas. There were about 195,000 ha of deepwater rice in 1992 or about 2% of the total planted area. Its output of 426,000 tons represents about 2% of the total output. The average yield shows a declining trend during the past 5 years; this is partly caused by the increase of environmental stresses and the limited research on deepwater rice varieties. An important problem for deepwater rice is its high labour requirement during harvesting. The increase in wage costs may make deepwater rice uneconomical in the near future.

C. Rice research in Thailand

The export orientation of Thai rice requires producers to remain competitive in the international market by the constant improvement of quality and lowering of production costs. During earlier periods, the main emphasis of rice research was on the improvement of rice quality, particularly the length of the grain and its cooking quality. A Thai rice variety, Pin Khaew, won the international award in Montreal in 1934 for the length of its grain.

The objective of rice research changed during the mid-1960s. The release of IR8, the miracle rice, captured the attention of the rice growers in tropical Asia. It increased the output of rice in most of the importing countries and reduced their demand for imports. With the availability of breeding materials and research funds, rice research in Thailand gravitated toward the pursuit of higher yields in order to lower the cost of production.

The breeding work which started in 1966 produced the first non-photo-period-sensitive rice variety in Thailand, RD1, to be used in irrigated areas. RD1 gave a 50% higher yield compared to the traditional varieties. However, RD-1 was not adopted due to the eating quality and additional input requirements such as fertilizers. The subsequent varieties RD7, RD15 and RD23 were developed and became dominant in the irrigated area.

There were low adoption rates of high-yielding varieties (HYVs) in Thailand because they need good water control, fertilizers and chemicals. While the limited irrigated area constrains the adoption of the HYVs, the higher cost of investment discouraged farmers who did not have access to credit. More importantly, HYVs did not taste as good as the traditional varieties. Most of the output of these HYVs are not consumed domestically. They are either used to make parboil rice for export or exported as lower-quality rice when prices are not attractive. While rice research contributed toward lowering the cost of rice production through increased yield during this period, it also lowered its quality and its price. The net benefit was not as much as expected.

In the 1980s, the decline in the international rice price put pressure on the Thai rice export market, especially with the emergence of Vietnam as a major rice exporter. The decline in international market price was substantial for lower-quality rice. In contrast, the price of higher-quality rice shows an upward trend. This fact triggered the Rice Research Institute of the Department of Agricultural Extension (DOAE) to revive its research effort on the quality of rice. Given the economic growth trend, the price of higher-quality rice was expected to be even higher in the future. This makes the return of research on rice quality higher.

High economic growth also put pressure on the cost of rice production, especially the cost of labour. Cost of labour is the most important cost component in the Thai rice production and affects the cost of rice production and the technology to be chosen.

III. Determination of Research Priorities

To determine priorities within rice research, two major approaches are used in this study: the yield constraints approach and the return on research approach.

The yield constraints approach seeks to identify the sources of yield loss and prioritize the research to address the most important constraints. The advantage of this approach is that it is easily understood by the rice research scientists because it is a measurable target. It was the approach that guided public research in the past. Although this approach is systematic, one of its weaknesses is that it does not incorporate other sources of reducing the cost of production such as post-harvest losses and increasing returns such as by improving the quality of rice.

The return on research approach investigates the benefit and cost of research and ranks the net benefit from the research. The priority is determined by the ranking of the net present value (NPV) of the benefit from a particular research activity. This approach takes into consideration the specific cost of solving the problem, the gain from the increase in output and the improvement in the quality of rice. The major problem with this approach is the reliability of information on the expected benefit from and cost of research.

This chapter uses both of these approaches in investigating research priorities for rice population in Thailand.

A. Three sources of information

Three sources of information were examined: secondary data, farmer responses and scientist responses. Each source has its own strengths and weaknesses.

Problems obtained from farm surveys and secondary data on damaged areas are quite similar, but the views of scientists are slightly different from the first two. This could be because of the bias in the scientists selected for the survey. Nevertheless, it serves to indicate that there is an information gap among farmers, extension workers and scientists. The gaps are not large, but exist for all the four ecosystems.

The disease problem is considered to be the most important among scientists because of the nature of the problem. Disease damage is less obvious to the eye of the non-scientist. From the farmers' point of view, both in the rainfed and irrigated areas, insects are the most important problem. Insect problems are given a relatively high priority because they are easily observed compared to diseases. This is the same as information from the secondary data collected by extension workers. The scientists were interested in most of the sources of yield loss, but some problems have been ignored. The reasons are: (i) the specific study area; (ii) the specialized scientists; (iii) the system of research planning; (iv) factors limiting farmers adopting new technology; and (v) environmental diversity.

B. Measurement of yield gaps

There are two yield gaps: the difference between potential yield and best-practice yield at the farm level (yield gap I) and the difference between best-practice yield and actual farmer yield (yield gap II). Yield gap I indicates the distance between scientific findings and its actual applicability. More basic and strategic research would be required to close this gap. Yield gap II can be caused by socioeconomic factors, such as the prices of outputs and inputs and government policies. Some of these factors can be corrected, e.g. biased government policies. However, there are many factors that may not be easily corrected, e.g. a declining world price of output and the overvaluation of foreign exchange.

To determine these yield gaps, the yield at farmer level, the maximum yield at farmer level and the maximum yield at experimental station level are required.

Actual yield

There are two sources of information on actual yield: secondary data and farm surveys. Since the secondary information is at the provincial level and there is no breakdown by production environment, the information of output and planted area in each production environment is calculated from the planted area and output of groups of provinces which have large proportions of land area under the different environments. This makes the yield slightly different from that obtained from the farm survey, which is more location specific.

The average rice yield in the irrigated area estimated from the farm surveys is 4 ton/ha^{-1} which is about 10% higher than the secondary data indicates. The yield for deepwater rice estimated from the farm surveys is only about 70% of the yield calculated from the secondary data. For consistency of comparison, farmers' yields will be used.

No-loss (best-practice) yield

Ideally, the yield level with no loss should be that obtained from the 'farm trial' result of the experiment. However, this information is not available; instead we calculated the no-loss yield level from a yield-loss equation estimated from the data for 73 provinces from 1986 to 1991.

Potential yield

The potential yield level is obtained from the rice experimental stations in the four production environments. In each station there are many varieties being tested. Table 13.4 shows some of the results. The ones marked with an asterisk (*) have been chosen to represent the best yield for the region, based on the popularity of that variety in the region, e.g. the use of KDML105 for the lower northeast region.

Table 13.5 shows that yield gap I is larger than yield gap II in irrigated and rainfed lowland rice areas. This result suggests that there are more scientific challenges in irrigated and rainfed lowland rice compared with rainfed dryland and deepwater rice. Rice research has remained limited for rainfed dryland and

Table 13.4. Potential (research) yield level results.

Varieties	Yield (tons ha^{-1})
Irrigated	
RD23	5.269
SPR90*	5.950
RD25	6.250
RD4	5.206
Rainfed dryland	
KDML105*	2.500
RD15	2.463
Rainfed lowland	
RD11	4.625
RD15	3.231
RD21*	4.888
RD23	4.481
PSL60	5.625
SPR90	5.131
Deepwater rice	
LMN*	2.288

Varieties marked with an asterisk (*) have been chosen to represent the best yield for each region.
Source: Rice Research Stations, Department of Agriculture.

deepwater rice. The lower level of potential yield for these two environments is evidence of this result.

For the irrigated area, yield gap II is about 0.52 ton ha^{-1} or about 10% of actual yield. Yield gap I is also large, about 1.12 ton ha^{-1} or about 20% of actual yield. On average, yield gap II is larger than yield gap I.

Table 13.5. Calculation of the yield gaps (tons ha^{-1}).

Varieties	Actual farm yield	Estimated no-loss yield	Research potential*	Yield gap I	Yield gap II
Irrigated	4.308	4.825	5.950	1.125	0.517
Rainfed dryland	1.835	2.236	2.500	0.264	0.401
Rainfed lowland	2.486	3.250	4.888	1.637	0.765
Deepwater	1.392	2.044	2.288	0.243	0.652

*Research potential best-practice yield: irrigated = Supan Buri 90 variety; rainfed dryland = Khao Dok Mali 105 variety; rainfed lowland = RD21 variety; deepwater = LMN variety.

C. Farm survey yields

The farm survey was used to identify farmers' problems in rice production. The questionnaire of the survey covers farmers' practices, inputs, outputs, on-farm problems which caused yield reduction, problem-solving practices, reasons underlying the farmers' practices, and farmers' needs. The information from this survey was also used for analysis of the yield-loss equation. Subsequently, the result of the survey was compared with the results of the secondary data and the view of the scientist.

The survey was conducted in three areas for irrigated rice, rainfed rice and deepwater rice/floating rice ecosystems. (The dryland rice ecosystem is not included in this survey because of its small cultivated area.)* For each sample province, a district was selected based on its agroclimate conditions and soil types. From these districts, a typical village was selected based on the cropping pattern, agroclimate conditions and soil types from the background information on the village from the district census of the DOAE. Farmers were randomly sampled and interviewed.

In each of the surveyed areas, extension workers were interviewed. They were requested to identify the sources of yield loss in the area, the damaged area and also the possible alternative to address the problem. Scientists doing research on rice were asked their opinion about the major rice production problems in each of the rice production environments. This provided a third result.

D. Sources of yield loss

The extent of yield loss can also be ascertained from secondary information on damage from biotic and abiotic factors. The data for damaged area were collected by 31 Plant Protection Surveillance Units (PPSUs in the DOAE), located around the country. The area is classified as a damaged area whenever the pest density is higher than the economic threshold (ET) level. A summary of the information from 1986 to 1991 is as follows. An average of 870,000 ha or about 8.35% of the total rice area were damaged yearly by abiotic factors (drought and flood). As shown in Table 13.6, the damaged areas from drought was 368,974 ha (about 4.57% of the total rice area), while flooding affected

---

*The provinces selected for the survey are: Suphan Buri and Phitsanulok for irrigated rice, Surion and Khon Kaen for rainfed rice and Prachin Buri for deepwater rice. Suphan Buri province was selected to represent irrigated rice in the central plain because it is the leading province in irrigated rice production. Phitsanulok province was chosen to represent the irrigated rice in the upper central plain and the lower north plain. Khon Kaen and Surion provinces were selected to represent rainfed rice in the rainfed dryland area. In the upper northeast, glutinous rice is used for home consumption. But in the lower northeast the conditions are suitable for non-glutinous rice, especially the aromatic KDML105. The deepwater area is concentrated in the central plain of Thailand. Prachin Buri province was selected because it is the most important deepwater rice-producing area in Thailand.

Table 13.6. Damaged area classified by production environments (average of 1986–1991).

Factors	Rainfed dryland (ha)	Rainfed lowland (ha)
Biotic factors		
Diseases	52,118	81,286
Insects	237,782	279,851
Other pests	102,090	335,372
Weeds	497	1,467
Total biotic factors	392,486	697,976
Abiotic factors		
Drought	368,974	143,561
Flood	166,096	190,568
Total abiotic factors	535,069	334,129
Total of rice area	927,555	1,032,105
Percent of rice area	13.82	53.70

Source: Plant Protection Surveillance Unit, Department of Agricultural Extension (1992).

166,096 ha (about 3.78% of the total rice area). The biotic factors can be divided into four groups: diseases, insects, other pests and weeds. The average damaged areas during 1986–1991 indicated that insects cause the most damage (about 517,000 ha or about 6% of the total rice area).

The causes of damage vary by ecological conditions. For irrigated rice, the damage from abiotic factors is small because of water control; however, its better vegetative growth induced higher damage from biotic factors. The damaged areas from individual constraints on irrigated and rainfed rice are compiled from the PPSUs and presented in Table 13.7.

Irrigated rice

Insects are a major problem in this area because of the use of fertilizers, pesticides and other modern farming practices. The main insects in this production environment are the brown planthopper, stemborer, rice leaf folder and white-backed planthopper. The rat is the second most important problem in this region. Major rice diseases in the irrigated areas are brown spot, narrow brown spot, tungro, ragged stunt virus (RSV) and bacterial leaf blight. Thrips, stemborer, rice leaf roller, cutworm and rice leaf folder are also important problems in the irrigated rice. However, scientists believe that in the future the rice leaf folder and white-backed planthopper will be the key pests.

Rainfed rice

Biotic factors are the most important sources of damage in the lowland area. The rat is the biggest production problem, while insects (especially the brown

Table 13.7. Average damaged area due to pests, 1986–1991 (ha).

Pests	Rainfed	Irrigated	Total
Diseases			
1 Brown spot	4,409	7,192	11,600
2 Narrow brown spot	628	712	1,340
3 Sheath rot	4,276	2,503	6,779
4 Sheath blight	1,780	3,213	4,993
5 Ragged stunt virus	17	321	338
6 Yellow orange leaf virus	870	6,730	7,600
7 Blast	27,300	46,915	74,215
8 Bacterial leaf blight	4,221	2,695	6,916
9 Bacterial leaf streak	17	321	338
10 Bakanae	12,238	230	12,468
11 Unclassified diseases	5,789	1,029	6,817
Insects			
12 Thrips	19,852	62,153	82,005
13 Mealy bug	5,370	1,682	7,052
14 Brown planthopper	64,571	181,066	245,637
15 White-backed planthopper	8,720	1,718	10,439
16 Green rice leafhopper	12,997	19,554	32,551
17 Zigzag rice leafhopper	10	142	151
18 Rice stemborer	57,853	16,318	74,172
19 Rice leaf roller	50,231	18,616	68,847
20 Rice leaf folder	12,101	32,657	44,758
21 Rice caseworm	10,410	15,753	26,163
22 Rice whorl maggot	2,235	582	2,817
23 Rice swarming caterpillar	39,337	12,537	51,874
24 Common army worm	14,652	12,569	27,220
25 Rice bug	6,994	1,058	8,052
26 Rice hispa	226	3,441	3,667
27 Rice gall midge	1,862	1,328	3,190
28 Rice black bug	4,649	617	5,266
29 Rice grasshopper	9,905	2,666	12,571
30 Unclassified insects	14,320	6,479	20,799
Other pests			
31 Rats	208,095	251,284	459,378
32 Crabs	14,255	5,163	19,418
33 Unclassified other pests	1,470	5,195	6,665
Other			
34 Weeds	683	1,280	1,963

Source: Department of Agricultural Extension (1992).

planthopper) are the second most important sources of production losses. Blast was the major rice disease which caused the highest damaged area, because of new races coming and inadequate screening. Among the abiotic factors in lowland regions, floods are relatively more important.

E. Statistical estimates of yield loss

The no-loss yield level is obtained from the yield-loss equation specified as follows:

$$Y = F(D_1, D_2, D_3, ..., D_n) \qquad (13.1)$$

where Y = rice yield (kg rai^{-1}); and D_i = damaged area caused by individual insects, diseases, pests or conditions like drought.

The data from the farmers' survey were used in this estimation. The number of constraints that gave statistically significant results was much smaller than reported by extension workers. The four estimated equations are statistically significant. The R^2 ranges from 0.32 to 0.7 (Table 13.8). The analysis of the yield-loss function shows that thrips and the brown planthopper are the two most important sources of damage.

The estimates reported in Table 13.9 show that most yield losses occur in rainfed dryland rice and that insects (thrips and brown planthopper) cause the greatest damage.

F. Other objectives of rice research

The emphasis of rice research in Thailand is not only to reduce the yield loss and increase rice production, but also to make the rice marketable. In order to make it marketable, the price must be attractive and the product must be of a desirable quality and conveniently available. Viewed from this angle, the important problems confronting rice production in Thailand are the increasing cost of production and the improvement of the quality of rice.

Mechanization

The rapid development of the Thai economy during the past decade, especially in the manufacturing and service sectors, led to a rapid increase in the rural wage rate. As labour costs account for about two-thirds of the total cost of rice production, the trend in rural wages plays an important role in determining the cost of rice production. This problem is aggravated by the seasonality of the demand for labour in rice production. On top of the increasing wage rate, during the period of peak demand for labour, the wage rate became even higher than the wage rate in the manufacturing sector.

This problem is acute for the irrigated area, especially at harvest. In order to capture the benefit of the available water and to maximize the cropping intensity, harvesting has to be done punctually to allow enough time to prepare

Table 13.8. Yield-loss equation for the four production environments in Thailand.

	Irrigated rice			Rainfed dryland rice			Rainfed lowland rice			Deepwater rice		
	Estimated coefficient	t statistic	Mean % DA	Estimated coefficient	t statistic	Mean % DA	Estimated coefficient	t statistic	Mean % DA	Estimated coefficient	t statistic	Mean % DA
Constant	772.031	45.765	689.290	337.005	15.480	508.000	626.338	41.251	501.465	329.942	28.295	222.778
Blast	−1.910	−1.327	3.498	0.218	0.284	4.694	−0.981	−0.927	3.553	−0.595	−1.193	5.737
Brown planthopper	−5.965	−10.247	13.057	−3.563	−4.532	4.495	−2.763	−4.420	8.129			
Crabs	1.045	0.372	0.417	1.726	1.069	1.844	−3.798	−1.560	1.151			
Submergence	−4.839	−3.079	2.124	−2.814	−5.586	4.958	−6.147	−13.623	9.905	−3.179	−15.104	31.071
Stemborer	8.888	1.753	0.809	−0.848	−0.814	3.668	−4.781	−3.313	2.755	−2.482	−1.619	1.141
Thrips	−5.073	−0.557	0.392	−1.465	−0.648	1.463	−5.490	−2.859	1.495	−0.328	−0.244	1.748
	0.403	0.314	3.640	−1.803	−3.762	9.098	−2.574	−3.999	6.867	−0.374	−0.505	4.242
Interaction												
Blast + drought	0.105	0.664	14.602	0.014	0.425	72.548	0.045	0.996	47.541			
Blast + submergence	0.035	1.388	90.646	0.045	2.374	38.908	0.041	2.594	91.131			
Blast + thrips	−0.489	−1.038	3.671	−0.157	−2.148	31.123	−0.081	−0.931	19.015			
Thrips + drought	0.251	1.577	8.244	0.119	2.008	43.645	0.026	0.705	41.620			
Dummy variable												
Irrigated rice				442.917	17.259	0.459						
Rainfed rice				35.134	1.218	0.339						
Adjusted R^2	0.321			0.679			0.320			0.699		
F	13.134**			50.743**			22.531-**			46.532**		
n	283			307			505			99		
Adjusted R_2	1.525			1.705			0.932			1.859		

Source: Setboonsarng et al., 1994.
F = F test value.

Table 13.9. Estimated output foregone due to various sources of yield loss estimated from yield-loss equations.

	Output foregone (tons)				
Sources	Irrigated	Rainfed dryland	Rainfed lowland	Total (tons)	Total output (%)
Thrips	3,992	206,826	44,975	255,793	1.25
Brown planthopper	6,699	206,985		213,684	1.05
Green leafhopper	9,794	133,340	6,707	116,840	0.57
Cutworm	17,167		81,330	98,497	0.48
Rice leaf roller	8,687	83,431		92,118	0.45
Crabs	9,518			9,518	0.03
Stemborer		118,250	111,744	229,994	0.03
Caseworm	5,451			5,451	0.03
Rats	4,527	14,401	40,974	50,848	0.00
Blast	6,707	47,702		115,241	0.56

Source: Setboonsarng et al., 1994.

the land and plant the next crop. The value of labour at this sharp peak is very high and farmers are willing to pay a high price for labour.

This high wage rate gives rise to the demand for labour-saving technology, especially farm mechanization. There has been an upward trend in the use of combine harvesters in the irrigated area of the central plain since 1991. In order to assess the economics of the combine harvester, a farm survey was conducted in the irrigated area. The cost computation shows that the break-even wage rate for a combine harvester is 82 baht (US$3.5) per day. In other words, a combine harvester will be economical if the wage rate is higher than 82 baht day^{-1}. For provinces adjacent to Bangkok, the wage rate of factory workers is about 90 baht (US$3.6) per day.

Considering the fact that the demand for harvesting and thrashing labour is seasonal, the wage cost would be even higher than the minimum wage in this area. Most of the farmers in this area had to hire workers from other regions and pay for transportation costs and some living allowances to attract these labourers. Some farmers had paid 150 baht day^{-1} for harvesting and thrashing.

The benefit of research on farm mechanization in the irrigated area was estimated based on the wage rate of 100 baht day^{-1}. For the rainfed areas, the break-even wage rate for a combine harvester is 45 baht (US$1.8) day and the market wage rate is 40 baht (US$1.6) per day. However, during the harvesting period, the wage rate is as high as 55 baht (US$2.2) per day. The NPV of the research on combine harvesters in the rainfed area is based on these parameters.

Rice quality

The ability to adjust the quality of rice to meet the requirements of the market is one of the key factors that enables Thailand to be the top exporter in the world rice market. The first effort of rice research in Thailand in the early 1900s aimed at increasing the length of rice grain thereby differentiating Thai rice from other rice. Subsequently, other post-harvest technology, such as thrashing and milling, was developed to ensure the post-harvest processing quality. The goodwill of Thai rice paid off handsomely for this research investment. This history gives rice research in Thailand a very special heredity culture.

During the past decade, the rapid increase in income in Thailand also led to an increase in the demand for higher-quality rice in the domestic market. The current emphasis of quality-related research is on the moisture content of the grain and the aroma of the rice. The moisture problem becomes more important with the expansion of irrigated rice harvested during the rainy season. In the export market, Thai aromatic rice has become more popular.

To reduce grain moisture, a dryer (farm mechanization) is the solution. The research on a medium-sized dryer was estimated to cost 2 million baht (US$80,000) for a period of 3 years. The price of dry grain (moisture less than 7%) is about 15% higher than the price of grain that has medium moisture (moisture of 10–15%). Given the cost of research and the potential pay-off, the NPV for dryer research was estimated.

For improving eating quality, two possible options were identified: breeding of an aromatic rice variety for the irrigated areas and research on post-harvest technology which preserves the aroma of rice. Currently, aromatic rice is only grown in the rainfed dryland in the lower northeastern region. By having aromatic rice grown in the irrigated area, it would be possible to greatly increase the output of aromatic rice. This could greatly increase the value of rice produced in the irrigated area.

Scientists have pointed out two options to improve eating quality: conventional plant breeding and biotechnology. Some scientists believe that conventional plant breeding can accomplish this task within 3–4 years. However, they all agreed that biotechnology would be a more viable alternative.

Aromatic rice fetches about a 15% higher price in the market. Scientists have estimated that the aromatic paddy will give a 5% lower yield (actual yield will be decreased from 3.925 tons ha^{-1} to 3.729 tons ha^{-1}). For the aromatic rice produced in the rainfed area, the research on how to maintain the aroma, cooking and eating quality of the existing production is very important. Post-harvest technology is important here. Since aromatic rice commands a premium of about 3 baht/kg^{-1} in the market, the potential benefit of this technology is the preservation of this value in the market. Given the budget of 2 million baht year^{-1} for a period of 3 years, the scientists expected that the probability of research success to be about 70%. The resulting information can be used to determine the packaging and method of storage that will reduce the rate of deterioration in the aroma of rice. Therefore, the benefit will not be the full 3 baht kg^{-1}. Given the packaging and storage cost, the net benefit is about 0.5 baht kg^{-1} or about 500 baht (US$25) per ton. This NPV is reported in the next section.

IV. Assessment of Benefits from Rice Research

The estimation of benefits of research investment take into account the extent of area in which the new technology is suitable, the expected productivity gain, the direct costs involved in using the new technology, the reduction in costs achieved and the probability of success. Scientists were asked to consider the alternative activities, addressing each constraint on rice yield, across three rice ecosystems: irrigated, rainfed and deepwater rice. For each option, scientists were asked to estimate the probability of research success (PR), the yield affected (kg rai^{-1}), the cost affected (baht rai^{-1}) and the rate of adoption. Except for the cost of production and rice quality discussed earlier, the scientists were requested to base their estimations on the same direct cost of research for each alternative option (US$400,000 year^{-1}) and the same period required to achieve success (5 years). The government bonds rate of 10.75% is used as the discount rate in calculating the expected net benefit because agricultural research is a public investment.

The benefit and cost of each research option are used for the calculation of expected net benefit. It is assumed that farmer adoption will reach the maximum level within 15 years after the new technology is available. The following formula was used in the estimation of the expected net benefit:

$$NPV = \sum_i \frac{B_i - C_i}{(1+r)^i} \qquad (13.2)$$

where B_i =[price(1 + %Δprice) × ΔY × area adopt$_i$ × adopt rate$_i$ × PR] and C_i = [Δcost × area adopt$_i$ × adopt rate$_i$ × PR + (research cost)], where B_i = benefit of research work, C_i = cost of research and farmer cost, r = government bonds rate = 0.1075, i = number of years, price = rice price (baht kg^{-1}), %Δprice = (%ΔQ) demand elasticity for rice, Y = affected yield (kg rai^{-1}), cost = affected cost (baht rai^{-1}), PR = probability of success of research work, area adopt = adopted area and adopt rate = adoption rate.

A price adjustment is required to take into consideration world demand elasticity for rice. This is needed because Thailand is a net exporter of rice; thus, any additional production would be exported. As one of the most important players in the international market, any additional export would affect the international market price. Based on previous estimates, demand elasticity on the international market is assumed to be − 4.0.

Expected net benefits were estimated for all the problems and their alternative research solutions for the three rice ecosystems. The alternative activities of each rice ecosystem were first compared and ranked, and then the alternative activities of the three rice ecosystems were ranked to establish the priorities. To prioritize the research options, the following steps were adopted:

1. The NPV for research of all the alternatives and problems were first ranked for each production environment.

2. The net NPVs were ranked for each problem in each production environment. The highest pay-off option was then chosen for each problem in each production environment.
3. Finally, the chosen option for each problem in all the production environments was then combined and ranked to determine the priority of rice research.

Irrigated rice ecosystem

For irrigated rice the results are reported in Table 13.10. The top three problems and their research options are: blast, rice quality and combine harvesters (cost of labour).

Among the top ten constraints, there are seven that farmers recognized ($>10\%$). These are blast, grain quality, green leafhopper, brown planthopper, stemborer, rice leaf folder and weeds. There are three constraints that are mentioned by scientists but not by farmers, i.e. bacterial leaf blight, tungro and ragged stunt virus. This could be because farmers already know how to deal with them and hence do not consider them as problems. The damaged areas due to bacterial leaf blight and tungro were very small (900 and 120 rai) in Suphan Buri (DOAE, 1991). There are six constraints that are recognized by both farmers ($>10\%$) and scientists. These are blast, grain quality, brown planthopper, stemborer, rice leaf folder and weeds.

The results show that the research on blast has the highest NPV. There are two alternative solutions: plant protection through integrated pest management (IPM) and variety resistance through conventional plant breeding. The research on IPM would yield a pay-off of US$37.5 million which is higher than investing in breeding a resistant variety (US$20.6 million). The high benefit of IPM comes from both the increase in yields and the reduction of production costs due to the optimum use of pesticides. A blast-resistant variety would have a lower yield but this is compensated by the decrease in the cost of pesticide. Therefore, IPM is chosen for the blast problem.

Research on combine harvesters ranks second for the irrigated area. It yields a net benefit of US$36.1 million. Most of the activity in research in this area is now carried out by private entrepreneurs. The high potential pay-off and the low level of private investment suggest that there is a constraint in the private sector; this could come from the risk involved. To close this gap, the government could do the initial research and, once the market has confidence, the government could sell the technology to recover its initial investment.

The third highest NPV is for research in rice quality to breed aromatic rice for irrigated areas. The perpetual flow pay-off is US$36.2 million. It should be noted here that this calculation assumes that the elasticity of demand for aromatic rice is -2.0, i.e. more inelastic than the world demand for rice (-4.0). This is because the market for aromatic rice is small and specialized.

Rainfed rice ecosystem

There are 13 problems for rainfed rice research. The alternative options to solve each of these problems and their NPVs are shown in Table 13.11. From the top five problems, there are two research activities indicated as important by the farmers: blast and stemborer. Three other problems are drought, rats and

Table 13.10. Net present value (NPV), present value (PV) of perpetual flow and payback period of return on rice research: irrigated rice.

Rank	Source	Alternative option	NPV start in 4th until 8th year (million US$)	PV of perpetual flow (million US$)	Pay-back period (years)
1	Harvester	Farm machinery	4.02	36.1	1.7
2	Blast	Plant protection	3.48	37.5	1.6
2	Grain quality	Post-harvest technology	3.29	36.2	1.6
4	Ragged stunt virus	Plant protection	2.44	29.7	2.0
5	Tungro	Plant protection	2.19	27.8	2.1
6	Nematodes	Plant protection	1.58	23.2	2.6
7	Stemborer	Plant protection	1.44	22.2	2.7
8	Brown planthopper	Plant protection	1.35	21.4	2.8
9	Rice leaf folder	Biology control	1.27	20.8	2.9
10	Ragged stunt virus	Conventional breeding	1.24	20.6	2.9
11	Bacterial leaf blight	Conventional breeding	1.24	20.6	2.9
12	Blast	Conventional breeding	1.24	20.6	2.9
13	Tungro	Conventional breeding	1.24	20.6	2.9
14	Aroma characteristic	Agricultural chemistry	1.22	20.5	2.9
15	Brown planthopper	Plant protection	0.92	18.2	3.3
16	Green leafhopper	Plant protection	0.66	16.2	3.7
17	Bacterial leaf blight	Plant protection	0.43	14.5	4.1
18	Weeds	Plant protection	0.37	14.1	4.2
19	Brown planthopper	Biology control	0.37	14.1	4.2
20	Weeds	Cropping pattern	0.21	12.9	4.6
21	Weeds	Water management	0.10	12.0	5.0
22	Brown planthopper	Conventional breeding	0.06	11.7	5.1
23	Stemborer	Biology control	−0.07	10.8	5.5
24	Pests	Forecasting model	−0.18	9.9	6.0
25	Acid sulphate soil	Conventional breeding	−0.19	9.8	6.1
26	Brown planthopper	Chemical control	−0.46	7.7	7.7
27	Weeds	Biology control	−0.52	7.3	8.1
28	Stemborer	Conventional breeding	−0.64	6.4	9.3
29	Weeds	Herbicide application	−1.46	0.2	336.7
30	Dryer	Dryer	−4.54	−23.1	−2.6

Source: Setboonsarng *et al.*, 1994.

weeds. The farmers did not consider grain quality as a problem because most of the farmers are satisfied with the market system in terms of rice price. There are three research activities suggested only by the scientists and not by the farmers. These are bacterial leaf blight (BLB), gall midge and grain quality. While the request for BLB and gall midge research came from the research community and extension agents, scientists received many requests from the rice exporters to do research on grain quality.

Table 13.11. Net present value (NPV), present value (PV) of perpetual flow and payback period of return on rice research: rainfed rice.

Rank	Source	Alternative option	NPV start in 4th until 8th years (million US$)	PV of perpetual flow (million US$)	Pay-back period (years)
1	Drought	Agronomy	2.80	35.9	1.8
2	Blast	Conventional breeding	4.05	46.4	1.4
3	Stemborer	Conventional breeding	3.36	40.6	1.6
4	Bacterial leaf blight	Conventional breeding	4.05	46.4	1.4
5	Rats	Plant protection	2.59	34.2	1.9
6	Blast	Plant protection	1.82	27.7	2.3
7	Bacterial leaf blight	Plant protection	5.06	54.9	1.2
8	Stemborer	Plant protection	3.81	44.4	1.5
9	Drought	Conventional breeding + biotechnology	2.66	34.8	1.9
10	Stemborer	Conventional breeding + biotechnology	5.43	58.0	1.1
11	Bacterial leaf blight	Conventional breeding + biotechnology	4.05	46.4	1.4
12	Blast	Conventional breeding + biotechnology	2.66	34.8	1.9
13	Gall midge	Conventional breeding	−0.23	10.5	6.2
14	Gall midge	Plant protection	3.61	42.8	1.5
15	Weeds	Cropping pattern	3.64	43.0	1.5
16	Gall midge	Conventional breeding + biotechnology	4.44	49.7	1.3
17	Thrips	Plant protection	1.59	25.8	2.5
18	Grain quality (appearance)	Post-harvest technology	4.64	51.4	1.3
19	Crabs	Plant protection	2.82	36.2	1.8
20	Weeds	Herbicide application	2.35	32.2	2.0
21	Drought	Soil management	−3.42	−16.2	−4.0
22	Low soil fertility	Soil management	1.50	25.1	2.6
23	Drought	Conventional breeding	0.09	13.3	4.9
24	Drought	Use own variety sources	0.88	19.9	3.3
25	Drought	Introduce variety from outside	1.48	24.9	2.6
26	Submergence	Conventional breeding	−0.22	10.6	6.1
27	Grain quality (aroma)	Agricultural chemistry	3.01	34.0	1.8
28	Harvester	Farm machinery	5.12	44.4	1.3

Source: Setboonsarng *et al.*, 1994.

The highest NPV is from the stemborer through conventional plant breeding plus biotechnology. This option dominates plant protection because plant protection would involve the use of other inputs such as chemical and labour while a resistant variety would save on these costs. The probability of success and time period for research of this option are enhanced by the use of biotechnology. The stemborer is the only problem identified by farmers and scientists. However, the reported damaged area is only average at about 62,000 ha for the whole country in 1986–1991. However, given its potential damage area in the future, farmers and scientists consider the stemborer as a possible problem in the future.

The second highest NPV is the research on BLB through plant protection. Plant protection dominates conventional plant breeding because of the variation of BLB strains across the rainfed area. It would be costly to do research addressing the diverse set of BLB. Therefore, plant protection research which is more general in application gives a higher investment pay-off. The difference of NPV between conventional plant breeding and conventional plant breeding plus biotechnology is negligible. While conventional plant breeding is already well established, the organization for the inclusion of biotechnology would take additional effort.

The third highest NPV is grain quality. The post-harvest technology which deals with storage and packaging problems yields higher pay-offs than the use of agricultural chemistry in maintaining the quality of rice. Better storage and packaging would allow control over the aroma and dehydration process of rice after it is milled. The dehydration process of rice affects the cooking and eating quality of rice. While newly harvested rice with higher moisture will be softer, the old rice will be harder. Although there is a question of whether post-harvest research should be public or private research, it is clear that initial steps should be taken by the public sector.

Deepwater rice ecosystem

There are ten problems identified for deepwater rice research. All ten research activities are suggested by scientists. There are seven research activities suggested by farmers. Seven research activities are mentioned by both farmers and scientists. The main problems are biotic factors – blast, rats and weeds.

There are three research activities mentioned by scientists that are not considered by farmers, i.e. the brown planthopper, BLB and nematodes. The damaged area due to BLB was only about 1000 ha in Prachin Buri in 1989–1991 (DOAE, 1991). BLB is not considered to be a problem by farmers because yield loss is limited. According to knowledgeable farmers, nematodes are not considered to be a problem by the average farmer.

Most of the options have negative NPVs when considered within 8 years. Part of the problem comes from the method used in determining the return on research. A fixed research budget of US$40,000 for 5 years was used as a basis for the estimation of the probability of success and the outcome of the research. The limited plant area of deepwater rice makes its potential return low. Research in blast through conventional plant breeding yields the highest pay-off for deepwater rice. This option dominates the use of plant protection because plant protection has a higher cost of application for the farmer.

Among the three rice ecosystems, the benefit of research activities for rainfed rice is the highest because of its large cultivated area. The benefit of research activities for deepwater rice is the lowest group, with negative NPVs.

There are many common research areas in the three rice production environments, notably stemborers, grain quality, blast and combine harvesters. If the spill-over of research between the rainfed and irrigated production environments for these four areas of research is assumed to be 100%, these four areas of research would be given the highest pay-off for rice research. In particular, grain quality and blast will be the two highest areas of research, followed by combine harvesters and the stemborer.

V. Conclusions

The expected return to rice research depends on the cultivated area and output of rice. The benefit of research activities for rainfed rice is the highest followed by irrigated rice. Research in deepwater rice has a small NPV.

Rainfed rice research is more profitable than the others. However, the allocation of the research budget at the Rice Research Institute of the Department of Agriculture for rainfed rice is less than that allocated for irrigated rice. In 1991, the number of research projects for rainfed rice was 165, and 298 for irrigated rice. The output of irrigated rice (15%) is less than rainfed rice. Therefore, the allocation of budget for rice research does not take into consideration the proportion of cultivated area and output, and may not be efficient.

Given the possible spill-over of research output, research on stemborer using conventional breeding and biotechnology has the highest pay-off. Farm machinery research on combine harvesters generates the second highest pay-off because of the rising wage cost. Plant protection research on bacterial leaf blight has the third highest pay-off. These research activities are all for rainfed rice.

Two important points to note are: (i) biotechnology research is a complement to the existing research, not a substitute; (ii) aside from the physical constraints, economic constraints, such as wage rate, are becoming more important for rice research in Thailand.

References

Department of Agricultural Extension (1991) Ministry of Agriculture 1991 Annual Report.
Department of Agricultural Extension (1992) Ministry of Agriculture 1992 Annual Report.

Office of Agricultural Economics (1990) Ministry of Agriculture 1990 Cost of Agriculture Survey.

Office of Agricultural Economics (1993) Ministry of Agriculture 1993 Cost of Agriculture Survey.

Setboonsarng, S., Suthad, A. and Satasart, S. (1994) *Priority Setting for Rice Research in Thailand*. Thailand Development Research Institute, Bangkok.

14 Constraints to Growth in Rice Production in the Philippines

M. Hossain[1], F.B. Gascon[1] and I.M. Revilla[2]

[1]Social Sciences Division, International Rice Research Institute, PO Box 933, 1099 Manila, Philippines; [2]Cross Ecosystem Program, Entomology and Plant Pathology Division, International Rice Research Institute, PO Box 933, 1099 Manila, Philippines

I. Introduction

The Philippines is one of few countries in Southeast Asia that faces food security problems. Nearly half the population lives in poverty (Balisacan, 1992). Rice is the most important staple food, and in terms of area planted to foodgrains it is second only to maize. Modern varieties cover 95% of the rice area, yet the yield is only 40% of the level reached in Japan and South Korea, and 60% that of Indonesia. The population is still growing at 2.2% per year (the highest in Southeast Asia) and the per capita consumption of rice may increase with economic prosperity because of the current widespread poverty (Balisacan, 1994). It is estimated that the demand for rice will grow at 2.5–3.0% per year if the government's target of economic growth is achieved, although rice production has remained almost stagnant since 1985. The World Bank projects that the Philippines will become a major rice-importing country in the world by the year 2000 if recent trends continue.

Few studies have looked into the factors behind the recent stagnation in rice production and the low level of yield despite widespread adoption of modern varieties. The study of constraint analysis conducted by the IRRI in the late 1970s provided valuable information, but as yields have increased considerably since then the importance of various problems may be different today (IRRI, 1979). A detailed study is needed to focus rice research on important problems and to support the government's efforts to achieve and sustain food security in the Philippines.

This chapter presents an overview of the growth in rice production and yield gaps under different rice-growing conditions using secondary data. It also reports farmers' perceptions of the yield loss in rice production and the broad factors contributing to the losses from in-depth village studies conducted by the Social Sciences Division at the IRRI during 1992–94.

II. Long-Term Trend in Production

The race between the growth in rice production and population growth is shown in Fig. 14.1. Rice output has kept pace with the growth of population over the last three decades, mostly due to the growth in rice yield. But there has been a cyclical movement in the performance of the rice sector. During most of the 1960s, rice production failed to keep pace with the growth in population. Then there was an impressive growth in rice yield during the early green revolution period (1968–1980) when rice production grew substantially faster than population growth and contributed to an impressive increase in per capita availability of rice. But there has been a drastic deceleration in the growth of rice yield since 1984, and by the end of the decade the per capita availability of rice slipped back to the level of early 1960s. In the 1960s the Philippines used to meet about 10–15% of its domestic requirement for rice through imports. It became self-sufficient by the end of 1970s and exported small amounts in the early 1980s. But during the last 3 years the average import of rice has been almost at the same level as three decades ago.

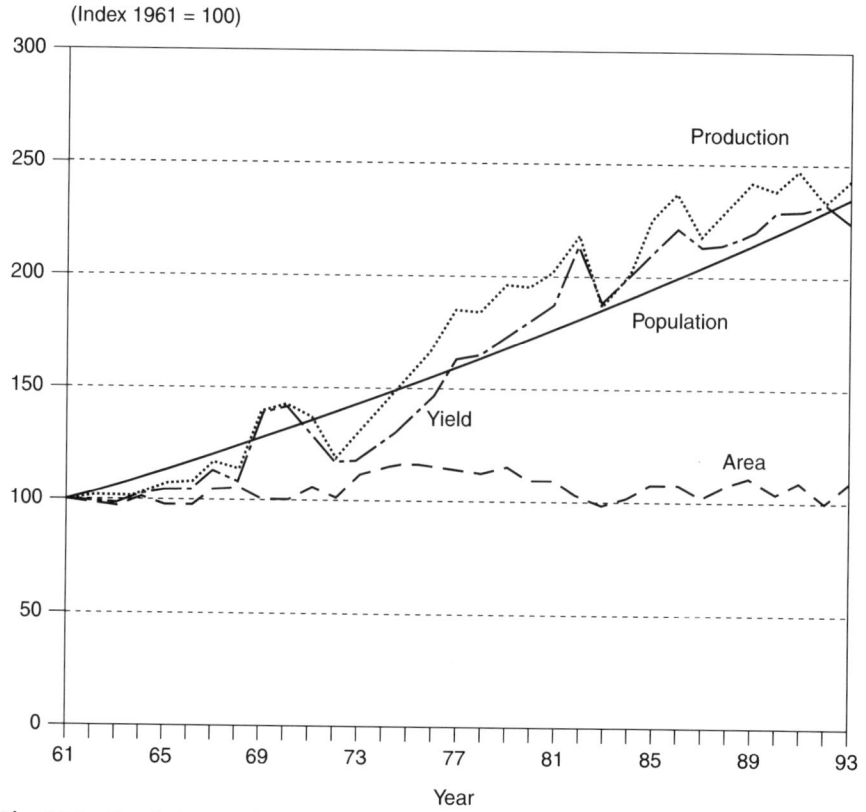

Fig. 14.1. Trends in population, rice area, production and yield, Philippines, 1961–1993.

The growth in rice production and the contribution of area and yield during the last three decades can be reviewed from Fig. 14.2. The contribution of the expansion of rice area to the growth in output was small, declined over the second decade and completely dried up during the 1980s. Rice yield increased at an impressive rate of about 4% per year during the 1970s but that declined to 1.8% during the 1980s. The growth in rice production during the 1980s did not even meet the population growth-induced demand for rice.

III. Sources of Growth

As improved rice varieties become available, an important source of growth in crop yield becomes the shifting of land from the traditional low-yielding varieties to the improved high-yielding ones. When this source of growth dries up, the only way that yield could be further increased is through more intensive use of inputs on the improved varieties, i.e. a movement along the same production function. In order to see the changes in the composition of the different elements of growth in rice production in the Philippines in the 1970s and 1980s we applied an additive method of decomposition of rice production,

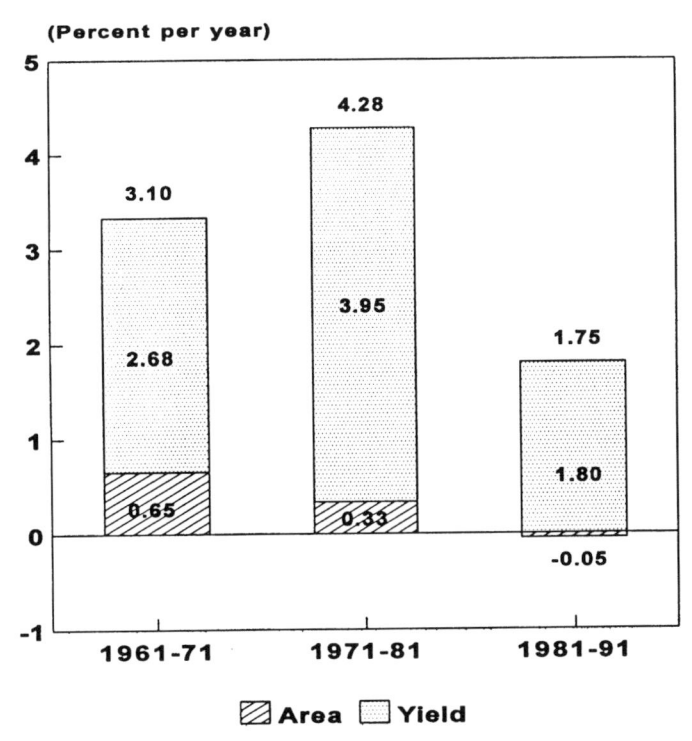

Fig. 14.2. Growth in rice production and productivity in the Philippines over the last three decades.

proposed by Minhas and Vidyanathan (1965). The analysis was done with time series data on the area and production of eight types of rice classified by varieties (modern and traditional), seasons (wet and dry) and water control (irrigated and rainfed). The method of decomposition is as follows:

$$(P_t - P_o)/P_o = \left(A_t \sum_{i=1}^{n} C_{it} Y_{it} - A_o \sum_{i=1}^{n} C_{io} Y_{io}\right)/P_o$$
$$= \left[(A_t - A_o) \sum C_{io} Y_{io}\right]/P_o + \left[A_t \sum C_{io}(Y_{it} - Y_{io})\right]/P_o$$
$$+ \left[A_t \sum Y_{io}(C_{it})\right]/P_o + \left[A_t \sum (Y_{it} - Y_{io})(C_{it} - C_{io})\right]/P_o$$

(14.1)

where P is the volume of rice output, A is the area under rice, Y_i is the yield of the variety i, and C_i is the proportion of rice area under the ith variety. Subscripts o and t refer to the base and terminal period, respectively.

The first term on the right-hand side of the equation measures the increase in crop acreage, i.e. by how much the production would have grown if the rice yield remained constant. The second term measures the pure yield effect, i.e. the increase in production due to more intensive use of inputs along a given production function. The third term measures the effect of the changes in cropping pattern, i.e. by how much the production would have increased if the yield of individual varieties remained stagnant and farmers had increased production by changing the allocation of land among varieties (movement across production functions). The fourth term is the effect of interaction between variety composition and crop yield.

The method was applied to the disaggregated data for the 1970s and 1980s for the eight groups of rice varieties in the Philippines. The unpublished official data were obtained through personal communications with the Bureau of Agricultural Statistics under the Department of Agriculture. In order to smooth out yearly fluctuations due to weather conditions, 3-yearly averages centred around the base and the terminal year was used. The results of the exercise are presented in Table 14.1.

The results show that during the 1970s nearly 62% of the growth in rice production arose from pure yield effect. The other major source of growth in 1970s was the interaction effect – land was shifted to varieties which had an upward trend in yield. This source of growth dried up in 1980s as the diffusion of land from traditional to modern varieties became almost complete. The growth from the movement along the production function has also decelerated substantially to only 1.0% per year, an indication of diminishing marginal returns on more intensive application of input along the production frontier. With rice area declining and the potential of raising yield through changes in land portfolio across varieties almost exploited, the intensity in input use will remain the only source for future growth in rice production in the Philippines unless there are further technological breakthroughs. With the present level of growth, the Philippines will be unable to meet the growing rice needs of its people.

Table 14.1. Relative contribution of different elements to the growth of rice output, in the Philippines 1971–1981 and 1981–1991.

	1971–1981		1981–91	
Elements of Growth	Changes in output due to the element (as a % of base output)	Increase per year (%)	Changes in output due to the element (as a % of base output)	Increase per year (%)
Area	5.0	0.49	0.0	0.0
Yield	30.1	2.67	11.0	1.04
Variety composition	6.3	0.61	7.8	0.75
Interaction of yield and variety of composition	10.7	1.02	0.3	0.03
Total	52.1	4.28	19.1	1.76

Source: Estimated using a growth decomposition method on 3-year moving average data for eight varieties of rice grown in different seasons and ecosystems.

IV. Yield Differences across Regions

Figure 14.3 shows the maximum and average yields achieved by farmers for modern and traditional varieties under different growing conditions in Central Luzon and Western Visayas, the two major rice bowls in the country. It may be noted that farmers have already achieved 4.6 tons of yield for modern varieties under irrigated conditions in the dry season. The yield is 25% lower for the same modern variety when grown in the wet season (this may partly be due to typhoons), and only about half for traditional varieties. The yield under the same growing conditions is lower in Western Visayas than in Central Luzon, which may be explained by erratic rainfall and inferior quality of irrigation. Visayas also grows rainfed rice in a substantial portion of land (63%) during the January–June period by taking advantage of the late rains. The average rice yield for this growing condition is typically less than 2.0 tons ha^{-1} because of frequent moisture stress and low use of inputs due to the uncertain weather. A large part of the yield gap in the Philippines may thus be due to abiotic stresses.

Table 14.2 shows the gap in rice yield for modern varieties between irrigated and rainfed conditions, and for irrigated conditions between the dry and the wet season. The availability of irrigation during the wet season increases rice yield by more than 1.0 ton in most of the regions. Since the crop is grown under heavy rains, irrigation presumably helps farmers save yield losses due to temporary droughts. The rice yield should be higher in the dry season because of longer day length, lower cloud cover and less build-up of pests, but in most of the regions in the Philippines the rice yield in the dry season is lower than in the wet season (except in Central Luzon, Southern Tagalog and Western Mindanao). This lower yield may be due to the erratic supply of irrigation water, as the water level in dams and irrigation channels is sometimes reduced

Central Luzon

Category	Avg	Max
WS Irrigated MV		3.44
WS Irrigated TV		2.93
WS Rainfed MV		3.58
WS Rainfed TV		2.43
DS Irrigated MV		4.61
DS Irrigated TV		3.41

Yield (tons ha^{-1})

☐ Avg ☐ Max

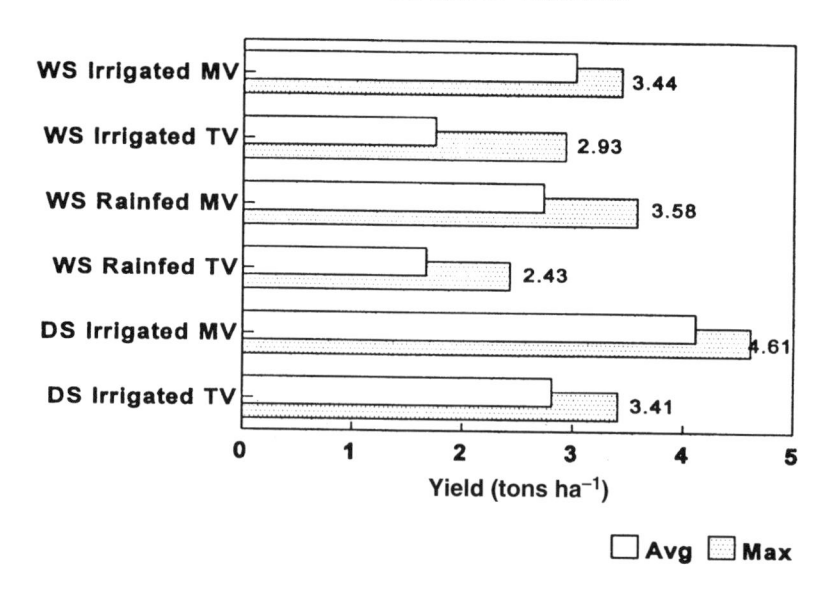

Western Visayas

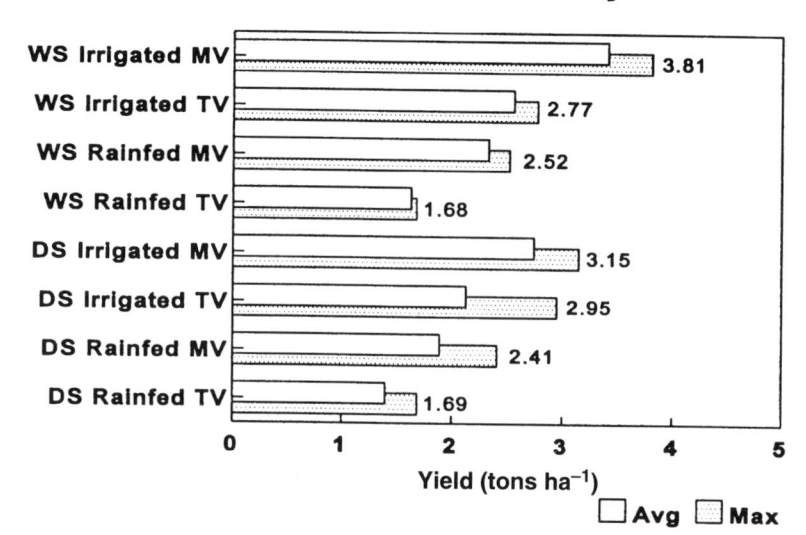

Category	Avg	Max
WS Irrigated MV		3.81
WS Irrigated TV		2.77
WS Rainfed MV		2.52
WS Rainfed TV		1.68
DS Irrigated MV		3.15
DS Irrigated TV		2.95
DS Rainfed MV		2.41
DS Rainfed TV		1.69

Yield (tons ha^{-1})

☐ Avg ☐ Max

Fig. 14.3. Gaps in yield achieved by average farmers in Central Luzon and Western Visayas, by season, water control and varieties, 1981–1991. DS, dry season; MV, modern variety; TV, traditional variety; WS, wet season; Avg, average yield; max, maximum yield during the 1981–1991 period.

Table 14.2. Highest rice yield of modern varieties achieved in different regions in the Philippines.

Regions	Wet season irrigated	Wet season rainfed	Dry season irrigated	Yield difference Irrigated over rainfed	Dry season over wet season
Ilocos	3.09	2.80	2.93	0.29	−0.16
Cagayan Valley	3.86	2.12	3.66	1.74	−0.20
Central Luzon	3.71	3.58	4.61	0.13	0.90
Southern Tagalog	3.23	2.60	3.70	0.64	0.46
Bicol	3.35	2.16	3.18	1.19	−0.17
Western Visayas	3.81	2.52	3.15	1.29	−0.66
Central Visayas	2.58	1.86	2.37	0.72	−0.21
Eastern Visayas	3.26	1.76	2.86	1.50	−0.40
Western Mindanao	3.66	2.62	4.14	1.04	0.48
Northern Mindanao	4.16	3.08	3.39	1.08	−0.77
Southern Mindanao	4.35	3.31	3.76	1.04	−0.59
Central Mindanao	3.84	3.06	3.84	0.79	0.00

Source: Bureau of Agricultural Statistics, undated.

by inadequate rainfall in the rainy season and prolonged droughts in the dry season. Development of cultivars resistance to temporary moisture stress could thus help increase rice yields substantially.

V. Estimates of Yield Gaps

The yield gap is the difference between potential yields and actual yields (Gomez et al., 1979a). The yield gap can be divided into two parts: yield gap I is the difference between experimental station yield and potential farm yield. It exists mainly because of environmental differences between experiment stations and the actual rice farms. The potential farm yield can be approximated by the yield obtained in on-farm experiments under non-limiting input conditions. Yield gap II is the difference between the potential farm yield and the actual farm yield. This gap reflects biological constraints, soil and water constraints and socioeconomic constraints that compel farmers to use inputs at a level below the technical optimum. A part of yield gap II could be reduced through research for developing resistances in rice cultivars against various biotic and abiotic stresses and through appropriate socioeconomic policies.

Table 14.3 presents information on rice yields obtained in research stations, on-farm experiments for 1993 and the average yield at the farm level for 1991–93. Because of the difference in environmental factors, the yield is substantially higher in the dry season than in the wet season both under on-farm experiments and at the farm level, when irrigation facilities are available. Under rainfed conditions, however, the yield is lower in the dry season than in the wet season

Table 14.3. Estimates of yield gaps between experimental station and farm-level yields, 1992/93

Variables	Wet season (tons ha^{-1})	Dry season (tons ha^{-1})
Highest experimental yield, 1993	6.22	8.12
Highest yield on-farm experiments, 1993	5.76	7.26
Average farm-level yield (1991–1993)		
Irrigated	3.31	4.21
Rainfed	2.28	1.71
Yield gap I	0.46	0.86
Yield gap II		
Irrigated	2.45	3.05
Rainfed	3.48	5.52

Source: Social Science and Policy Research Survey, Philippine Rice Research Institute, Nueva Ecija, and the Bureau of Agricultural Statistics, Quezon City.

because of the lower intensity and erratic nature of rainfall. This again shows the importance of water stress as a constraint to increase in rice yields. Yield gap I is estimated at 0.46 ton ha^{-1} for the wet season and 0.86 ton ha^{-1} for the dry season. Yield gap II is, however, substantial even when water is non-limiting. It is estimated at 2.45 tons ha^{-1} for the wet season and 3.05 tons ha^{-1} for the dry season, more than 70% of the yields that have been achieved by farmers. Thus, even with existing varieties, the Philippines appears to have a vast potential for increasing rice production if rice research could help reduce the yield gap.

VI. Constraints to Higher Yields

Why are farmers unable to achieve the full potential yields of the technologies already available? If the rice plants are provided with all their biological needs, and if they are adequately protected from damage caused by insects, diseases and climatic variations they should give high yields. Obviously there are a combination of factors – physical, economic and social – that prevent farmers from applying inputs at technically optimum levels and getting the maximum potential yields possible with the new varieties.

To study the constraints to higher yields, the IRRI carried out farm-level experiments in four areas in the Philippines during the 1975–1977 period with two different input management packages with respect to three factors – fertilizer use, insect control and weed control – while keeping all other factors constant (De Datta et al., 1979a,b,c; Gomez et al., 1979b). Potential farm-level yield was estimated with the high-input package that was expected to give maximum profits under non-limiting moisture availability conditions. Actual farm-level yield was estimated with a low input package that averaged farmers' prevailing practice. Yield gaps contributed by these three factors were estimated by comparing the yields obtained from these two different input packages. The results obtained from the study are summarized in Table 14.4. The yields were higher in Central Luzon (Laguna and Nueva Ecija) than in Visayas (Carmarines Sur and Iloilo), and for the dry season than for the wet season, again demonstrating differences in physical and environmental factors across regions and between seasons. These differences in factors that are beyond farmers' control also cause risks in rice cultivation and affect farmers' decisions regarding the level of input use. The three factors accounted for a yield gap varying from 0.9 to 1.8 tons ha^{-1} for the wet season (depending on the region), and from 1.6 to 2.2 tons ha^{-1} for the dry season (see Table 14.4). The use of chemical fertilizer contributed to the major portion of the gap, followed by insect control, while weeds accounted for a small fraction of the gap.

VII. Yield Losses

A recent household survey conducted by the Social Sciences Division at the IRRI collected information from farmers on the amount of crop loss in four consecutive rice-growing seasons during the 1992–1994 period. The respondents were asked to report the yield obtained at harvest and the expected normal yield, and estimate the amount of yield loss from major factors contributing to the loss. The information was collected at the parcel level from all farm households in the villages under study: an irrigated village in Laguna, a partially irrigated (with tubewells) village in Central Luzon, and two villages from Visayas – one irrigated village and one rainfed village with poor infrastructural facilities. The results from the survey are presented in Table 14.5. All farmers grew rice in the wet season and 60% of them raised a second rice crop during the dry season; the availability of irrigation facilities was the most important factor determining this choice. The average size of rice farm was 1.31 ha for the wet season and 0.98 ha for the dry season.

Table 14.4. Contribution of insect control, fertilizer and weed control to increased rice yield (tons ha^{-1}): estimates from farm-level experiments, 1975–1977.

Variable	Laguna Wet season	Laguna Dry season	Nueva Ecija Wet season	Nueva Ecija Dry season	Camarines Sur Wet season	Camarines Sur Dry season	Iloilo Wet season	Iloilo Dry season
Potential farm yield	5.41	6.48	6.3	6.8	4.1	5.6	4.8	5.4
Actual farm yield	3.59	4.36	4.8	4.7	3.2	3.5	3.5	3.8
Yield gap	1.82	2.18	1.5	2.1	0.9	2.1	1.3	1.6
Insect control	0.96	0.91	0.7	0.9	0.4	0.7	0.4	0.3
Fertilizer	0.62	1.04	0.5	1.0	0.4	1.2	0.6	1.3
Weed control	0.25	0.15	0.1	0.3	0.1	0.2	0.3	0.2

Source: IRRI, 1979.

Table 14.5. Yield loss reported by farmers from a household survey, 1992–94.

Characteristics	Wet season	Dry season
Number of sample farmers	411	246
Average size of rice farm (ha)	1.31	0.98
Rice yield at harvest (kg ha^{-1})	3270	3822
Production losses (kg ha^{-1})		
Drought/irregular supply of irrigation	198	759
Typhoon/strong wind	358	253
Submergence/excess water	49	27
Insects and diseases	250	206
Inferior variety	25	11
Limited capital	48	42
Others	17	0
Expected normal yield	4215	5120
Loss as percent of harvest	28.9	34.0

Source: Social Sciences Division, IRRI.

The yield loss was found to be substantial – 29% of the yield for the wet season, and 34% of the yield for the dry season. If the losses were eliminated, the yield could be raised to 4.2 tons ha^{-1} in the wet season and 5.1 tons ha^{-1} for the dry season. With the existing rice area at the national level, this would increase production to 15.4 million tons – about 50% higher than the present production level. Insects and diseases account for a smaller fraction of the yield loss than the abiotic stresses such as typhoons, droughts and poor quality of irrigation. The yield loss due to insects and diseases was reported at 26% of the total loss for the wet season and 16% for the dry season. The major pests were reported as the golden snail, stemborer, tungro, brown planthopper and rice bugs. In the wet season, typhoons and bad weather were reported as the most important factor contributing to the yield loss, estimated at 0.36 tons ha^{-1}, accounting for 38% of the total loss. For the dry season, prolonged drought and irregular supply of irrigation water were reported as the most important constraints to an increase in yield, contributing to 58% of the production loss. Even in the wet season, moisture stress due to prolonged drought reduced rice yield by 0.20 tons ha^{-1}.

VIII. Conclusions

In the Philippines, the growth in rice production has substantially decelerated during the last decade. The main factor behind the deceleration of growth is that the exploitation of the growth potential through reallocation of land from low-yielding traditional varieties to high-yielding modern varieties has almost been completed. The growth in rice yield through more intensive use of inputs with the same variety has also decelerated in the 1980s, when it was only 1.0% per year. The low average rice yield at the national level is due to diverse growing conditions. In some regions and varieties, farmers have already achieved a yield of 4.5 tons ha^{-1}. In a household survey of four villages representing different rice ecosystems, farmers reported yield a loss of 0.95 tons ha^{-1} in the wet season and 1.3 tons ha^{-1} in the dry season, mostly due to abiotic stresses of which typhoons and prolonged droughts and irregular supply of irrigation are important. Insects and diseases account for less than 25% of total losses. It will be difficult for the Philippines to meet the demand for rice from the fast-growing population, unless rice research succeeds in helping farmers to recoup these losses, through research on developing resistance of rice varieties against abiotic stresses.

References

Balisacan, A.M. (1992) Rural poverty in the Philippines: incidence, determinants and policies. *Asian Development Review* 10(1), 126–163.

Balisacan, A.M. (1994) Demand for food in the Philippines: responses to price and income changes. Paper presented to the Third Workshop on Projection and Policy Implications of Medium and Long-Term Rice Supply Demand, IRRI, IFPRI, held in Thailand, 24–26 January, 1994.

Bureau of Agricultural Statistics (Undated) *Rice Statistics Handbook*. Department of Agriculture, Government of the Philippines, Quezon City.

De Datta, S.K., Garcia, F.V., Abilay, W.P., Alcantara, J.M., Mandac, A. and Marciano, V.P. (1979a) Constraints to high rice yields, Nueva ecija, Philippines. In: IRRI, *Farm Level Constraints to High Rice Yields in Asia: 1974–77*. International Rice Research Institute, Los Baños, Philippines.

De Datta, S.K., Chatterjee, A.K., Cia, B.S. and Musicat, P.B. (1979b) Constraints to high rice yields, Camarines Sur, Philippines. In: IRRI, *Farm Level Constraints to High Rice Yields in Asia: 1974–77*. International Rice Research Institute, Los Baños.

De Datta, S.K., Cia, B.S., Jereza, H.C. and Musicat, P.B. (1979c) Constraints to high rice yields, Iloilo, Philippines. In: IRRI, *Farm Level Constraints to High Rice Yields in Asia: 1974–77*. International Rice Research Institute, Los Baños, Philippines.

Gomez, K.A., Herdt, R.W., Barker, R. and De Datta, S.K. (1979a) A methodology of identifying constraints to high rice yields on farmers' fields. In: *Farm-Level Constraints to High Yields in Asia: 1974–77*. International Rice Research Institute, Los Baños, Philippines.

Gomez, K.A., Lopez, L., Novenario, M.J., Herdt, R.W. and Marciano, V.P. (1979b) Constraints to high yield in Laguna, Philippines. In: *Farm-level Constraints to High Yields in Asia: 1974–77*. International Rice Research Institute, Los Baños, Philippines.

International Rice Research Institute (IRRI) (1979) *Farm-Level Constraints to High Yields in Asia: 1974–77*. IRRI, Los Baños, Philippines.

Minhas, B.S. and Vidyanathan, P. (1965) Growth of crop output in India 1951–54 to 1958–61. *Journal of the Indian Society of Agricultural Statistics*, 17(2).

15 Rice Production Losses from Pests in Indonesia

T. Jatileksono
Department of Agricultural Economics, Faculty of Agriculture, Gadja Mada University, Yogkakarta, Indonesia

I. Introduction

Rice has been the most important agricultural product in Indonesia, comprising about 47% of the food crops added value, or about 27% of the total agricultural added value in recent years (CBS, 1991). Rice is also the major staple food in the country, contributing around 53 and 45% of the total calorie and protein intakes, respectively. The share of rice in the consumption expenditure of the Indonesian household was around 17%, or 28% of the expenditure on food in 1990. Besides, rice production continues to be one of the most important sources of livelihood in the rural areas.

Increasing rice production to achieve self-sufficiency has been given high priority in Indonesia's economic development planning during the last two decades. Dissemination of modern varieties, investment in irrigation, fertilizer subsidies, credit facilities with subsidized interest rates and the establishment of intensive extension services have been the major instruments of the government's rice program. As a result, rice production increased rapidly during the last two decades, and rice self-sufficiency was achieved by 1985. Whether Indonesia will be able to maintain self-sufficiency in rice in the future is a critical policy issue.

The significant contribution of modern varieties (MVs) to increase rice production in Indonesia has been widely recognized and well documented. However, a levelling-off of rice yields has been observed with almost complete adoption of MVs in the lowland areas in recent years. This indicates that a new generation of MVs with a higher potential yield is required to maintain self-sufficiency in the future. The development of biotechnology research and its application to rice farming may be a major opportunity for a further increase in yields.

The objectives of this chapter are (i) to estimate rice production losses due to various insects and diseases at the national level; (ii) to examine the yield loss at the farm level; (iii) to evaluate the contribution of varietal improvement on yield losses; and (iv) to shed light on the effectiveness of the integrated pest management (IPM) program in Indonesia. Section II presents the progress of modern rice varietal improvement and its impact on fertilizer use and land productivity. The results of rice production loss estimation are presented in Section III, and the main findings of the study in Section IV.

II. Varietal Improvement, Fertilizer Use and Productivity

Early MVs of rice, such as IR5 and IR8, were highly susceptible to pests and diseases and major losses occurred occasionally due to epidemic outbreaks. To reduce such production losses, the breeding programs in the IRRI, as well as the national research institutions, have been designed to develop improved MVs resistant to pests and diseases (Khush, 1987). The green revolution was characterized by widespread adoption of early MVs followed by even wider dissemination of the later MVs characterized by multiple pest and disease resistance. Meanwhile, the shorter growth duration of later varieties seems to have brought about higher rice-cropping intensities. Thus, the higher potential yield, lower yield instability and short growth duration are all important components of the green revolution that contributed significantly to the growth of rice production (Otsuka and Gascon, 1992).

In Indonesia, adoption of MVs in the lowlands has increased significantly from 11% of the rice area in 1969/70 to 66% in 1979/80 and to 84% in 1989/90. Expansion and rehabilitation of irrigation facilities, as well as improvements in the characteristics of newer MVs in terms of greater pest resistance, better grain quality, shorter growth duration and more tolerance to adverse physical conditions, have led to a wider diffusion of MVs.

MVs bred at both the IRRI and the Indonesian Rice Research System are currently planted in the country. The improved Indonesian varieties are classified as IMV1 and IMV2. The cross-bred varieties between IR5 and Syntha, a national improved variety, are classified as IMV1, (i.e. Pelita I-1 and Pelita I-2). Cross-bred varieties between IMV1 and other improved varieties are classified as IMV2 (e.g. Cisadane, Cimandiri, Cipunegara, Citarum, Citandui, Krueng Aceh, Semeru and Sadang).

The first-generation IRRI-bred varieties (IR5, IR8, and C4-63) were introduced in the late 1960s and disseminated on up to 17% of the rice area in 1975/76. IMV1, which was released in the early 1970s, spread over a larger area and reached its peak in 1975/76 covering almost 22% of the rice area. IRRI-bred varieties released after IR8, but before IR36, (i.e. IR20 to IR34), were widely grown in the late 1970s – primarily to overcome the outbreak of brown planthopper. In the 1980s, however, all of these varieties were almost completely replaced by the second-generation (IR36) and third-generation (IR64) IRRI-bred varieties, as well as IMV2. IR36 was first introduced in the 1976/77 wet season, and it was rapidly and more widely adopted to cover 37% of the rice area in 1980/81, and continued to be planted over wide areas even up to the mid-1980s. IRRI varieties released after IR36, but before IR64 (i.e. IR38 to IR54), accounted for around 12% of the rice area, and were adopted mostly outside Java. IMV2, which was introduced in the late 1970s, initially spread slowly, but by the early 1980s the adoption rate increased sharply, surpassing the area planted to IR36 by a wide margin. IR64, which was released in 1986/87, appeared to be the most popular of the varieties that have replaced IR36 in recent years. It should be noted that both IR64 and IMV2 not only replaced IR36, but also significantly reduced the area planted with traditional varieties (TVs).

The adoption of IR5, IR8, C4–63 and IMV1 in the early 1970s was complemented by around 90 kg ha^{-1} fertilizer use (urea + TSP). Fertilizer use increased significantly in the late 1970s along with the release of IR36, and it continued to increase sharply as IMV2 and IR64 were more widely adopted in the 1980s. It is important to note that fertilizer use on upland rice had also increased significantly since the early 1980s, though its level was much lower than that used on the lowland rice.

Yields of lowland rice were stagnant at around 1.8 tons ha^{-1} during the 1950s, and at around 1.95 tons ha^{-1} in the early to mid-1960s. With the adoption of first-generation MVs, yields of lowland rice increased to slightly above 2 tons ha^{-1} in the late 1960s. The introduction of IMV1 in the early 1970s raised yields to more than 2.5 tons ha^{-1}, and the release of IR36 stimulated yields to achieve more than 3 tons ha^{-1} in the late 1970s. When IMV2 was introduced in the early 1980s, rice yields sharply increased to above 4 tons ha^{-1}. However, the widespread adoption of IR64 in recent years has not increased yields significantly, and seems to be levelling off at around 4.5 tons ha^{-1}.

Yields of upland rice were also stagnant in the 1950s and 1960s at slightly above 1 ton ha^{1}. Because no appropriate MVs have been developed for upland rice, upland areas are still mostly planted with traditional rice varieties. It should be noted, however, that yields of upland rice rose from around 1.1 tons ha^{-1} in the early 1970s to about 2 tons ha^{-1} in recent years. Observed increases in yields in the lowland areas, therefore, have not been due solely to the adoption of MVs, but also to other factors, such as the decline in fertilizer–rice price ratios as substantial subsidies were provided over time (Jatileksono, 1987; Timmer, 1989; Heytens, 1991).

The increases in rice yields were also due to the more intensive use of labour as MV adoption itself occasionally increased labour use per hectare by increasing labour requirements for crop care and harvesting (Barker and Cordova, 1978). In fact, higher-yielding varieties, the application of large amounts of fertilizer, irrigation, drainage and improved cultivation practices all belong to labour-using 'technology' which has yield-increasing properties at the same time (Ishikawa, 1978).

III. Rice Production Losses

Rice production losses were estimated as the product of damage intensity, area infected by various insects and diseases, and rice yields for each district. The results were aggregated and adjusted to the national level. This study utilizes the annual agricultural survey data published by the Central Bureau of Statistics which is now available for the period 1976–1990, i.e. data on the production of food crops and data on the area and damage intensity caused by insects, diseases and other calamities on paddy areas. Farm-level surveys were also conducted to explore the farmers' estimates of rice yield losses in 1990/91 by interviewing 640 paddy farm households selected from 32 villages, i.e. ten villages in West Java, ten villages in Central Java, ten villages in East Java and two villages in the Yogyakarta Special Region.

Table 15.1 presents the estimates of rice production losses compared to the actual rice production for the period 1976–1990. Total rice production loss was quite high during the 1976–1979 period, averaging 1.34 million tons of paddy per annum, or about 5.5% of total production. The average annual loss of lowland rice was around 1.32 million tons of paddy in 1976–1979. As shown in Table 15.2, this loss was due to attacks of the brown planthopper (40%), rats (17.6%), gall midge (14.6%), stemborer (9.7%), stink bug (8.1%), leaf folder/roller (3.3%) and armyworm (1.4%).

Rice production losses declined substantially as improved varieties spread out in the 1980s, especially IR36, Cisadane, Krueng Aceh and IR64. Estimates for 1986–1990 show that the annual loss of lowland rice was about 287,000 tons of paddy, or only 0.7% of total production. This loss was due to crop damage by rats (33.5%), stemborer (23.7%), leaf folder/roller (12.1%), brown planthopper (7.3%), gall midge (4.5%), stink bug (4.2%), armyworm (2.8%), leaf blight (2.3%), blast (1.5%) and tungro (1.3%).

Table 15.1. Rice production and losses in Indonesia (thousand tons of paddy), 1976–1990.

Year	Rice production Lowland	Upland	Total	Production loss Lowland	Upland	Total	Loss as percent of total production
1976	21,852	1,449	23,301	960	11	971	4.17
1977	21,808	1,539	23,347	1,842	10	1,852	7.93
1978	24,172	1,599	25,772	1,030	22	1,051	4.08
1979	24,732	1,551	26,283	1,442	57	1,499	5.70
1980	27,993	1,659	29,652	428	10	438	1.48
1981	30,989	1,785	32,774	484	10	494	1.51
1982	31,776	1,808	33,584	444	7	450	1.34
1983	33,294	2,009	35,303	421	26	448	1.27
1984	36,017	2,119	38,136	293	15	309	0.81
1985	37,027	2,006	39,033	304	3	307	0.79
1986	37,740	1,987	39,727	231	4	235	0.59
1987	37,970	2,109	40,079	198	16	214	0.53
1988	39,316	2,360	41,676	279	8	286	0.69
1989	42,371	2,354	44,725	296	10	305	0.68
1990	42,825	2,353	45,178	390	8	398	0.88

Source of data: Central Bureau of Statistics, Indonesia.

Rice Production Losses from Pests in Indonesia

Table 15.2. Estimated production loss for the lowlands, 1976–1990.

Constraints	Average for 1976–1979 Production loss (thousand tons)	Percent of total loss	Average for 1986–1990 Production loss (thousand tons)	Percent of total loss
Brown planthopper	527.4	40.0	20.4	7.3
Rodents	231.4	17.6	93.4	33.5
Stemborer	127.4	9.7	66.1	23.7
Gall midge	192.1	14.6	12.5	4.5
Stink bug	106.6	8.1	11.8	4.2
Leaf folder	43.7	3.3	33.7	12.1
Armyworm	18.5	1.4	7.8	2.8
Tungro	8.7	0.7	3.7	1.3
Blast	4.6	3.5	4.1	1.5
Brown spot	8.9	6.7	2.9	1.0
Rice bug	11.5	8.7	2.2	0.8
Bacterial leaf blight	4.2	3.2	6.5	2.3
Sheath blight	3.5	2.7	2.4	0.9
Grassy stunt	9.9	0.8	0.0	0.3
Others	20.6	1.6	10.3	3.7
Total	1,319	100.0	279	100.0

Source: Central Bureau of Statistics, Indonesia.

It should be noted that upland rice production losses were very small (see Table 15.1). The top five major problems that caused 73% of the production losses in upland rice in Indonesia during the 1976–1990 period were the brown planthopper (34%), rodents (13%), stink bug (9%), wild pig (9%) and stemborer (8%).

It may be tempting to relate the substantial decline in production losses in recent years to the IPM program. In November 1986, the IPM program was officially launched in Indonesia. This program has been implemented by: (i) scheduling cropping patterns with rotation of rice varieties; (ii) growing high-yield varieties resistant to pests and diseases; (iii) eradication and sanitation of crop damage; (iv) judicious use of insecticides; (v) pest monitoring; (vi) supervised pest management; and (vii) the implementation of coordinated efforts at the national and regional levels. The most remarkable starting point with IPM was to ban 57 brands of pesticides used for rice. Rice yields have continued to increase slightly in recent years despite the fact that pesticide use has been reduced significantly. But the reduction in production losses from insects and diseases cannot be credited entirely to IPM. It will be noted from Table 15.1 that the reported production losses dropped substantially in 1980, long before the introduction of the IPM. Since then the downward trend in production losses has been only marginal.

This leads one to believe that official reports seriously underestimate the rice production losses in Indonesia. The estimated production loss reported by the Central Bureau of Statistics for the 1980s was only about 350,000 tons, or about 37 kg ha^{-1} of land, which is about one-tenth of the figures estimated for other countries reported in this volume.

The results from the farm-level survey presented in Table 15.3 clearly indicate that farmers' estimates of rice production losses in all the sample villages were much higher than what we have learned from the official statistics. The loss in rice yield due to pests is reported at 18% for the irrigated area, 21% for the uplands and 12% for the rainfed lowlands. Rice research which eliminated or reduced these losses could, therefore substantially increase farmers' yields. The factors behind the yield losses differ across rice ecosystems. The major constraints for the irrigated ecosystems are the stemborer and rodents, which account for 84% of the production losses. In the upland areas, the most serious constraints are grubs, brown planthopper and bacterial leaf blights; these three problems account for 85% of the reported production losses. In the rainfed lowlands most of the production losses are from rodents and the stink bug.

The fact that Indonesian rice farmers have been approaching their potential yields and that rice production losses are at low levels in recent years, suggests that rice research should emphasize the use of biotechnology to increase the potential yield, besides continuing the use of conventional techniques for maintaining resistance in MVs against major pests such as the stemborer, brown planthopper, gall midge and stink bug. The government should take steps to institutionalize infrastructures for the use of advanced research tools for rice research, and train scientists and allocate appropriate resources for this purpose. Chavas and Cox (1992) report that it takes 30 years to fully capture the effects of public research expenditure but the internal rate of return on research investment for increasing agricultural productivity is very high.

Table 15.3. Estimates of rice production losses based on farmers' assessments, 1990/91 (figures in percent of farm-level yield).

Constraints	Irrigated ecosystem	Rainfed lowland ecosystem	Rainfed upland ecosystem
Stemborer	9.9	0.8	0.3
Rodents	5.4	6.1	0.1
Brown planthopper	0.7	0.2	5.9
Stink bug	0.8	2.4	0.9
Bacterial leaf blight	0.7	0.0	1.6
Grub	0.0	0.0	10.2
Others	0.7	2.3	1.9
Total	18.3	11.8	20.9

Source: Survey of 640 households from 32 villages carried out by the author.

IV. Conclusions

The experience of technological change led by varietal improvement in Indonesia has significantly contributed to the growth of rice production and improved grain quality as well. Varietal improvement complemented by government policy to heavily subsidize fertilizers and irrigation have contributed to remarkable rice yield increases with limited production losses, allowing Indonesian farmers to approach the potential yield in most irrigated lowlands. The IPM program started in 1987 seems to have been successful in reducing pesticide use without significant adverse impacts on rice yield. But, some preventive measures should be further developed as rice production loss due to the stemborer has significantly increased in recent years.

Presently, Indonesian rice farmers are really waiting for the next technological breakthrough since the potential yield has been approached, and rice production losses are at low levels in recent years. However, a combined effort might be more beneficial. The most damaging insects and diseases contributing to rice production losses in recent years were rats and the stemborer, followed by the leaf folder/roller, brown planthopper, gall midge, stink bug, armyworm, leaf blight, blast and tungro. It is suggested that biotechnology research on rice growing in Indonesia should give priority to increasing the potential yield, as well as continuing conventional research methods for yield loss prevention.

References

Barker, R. and Cordova, V. (1978) Labor utilization in rice production. In: *Economic Consequences of New Rice Technology.* IRRI, Los Baños, Philippines.

Central Bureau of Statistics (CBS) (1991) *Statistik Indonesia 1991 (Statistical Yearbook of Indonesia 1991).* Jakarta, Indonesia.

Chavas, J-P. and Cox, T.L. (1992) A nonparametric analysis of the influence of research on agricultural productivity, *American Journal of Agricultural Economics,* 74(3), 583–591.

Heytens, P. (1991) Technical change in wetland rice agriculture. In: Pearson, S., Falcon, W., Heytens, P., Monke, E. and Naylor, R. (eds) *Rice Policy in Indonesia.* Cornell University Press, Ithaca, New York.

Ishikawa, S. (1978) *Labor absorption in Asian Agriculture: An Issue.* ILO–ARTEP, Bangkok.

Jatileksono, T. (1987) *Equity Achievement in the Indonesian Rice Economy.* Gadjah Mada University Press, Yogyakarta.

Khush, G.S. (1987) Rice breeding: past, present, and future. *Journal of Genetics,* 66, 195–216.

Otsuka, K. and Gascon, F. (1992) Two decades of green revolution in Central Luzon: A study of technology adoption and productivity changes. *Southeast Asian Journal of Agricultural Economics,* 1(1), 45–62.

Timmer, C.P. (1989) Indonesia: transition from food importer to exporter. In: Sicular, T. (ed.) *Food Price Policy in Asia.* Cornell University Press, Ithaca, New York.

III Crop-Loss Studies

16 Technical Issues in Using Crop-Loss Data for Research Prioritization

P.S. Teng and I.M. Revilla
Cross Ecosystem Program, Entomology and Plant Pathology Division, International Rice Research Institute, PO Box 933, 1099 Manila, Philippines

I. Introduction

Research prioritization must inevitably take into account the magnitude and frequency of factors that limit productivity, among which crop losses caused by biotic and abiotic stresses are the most common. The techniques for crop-loss assessment have themselves been the subject for much research investment, although this has not always resulted in more accurate or extensive estimates of losses. There is a gap between the availability of loss-assessment techniques and the actual application of these to determine regional-level losses. In the rice-growing countries of Asia, research institutions are woefully ill-equipped to conduct the surveys needed to determine regional pest prevalence and loss, while the extension and/or development agencies with capacity to do so have not considered this a priority activity. Accurate loss estimates furthermore require much investment of effort for data to be collected on a regular basis. Nevertheless, there is as yet no better basis for priority setting than crop losses determined under real-world situations.

The methodology for loss assessment has been previously reviewed (IRRI, 1990) and the intention in this chapter is to: (i) provide a framework in which to interpret crop-loss estimates from the literature and those generated in the country studies reported in Chapters 7–15 in this volume; (ii) to provide baseline information on the variability in crop losses reported for different pests; and (iii) review the rankings of importance of key pests found in Asia.

II. Interpretation of Crop-Loss Data

A. A conceptual framework for using crop-loss data

Crop loss is always estimated relative to a reference yield and it is therefore important that for studies to provide comparable results, this reference yield be clearly determined. Biological scientists have commonly defined four yield levels when considering constraints – potential (maximum attainable, theoretical), attainable, economic and actual (Fig. 16.1) (Zadoks, 1990; Nutter *et al.*, 1993).

Potential yield is the highest possible, unlimited yield when crops are grown constraint-free and is the fullest expression of a specific crop genotype for a particular environment. We have considered this, for practical purposes, to be equivalent to the maximum attainable, in which crops are grown under optimal conditions.

Attainable yield is obtained using all available technology to minimize biotic stress under experimental conditions and with no consideration of costs. Attainable yield is commonly less than, or at best equal to, potential yield.

Economic yield is obtained at affordable management costs, in which the farmer implicitly attempts to optimize his output/input ratio. It is commonly lower than and at best equal to, attainable yield.

Actual yield is that obtained by farmers at the level of management and technology available, and in the presence of stresses.

In the context of the estimates used in country studies in this volume, yield level 1 is equivalent to actual yield, yield level 2 is equivalent to attainable yield and yield level 3 to potential yield (see Fig. 16.1).

Crop loss is commonly calculated by researchers as the difference between attainable yield and actual yield. This gap is represented by the effects of loss-reducing factors such as biotic and abiotic stresses. Under farming conditions, crop loss is more realistically the difference between economic yield and actual yield since the economic yield is the attainable yield adjusted for profit maximization given the output and input prices, and also for economic risks in rice cultivation due to the biotic and abiotic stresses. The agreement between these

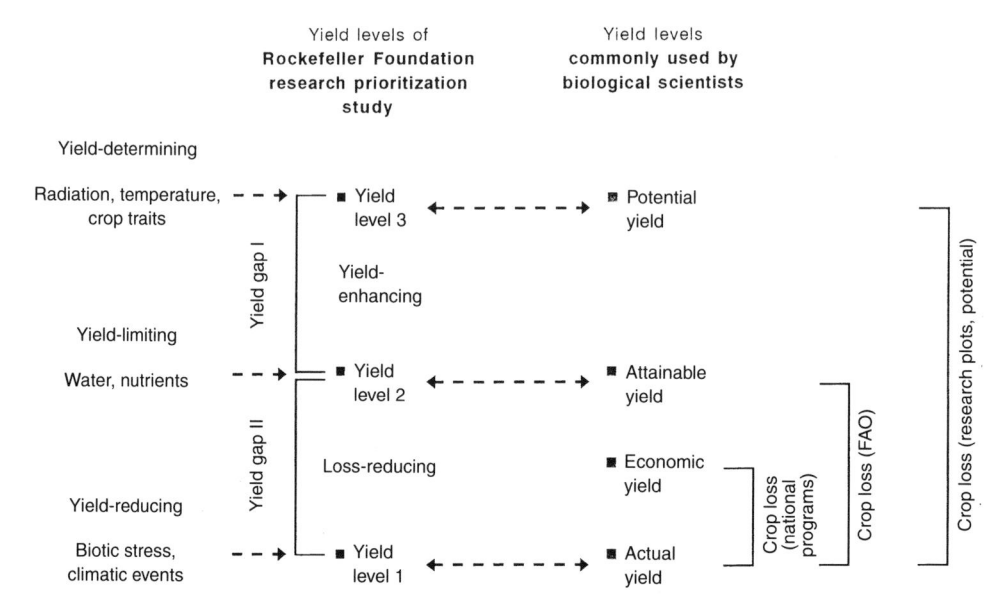

Fig. 16.1. Schematic diagram of different yield and loss concepts.

definitions of yield gaps and those understood by agroecologists is best summed up in the system proposed by de Wit (1989), in which factors of production are categorized as yield-reducing (biotic, abiotic stress), yield-limiting (water, nutrients) and yield-determining (crop genotype, radiation, temperature). The difference between yield level 1 and yield level 2 then becomes the crop loss gap, or yield gap II, and is accounted for by loss-reducing factors (see Fig. 16.1). Yield-limiting factors in yield gap I determine the attainable yield. Computer simulation of loss caused by specific pests tends to produce loss figures which are the difference between the potential and actual yields. Based on the above, it would therefore be expected that for the same pest, potential loss (the difference between potential and actual yields) is higher than attainable loss (the difference between attainable and actual yields) which in turn is higher than economic loss (the difference between economic and actual yields).

Most of the data reported by extension agencies in national programs are economic loss figures while researchers have generally published attainable losses. This distinction is important because of the role that prevailing practices and technologies have on actual and economic yield. Many loss estimates from farmers' fields occur in spite of the presence of host plant resistance or pesticides and, therefore, may lead to an underestimate of the importance of a pest.

B. Issues affecting interpretation of crop-loss data

Limited time periods

The ideal set of crop-loss data would be time series data over, at best, 10 years, in which all major biotic and abiotic stresses are quantified into production constraints and partitioned among the major rice ecosystems in a country. Such a dataset does not currently exist for any rice-growing country. Most of the data available are those reported for one season, 1 year or at best up to 3 years. The positive side, however, is that while crop-loss data *per se* do not exist, there is a lot of data available on area affected over relatively long time periods in several countries. This could be used to corroborate the shorter-period crop-loss data.

Limited area covered

Crop-loss data are often not available over a large, representative area to enable national or regional averaging. Furthermore, some pests only cause localized damage, which although it may amount to 100% loss, becomes relatively unimportant when compared to national rice production. This is the case with pest damage such as that caused by tungro. Research prioritization needs to recognize that there are losses which occur over large areas at low levels every year as well as losses which occur infrequently in small areas at high levels.

Limited scale data

Crop losses are often calculated using empirical damage functions derived from plot experiments on research stations. Extrapolating their use to farmers' fields requires that proper accounting be done for scale-dependent phenomena such as interplant compensation, interhill compensation, etc. (Teng, 1990). As indicated previously, the reference yield for calculating losses may also differ between researcher plots and farmers' fields and there is a need to equilibrate these.

Externalities and internalities

Crop losses in farmers' fields occur under a set of technological, environmental and management conditions. For any period, deployment of host plant resistance and pesticides exerts a strong influence on the magnitude of losses. Crop losses estimated for one period and used to interpret trends in another period further assume that the variability in environmental conditions for both periods is the same, e.g. there is no significant difference in the weather or soil. Similar assumptions are made about pests and beneficial insect populations. A major change in management in some localities, such as in the use of integrated pest management, may dramatically change the level of losses in that locality. There is, therefore, a need to qualify each set of crop-loss data with the prevailing conditions in the ecosystem.

III. General Status of Crop Losses

It is necessary to distinguish between the potential losses that a pest can cause and the actual field losses that are observed. The first represents the maximum losses any pest is capable of causing, given the optimum conditions for its increase. The second represents the losses under different scenarios of technology such as presence of host plant resistance, use of pesticides and use of other pest management techniques.

Few current estimates of actual field losses caused by pests in farmers' fields have been obtained through direct field measurement surveys using scientific experiments. Most data have been generated indirectly using the techniques described by Zadoks and Schein (1979), the most common of which have been survey/interviews of local experts and the use of pesticide trials.

A. *Potential losses estimated empirically*

A pioneering exercise to derive global estimates was done by Cramer (1967), who found losses of the following magnitude: Potential production lost before harvest = 55.1%, consisting of:

- insects 34.4%
- diseases 9.9%
- weeds 10.8%

These are potential losses and indicate a maximum at the locations used. Given the current knowledge on how broad-spectrum insecticides may induce higher levels of pest infestations, it is not surprising that such high levels of potential loss for insects were determined during the period of the study. The estimates for diseases may actually be lower than the current situation as there is now more use of genetically uniform material and increased fertilizer rates.

Cramer's (1967) estimates must, by default, remain the most authoritative, generalized estimates of potential loss, until an attempt is made to involve national crop protection programs in tropical Asia to collect data in a more systematic manner. In a more recent analysis, Ahrens *et al.* (1983) found that, using data from pesticide evaluation trials over twelve years, the losses due to insects in East and Southeast Asia were 23.7%. Litsinger *et al.* (1987) have also estimated that losses due to chronic pests (i.e. non-outbreak levels) were 18.3% of potential production in the Central Luzon region of the Philippines. Again, it is important to note that insecticide-induced losses may in actual practice be low due to increasing acceptance of the integrated pest management concept among rice producers.

B. Potential losses estimated using crop physiological models

The effect of pest infestations on different rice cultivars, grown in different soil types and sites, and under different management practices, may be estimated using physiological crop simulation models to predict pest-free yield and yield under different degrees of pest attack. Models such as the IBSNAT CERES/RICE model and ORYZA1 (Kropff *et al.*, 1994) facilitate the coupling of stress factors, such as pests, to the crop.

For example, Teng (1990) used two approaches to represent leaf defoliation in CERES/RICE. In the first method, the 1–99% severity levels were introduced one at a time throughout all crop stages and the estimated biomass losses of IR50 were in the range of 0.04 to 11.02 millions tons ha^{-1} and grain yield losses of 0 to 4.4 million tons ha^{-1}. (Simulations were done using 1987 IRRI weather data and IRRI pedon soil.) With no pests, the predicted yields of biomass and grain were 11.05 and 4.4 million tons ha^{-1}, respectively, and 10.1 and 4.5 million tons ha^{-1} for the observed yields (data taken from the studies of blast effects on yield components, Plant Pathology Department, IRRI, 1987 WS).

The authors used a second approach, in which a sigmoid curve for disease increase was introduced throughout crop growth, and the resulting crop loss was then estimated. Simulation of the variety IR50 at different rates of disease increase showed that at a given onset time of defoliation, a higher disease progress curve caused a decrease in yield, which was represented by a decrease in panicle weight per unit area. Generally, the onset time of a disease, its rate of development and its duration are needed before yield loss can be estimated. The authors' results (Teng, 1990) also showed that early season defoliation (up to 40 days after sowing (DAS)) could not result in any yield loss. This rule has now been empirically evaluated by entomologists and is the basis for a no-spray recommendation during this period.

It is anticipated that the direct effects of specific pests such as the yellow stemborer, brown planthopper, sheath blight, blast and tungro will be quantifiable in the near future using crop models.

C. Actual farm-level losses

Some reported loss estimates of tropical rice are, for insects, 35–44% worldwide (Pathak and Dhaliwal, 1981), 35% in India (Way, 1976), and 16–30% in the Philippines (Way, 1976). Alam (1961) was of the opinion that in Bangladesh during 1951, 6% of the total rice crop was lost to insects. Fernando (1966) estimated average annual losses due to insects in Sri Lanka at 10–20%. Weed losses in the Philippines were estimated by Moody (1982) to range from 11–15% to 65%. No recent estimates of total losses due to diseases are available.

Loss estimates in farmers' fields for pests that have been reported were recently reviewed by Teng (1990), Litsinger (1991) and Teng and Revilla (1994) (see Appendix 1); a summary is provided here. These estimates, of course, relate to fields and regions of severe infestation or outbreaks rather than for all rice areas in any specific country. There is no case of any country suffering a 95% or even a 33% production loss in one year compared to an earlier year.

Stemborers

The rice stemborers have been considered by many entomologists to be the most serious of the insect pests (Barr *et al.*, 1975), and consist of a group of some five species. Some loss estimates are as follows:

Bangladesh	
(Outbreak year)	30–70% (Alam *et al.*, 1972);
(Non-outbreak year)	3–20% (Alam, 1967);
India	3–95% (Ghose *et al.*, 1960);
Indonesia	Up to 95% (Soenardi, 1967).
Malaysia (North Krian district)	33% (Wyatt, 1957).

Leaf and plant hoppers

It is difficult to separate the losses caused by the leafhoppers alone from the losses caused by the viruses that they transmit. Some estimates of losses are:

Bangladesh (leafhoppers)	50–80% (Alam, 1967);
Malaysia (brown planthopper)	M$10 million (Lim *et al.*, 1980);
India (brown planthopper)	1.1–32.5% (Jayaraj *et al.*, 1974).

Other insects

Apart from the stemborers and hoppers, little consistent data exist to give average losses from other insect pests. Rice bugs (*Leptocorisa* spp.) were reported in India to have caused a 10% loss in some 3 million ha during 1952 (Pruthi, 1953). According to Reddy (1967), larvae of the gall midge (*Pachydiplosis oryzae*) have occurred in outbreak levels in some years, causing losses of 12–35% in India (1934), 50–100% in Vietnam (1922) and severe losses in Sri Lanka (1951) and Burma (1934).

The rice hispa (*Dicladispa armigera*) has been reported to cause losses of 10–65% in Bangladesh, and about 10,000 ha in Bihar, India commonly suffer up to 50% loss (Barr *et al.*, 1975). Of the remaining rice pests, leaf folders can cause field losses of up to 50% (Balasubramanian *et al.*, 1973). Armyworms were reported to have devastated about 10,000 ha of rice in Malaysia in 1967 (Dunsmore, 1970).

Blast

Although blast is generally considered an important disease, capable of causing very severe losses of up to 100%, little information exists on the extent and intensity of actual losses in farmers' fields. According to Padmanabhan (1965), during 1960/61, some states in India suffered a 1% overall loss, with a range from 5 to 10%. In temperate rice environments such as Japan and Korea, losses due to blast have reportedly been 3% for Japan (1953–1960) and in epidemic levels in Korea (mid-1970s), in spite of extensive fungicide use. In the Peoples' Republic of China, losses due to blast in 1980 and 1981 were estimated at 8.4% and 14.0% respectively (Teng, 1986). The yield loss was estimated ranging from 50 to 60% in several thousand hectares of land planted to Peta in the Leyte and Southern Leyte provinces in 1963. In another blast outbreak in 1969–1970 in Laguna and Quezon provinces, yield loss on cultivars BPI-76 and C4-63 ranged from 70 to 85%.

Brown spot

The 'Great Bengal Famine' of 1942/43 in India was attributed to an epidemic of this disease, in which up to 80% yield reductions were probably common (Padmanabhan, 1973). More recent epidemics in India have resulted in 14–41% losses in high-yielding varieties (Vidhyasekaran and Ramadoss, 1973). Because the disease is associated with adverse soil conditions, it is not always possible to partition yield losses due to the disease from other factors which reduce yield.

Tungro

In recent years, rice tungro virus (RTV) has become a problem of concern in many tropical environments because of the potential of the disease to cause total loss, and the lack of effective corrective measures once symptoms are observed by farmers. Following outbreaks of the disease in Malaysia, surveys showed that the disease caused only localized damage, and when losses were averaged over production regions the average loss was less than 1% during 1981–1984 (Heong and Ho, 1986). Moi and Timin reported that from 1986 to 1989, the area affected with RTV was between 330 and 740 ha. However, in 1990, 1800 ha was sporadically infected with this disease and this warranted quick action. Subsequently, the hectarage was greatly reduced to only 67 ha in 1992. Chang *et al.* (1985) estimated RTV-induced losses at Malaysian (M)$ 21.6 million from an affected area of 17,628 ha in 1982. RTV has caused total crop losses in parts of Indonesia (244,904 ha in 1969–1992), 40–60% losses in Bangladesh and about 50% in parts of Thailand (Wathanakul and Weerapat, 1969; Reddy, 1973). In the Philippines 'aksip na pula' or 'red disease' (Serrano, 1957), probably RTV, caused annual losses of about 30% – an equivalent to 1.4 million tons of rough rice every year during the 1940s. In 1971, yield losses due to tungro were estimated at 456,000 tons of rough rice.

Bacterial blight

Bacterial blight is one of the major disease of rice in many rice-growing areas of the world. The losses induced by the disease have been related to the increased use of nitrogen-responsive and high-yielding varieties in some countries. In Japan, 300,000 to 400,000 ha of rice land have been affected annually in recent years. Losses in severely infected fields range from 20 to 30% and, occasionally, 50%. In tropical Asia such as India, Indonesia, the Philippines and others, losses are higher than in Japan. However, figures on yield losses are very scanty. Losses in yield vary from 6 to 60% in some states of India (Srivastava, 1972). Losses in yield range from 6 to 7% for IR20, 58% for TN1 and 74% for Bala (Rao and Kauffman, 1977). In mainland China, losses due to the disease were estimated at 6% in 1980 and 4.9% in 1981 (Teng, 1986).

Sheath blight

Sheath blight has assumed economic importance in the last two decades when modern, semidwarf nitrogen-responsive cultivars have been commercially grown. Information on the actual yield losses induced by the disease in farmer's fields are very few. A loss of 24,000–38,000 tons of rice annually in Japan was estimated by the National Institute of Agricultural Sciences (1954). Yield reductions equivalent to 20% (Mizuta, 1956) or 25% (Hori, 1969) may be incurred if the disease develops or reaches the uppermost flag leaves. Losses of 7.5–22.7% in high nitrogen plots have been reported (Ou and Bandong, 1976) in a susceptible variety and 0.4–8.8% and 2.5–13.2% losses, respectively, reported for moderately resistant varieties. Field prevalence of sheath blight was estimated at about 10% of rice tillers in one district of Sri Lanka (Abeygunawardena, 1966). Data from mainland China suggest that average losses due to sheath blight were 12.6% in 1980 and 9.1% in 1981 (Teng, 1986).

Other Diseases

A lot less data are available on field losses caused by other diseases, although it can be generalized that most pathogens, given favourable conditions, have the potential to cause severe losses. Of the remaining important diseases in tropical rice, stem rot has been reported to cause average losses of 5–10% annually in parts of India (Chauhan et al., 1968). In the Philippines, losses have been estimated at 30–80% in Tarlac province (Hernandez, 1973). In Arkansas, USA the annual average yield loss was about 16,000–35,000 tons (Ou, 1985).

Nematodes

Literature on yield losses due to nematodes is scanty. Losses due to white tip in Japan have been estimated at 30–35% (Yoshii and Yamamoto, 1950) and 40–50% on artificially inoculated susceptible cultivars in the USA (Atkins and Todd, 1959). Reduction in grain yields in Taiwan range from 29 to 46% in ten cultivars surveyed (Hung, 1959).

Yield losses due to ufra or stem nematodes in limited areas have been estimated at 50% in Uttar Pradesh, India (Singh, 1953), 20–90% in Thailand (Hashioke, 1963) and 30% in West Bengal, India (Pal, 1970).

Weeds

Almost no information exists on the losses caused by individual weed species.

D. Relative importance of pests

The relative importance of pests changes over time in response to many factors (Geddes, 1990). Current assessment of pest importance in Indonesia as a whole has shown that rats are the most important pre-harvest pest. The most important non-weed pests of rice were rats, stemborers, bacterial leaf blight and brown planthopper while the most important weeds were *Echinochloa crus-galli* and *Monochoria vaginalis* (Geddes, 1992). In South Asia, ranking estimates showed that the most serious rice pests were rice blast, yellow stemborer, bacterial leaf blight and brown planthopper (Geddes and Iles, 1991).

IV. Conclusions

Crop-loss data that have been collected using acceptable methodology are essential to guide decisions at different levels of government, from the national policy level to the village level. The justification of, and evaluation of, plant protection programs often relies on availability of loss data, yet little effort has been spent to ensure that these data are adequately collected. Apart from their usefulness in decisionmaking, the study of pest effects on crop yield should be a component of pest management programs, since many control actions are based on knowing the pest intensity–yield loss relationship. Furthermore, developments in pest and crop modelling offer new opportunities for understanding pest–loss effects, and for estimating locality-dependent crop responses to pest infestations.

References

Abeygunawardena, D.V.W. (1966) Recent investigations in the control of rice diseases. In: Abeygunawardena, D.V.W. (ed.) *Proceedings of a Symposium on Research and Production of Rice in Ceylon*. Ceylon Association of Advanced Science, Colombo, pp. 121–131.

Ahrens, C., Cramer, H.H., Mogk, M. and Peschel, H. (1983) Economic impact of crop losses. In: *Proceedings of 10th International Congress of Plant Protection*. British Crop Protection Council, Croydon, UK. pp. 65–73.

Alam, M.Z. (1961) *Insect Pests of Rice in East Pakistan and Their Control*. East Pakistan Department of Agriculture, Dacca, Pakistan.

Alam, M.Z. (1967) Insect pests of rice in Pakistan. In: Department of Ariculture, Dacca, Pakistan. *The Major Insect? Pests of the Rice Plant.* IRRI, Manila, Philippines, pp. 634–655.

Alam, M.Z., Alam, M.S. and Abbas, M. (1972) Status of different stem borers as pests of rice in Bangladesh. *International Rice Community Newsletter,* 21(2): 729.

Atkins, J.G. and Todd, H.E. (1959) White tip disease of rice. II. Yield tests and varietal resistance. *Journal of Plant Protection* 49, 189–191.

Balasubramanian, G., Saravanabhavanandam, M. and Subramanian, T.R. (1973) Control of the rice leaf roller *Cnaphalocrocis medinalis* Guenee. *Madras Agricultural Journal* 58, 717–718.

Barr, B.A., Koehler, C.S. and Smith, R.F. (1975) *Crop Losses – Rice: Field Losses to Insects, Diseases, Weeds, and Other Pests.* UC/AID Pest Management and Related Environmental Protection Project, University of California, Berkeley.

Chang, P.M., Hashim, H., Omar, O., Abdullah, S. and Amin S. Mohd. (1985) Penyakit merah virus disease of paddy – the problem and strategy for disease control in Malaysia. In: Lee, B.S., Loke, W.H. and Heong, K.L. (eds) *Integrated Pest Management in Malaysia.* Malaysian Plant Protection Society (MAPPS) (MARDI), pp. 159–182.

Chauhan, L.S., Verman, S.C. and Bajpai, G.K. (1968) Assessment of losses due to stem rot of rice caused by *Sclerotium oryzae. Plant Disease Reporter* 52, 963–965.

Cramer, H.H. (1967) *Plant Protection and World Crop Production.* Pflanzenschutz-Nachricten. Crop Protection Advisory Department of Farbenfabriken Bayer AG-Leverkusen. 524 pp.

deWit, C.T. and Rabbinge, R. (1989) In: Rabbinge, R., Ward, S.A. and van Laar, H.H. (eds) *Simulation and Systems Management in Crop Protection.* Wageningen, Pudoc. pp. 3–15.

Dunsmore, J.R. (1970) Investigations on the varieties, pests and diseases of upland rice (hill padi) in Sarawak, Malaysia. *International Rice Community Newsletter* 19(1), 29–35.

Fernando, H.E. (1966) Recent developments in paddy pest control. In: Abeygunawardena, D.V.W. (ed.) *Proceedings of a Symposium on Research and Production of Rice in Ceylon.* Ceylon Association of Advanced Science. pp. 103–120.

Geddes, A.M.W. (1990) *The Relative Importance of Crop Pests in Sub-Saharan Africa.* Bulletin no. 36, Natural Resources Institute, Chatham, UK, 68 pp.

Geddes, A.M.W. (1992) *The Relative Importance of Pre-harvest Crop Pests in Indonesia.* Bulletin no. 47, Natural Resources Institute, Chatham, UK, 70 pp.

Geddes, A.M.W. and Iles, M. (1991) *The Relative Importance of Crop Pests in South Asia.* Bulletin no. 39, Natural Resources Institute, Chatham, UK, 102 + vi pp.

Ghose, R.L.M., Ghatge, M.B. and Subrahmanyan, V. (1960) *Rice in India.* India Council of Agricultural Research, New Delhi.

Goto, K. (1963) Estimating losses from rice blast in Japan. In The Rice Blast Disease. The John Hopkins Press, Baltimore, Maryland, pp. 195–202.

Hashioke, Y. (1963) The rice stem nematode *Ditylenchus angustus* in Thailand. *FAO Plant Protection Bulletin* 11, 97–102.

Heong, K.L. and Ho, N.K. (1986) Farmers' perceptions of the rice tungro virus problem in the Muda irrigation scheme, Malaysia. In: Tait, J., Bottrell, D. and Napompeth, B. (eds) *Integrated Pest Management: Farmers' Perceptions and Practices.* Westview Press, Boulder, Colorado.

Hernandez, A. (1973) *Report of Plant Disease Section.* Report No. 23. Philippines Bureau of Agriculture, Quezon City, Philippines.

Hori, M. (1969) On forecasting the damage due to sheath blight of rice plants and the critical point for judging the necessity of chemical control of the disease. *Review of Plant Protection Research* 2, 70–73.

Hung, Y.P. (1959) White tip disease of rice in Taiwan. *Plant Protection Bulletin, Taiwan* 1(4), 1–4.

International Rice Research Institute (IRRI) (1990) *Crop Loss Assessment in Rice.* IRRI, Manila, Philippines.

Jatileksono, T. (1991) Rice variety improvement and yield loss assessment: the case of Indonesia. Paper presented at the Workshop on Rice Research Prioritization held at the International Rice Research Institute, Philippines, 13–15 August 1991.

Jayaraj, S., Velayutham, B., Rathinasamy, C. and Regupathy, A. (1974) Damage potential of the brown planthopper on rice. *Madras Agricultural Journal* 61, 144–148.

Karim, A.N.M.R. (1987) The hispa episode. In: Proceedings of a Workshop on Experience with Modern Rice Cultivation in Bangladesh held at the Bangladesh Rice Research Institute, Joydebphur, Gazipur, 5–7 April 1987.

Karim, A.N.M.R. (1991) Insect pests and diseases: the sources of risks in crop production. Paper presented at the National Workshop on Risk Management in Bangladesh Agricultural Research Council, Dhaka, 24–27 August 1991.

Kropff, M.J., van Laar, H.H. and Mathews, R.B. (1994) ORYZA1: an ecophysiological model for irrigated rice production. Research Institute for Agrobiology and Soil Fertility, Wageningen Department of Theoretical Production Ecology, IRRI.

Lim, G.S., Ooi, A.C.P. and Koh, A.K. (1980) Brown planthopper outbreaks and associated yield losses in Malaysia. *International Rice Research Institute Newsletter* 5, 15–16.

Lin, J.Y. (1991) Rice production constraints in Zhejiang province: a pilot study for agricultural research priority in China. Paper presented at the workshop on Rice Research Prioritization held at the International Rice Research Institute, Philippines, 13–15 August 1991.

Litsinger, J.L. (1991) Crop loss assessment in rice. In: Heinrichs, E.A. and Miller, T.A. (eds) *Rice Insects: Management Strategies.* Springer Verlag, New York, pp. 1–65.

Litsinger, J.A., Canapi, B.L., Bandong, J.P., dela Cruz, C.G., Apostol, RF, Pantua, P.C., Lumaban, M.D., Alviola, A.L., Raymundo, F., Libetario, E.M., Loevinsohn, M.E. and Joshi, R.C. (1987) *Rice Crop Loss from Insect Pests in Wetland and Dryland Environments of Asia with Emphasis on the Philippines.* Entomology Department, International Rice Research Institute, Los Baños, Philippines (mimeographed).

Mizuta, H. (1956) On the relation between yield and inoculation times of sheath blight, *Corticium sasakii* in the earlier planted paddy rice. *Association of Plant Protection, Kyushu.* 2, 100–102.

Mogi, Z. (1992) Bacterial red stripe disease in Indonesia. Paper presented at the International Rice Research Conference held at the International Rice Research Institute, Philippines, 21–25 April 1992.

Moody, K. (1982) The status of weed control in rice in Asia. *FAO Plant Protection Bulletin* 30, 119–123.

National Institute of Agricultural Sciences, Japan (1954) Insects and diseases of rice plants in Japan. Paper presented at the 4th Session, International Rice Commission, Tokyo.

Ou, S.H. (1985) *Rice Diseases*, 2nd edn. Commonwealth Mycological Institute, Kew, Surrey, UK.

Ou, S.H. and Bandong, J.M. (1976) Yield losses due to sheath blight of rice. *International Rice Research Newsletter* (76), 14.

Padmanabhan, S.Y. (1963) Estimating losses from rice blast in India. In: *The Rice Blast Disease*. The John Hopkins Press, Baltimore, Maryland. pp. 203–221.

Padmanabhan, S.Y. (1965) Estimating losses from rice blast in India. In: *The Rice Blast Disease. Proceedings of a Symposium at the International Rice Research Institute, July 1963*. Johns Hopkins Press, Baltimore, pp. 203–221.

Padmanabhan, S.Y. (1973) The great Bengal famine. *Annual Review of Phytopathology* 11, 11–26.

Pal, A.K. (1970) Survey on nematodes of paddy and successful treatment with hexadrin. In: *Proceedings of Indian Sciences Congress Association* 57(III). p. 496.

Pathak, P.K. and Dhaliwal, G.S. (1981) Trends and strategies for rice pest problems in tropical Asia. *IRRI Research Paper Series* 64, 1–15.

Pruthi, H.S. (1953) An epidemic of rice bug in India. *FAO Plant Protection Bulletin* 1, 87–88.

Rao, P.S. and Kauffman, H.E. (1977) Potential yield losses in dwarf rice varieties due to bacterial blight in India. *Phytopathologische Zeitschrift* 90, 281–284.

Reddy, A.P.K., MacKenzie, D.R., Rouse, D.I. and Rao, A.V. (1979) Relationship of bacterial leaf blight severity to grain yield of rice. *Phytopathology* 69, 967–969.

Reddy, D.B. (1967) The rice gall midge *Pachydiplosis oryzae* (Wood-Mason). In: *The Major Insect Pests of the Rice Plant*. Proceedings of a Symposium at the International Rice Research Institute, September 1964. The Johns Hopkins Press, Baltimore, pp. 457–491.

Reddy, D.B. (1973) High yielding varieties and special plant protection problems with particular reference to tungro virus of rice. *International Rice Community Newsletter* 22(2), 34–42.

Reyes, T.T. (1989) Selected economically important diseases of some major crops in the Philippines. In: *Crop Losses Due to Disease Outbreaks in the Tropics and Countermeasures*. Tropical Agriculture Research Series No. 22. Tropical Agriculture Research Center, Ministry of Agriculture, Forestry and Fisheries, Tsukuba, Ibaraki, Japan. pp. 11–20.

Serrano, F.B. (1957) Rice 'aksip na pula' or stunt disease – a serious menace to the Philippine rice industry. *Philippine Journal of Science* 86, 203–230.

Shahjahan, A.K.M. (1993) *Practical Approaches to Crop Pest and Disease Management in Bangladesh*. Bangladesh Agricultural Research Council, Dhaka, 168 pp.

Shen, M. and Lin, J.Y. (1994) The economic impact of rice blast disease in China. Proceedings of the International Blast Symposium, Wisconsin. (mimeographed)

Singh, D., Sardan, M.G. and Khosia, R.K. (1971) Estimates of incidence of diseases and consequent field losses in yield of paddy crop. *Indian Phytopathology* 24, 446–456.

Singh, L.N. (1953) Some important diseases of paddy. *Agriculture and Animal Husbandry in India* 3(10–12), 27–30.

Soenardi, I. (1967) Insect pests of rice in Indonesia. In: *The Major Insect Pests of the Rice Plant*. Proceedings of a Symposium at the International Rice Research Institute, September 1964. Johns Hopkins Press, Baltimore, pp. 675–683.

Srivastava, D.N. (1972) Bacterial blight of rice. *Indian Phytopathology* 25, 1–16.

Teng, P.S. (1986) Crop loss appraisals in the tropics. *Journal of Plant Protection in the Tropics*, 3(39).

Teng, P.S. (1990) Crop loss assessment: a review of representative approaches and current technology. In: *Crop Loss Assessment in Rice*. International Rice Research Institute, Los Baños, Philippines, pp. 19–38.

Teng, P.S. and Revilla, I.M. (1994) Discussion notes: crop losses due to diseases and insects. Paper presented at the Workshop on Rice Research Prioritization in Asia held at the International Rice Research Institute, Philippines, 21–22 February 1994.

Vidhyasekaran, P. and Ramadoss, N. (1973) Quantitative and qualitative losses in paddy due to helminthosporiose epidemic. *Indian Phytopathology* 26, 479–484.

Waibel, H. (1986) *The Economics of Integrated Pest Control in Irrigated Rice: a Case Study from the Philippines*. Springer-Verlag, Berlin, 196 pp.

Wathanakul, L. and Weerapat, P. (1969) Virus diseases of rice in Thailand. In: *The Virus Diseases of the Rice Plant*. Proceedings of a Symposium at the International Rice Research Institute, April 1967. Johns Hopkins Press, Baltimore, pp. 79–85.

Way, M.J. (1976) Entomology and the world food situation. *Bulletin of the Entomological Society of America* 22, 125–129.

Wyatt, I.J. (1957) *Field Investigations of Padi Stem-borers, 1955–1956*. Department of Agriculture, Federation of Malaya, Kuala Lumpur.

Yoshii, H. and Yamamoto, S. (1950) A rice nematode disease 'Senchu Singare Biyo'. I. Symptoms and pathogenic nematode. II. Hibernation of *Aphelenchoides oryzae*. III. Infection course of the present disease. IV. Prevention of the present disease. *Journal of the Faculty of Agriculture, Kyushu University* 9, 209–222 223–233 289–292 293–310.

Zadoks, J.C. and Schein, R.D. (1979) *Epidemiology and Plant Disease*. Oxford University Press, Oxford.

Appendix 1. Summary of Major Rice Insect and Disease Losses in the Field Reported from Various Sources

Country	Reference	Year		Insect/disease	Loss	Unit
China	Lin, 1991	1980–90				
			Early rice	Striped stemborer	17.09	kg ha^{-1}
				Whitebacked planthopper	15.69	kg ha^{-1}
			Late rice	Brown planthopper	34.06	kg ha^{-1}
				Whitebacked planthopper	23.86	kg ha^{-1}
			Early rice	Sheath blight	146.79	kg ha^{-1}
				Blast	29.86	kg ha^{-1}
				Bacterial leaf blight	12.14	kg ha^{-1}
			Late rice	Sheath blight	65.04	kg ha^{-1}
				Blast	38.22	kg ha^{-1}
				Bacterial leaf blight	25.18	kg ha^{-1}
	Shen and Lin	1980–90		Sclerotial blight	35.55	kg ha^{-1}
				Blast	23.01	kg ha^{-1}
Indonesia	Jatileksono, 1991	1976–89				
			Lowland rice	Brown planthopper	20.1	kg ha^{-1}
				Stemborer	8.9	kg ha^{-1}
				Gall midge	7	kg ha^{-1}
			Upland rice	Brown planthopper	4.4	kg ha^{-1}
			Lowland rice	Tungro	<1	kg ha^{-1}
				Blast	<1	kg ha^{-1}
			Upland rice	Blast	<1	kg ha^{-1}
				Brown spot	<1	kg ha^{-1}
	Mogi, 1992	1987		Bacterial red stripe	16–72	%

Country	Reference	Year	Insect/disease	Loss	Unit
Philippines	Reyes, 1989	1940	Tungro	30	%
	Waibel, 1986	1971–81	Tungro	36	%
India	Karim and Saxena, 1990				
		1914	*Nephotettix* sp.	25.40	%
		1952	*Nephotettix* sp.	25	%
	Padmanabhan, 1963	1960–61	Blast	0.80	%
	Vidhayasekaran and Ramadoss, 1973	1970–71	Helminthosporiose	14.2–40.6	%
	Singh et al., 1971	1959–62	Helminthosporiose	12.9	%
	Reddy et al., 1979		Bacterial leaf blight	2–74	%
Bangladesh	Shahjahan, 1993		Stemborer	13–26 (30–70% in epidemic year)	%
			Brown planthopper	20–43	%
			Tungro	up to 70	%
			Bacterial leaf blight	10–30	%
			Blast	up to 80	%
			Ufra	up to 100	%
			Sheath blight	15–35	%
			Sheath rot	up to 70	%
			Stem rot	10–25	%
			Leaf scald	10–20	%
			Hispa	14–62	%
	Year Book of Agricultural Statistics, Bangladesh (various issues)	1989–90	All insects	15.8	%
			All diseases	9.9	%
	Alam et al., 1985	Boro season	Stemborer	41–85	%
	Karim, 1986	Boro season	Hispa	14–62	%
	Karim, 1991	1977–79			
		Boro season	Major insect pests	13	%
		Aus season		24	%
		T. Aman season		18	%
		T. Aman season		22–26	%
		Aus season	Stemborer & other minor pests	15	%
		Boro season	Hispa	14–62	%
		Boro season	Brown planthopper	20–44	%
	Alam and Karim, 1981		Mealybug	25–30	%
	Karim, 1991	Aus/T. Aman season	Sheath blight	20	%
		All seasons	Stem rot	29.5	%
		T. Aman season	Tungro	55	%
		Aus season	Bacterial leaf blight	7–26	%
		Boro season	Blast	22	%
		Boro season	Seedling blight	98	%
		Boro season	Damping-off	45	%
		T. Aman season	Sheath rot	19–46	%
		Aus season	White-tip	60	%
		Boro season	Backanae	21–34	%
		Aus season	Bacterial leaf streak	13	%

Country	Reference	Year	Insect/disease	Loss	Unit
Japan	Goto, 1963	1953–1960	Stemborer	1.15	%
			Blast	7.3	%
Korea	Anonymous	1985–1987	Blast	36.67	kg ha^{-1}
			Sheath blight	1166.67	kg ha^{-1}
			Bacterial leaf blight	200	kg ha^{-1}
			Viral diseases	43.34	kg ha^{-1}

Source: Teng and Revilla (1994).

17 Priorities for Weed Science Research

K. Moody

Agronomy, Plant Physiology and Agroecology Division, International Rice Research Institute, P.O. Box 933, Manila, Philippines

I. Introduction

Despite the most sophisticated efforts at pest management of modern times, which includes plant breeding and genetic resistance, powerful pesticides, machinery to remove or incorporate infected crop residues, an arsenal of other control measures, and the ability to utilize several techniques simultaneously, pest problems have not been eliminated or brought under complete control. New pest problems continually arise as a result of evolutionary developments in many pest organisms, introduction of new pests to an area, changing cultural practices or crop intensification (Stoskopf, 1985).

While agronomic problems change from country to country, the proliferation of weeds remain a constant problem (Anon., 1988). Weed losses which tend to be more consistent and predictable than losses from insects and diseases must be addressed on an ongoing basis to achieve a profitable yield (Stoskopf, 1985).

Throughout much of Asia, it is impossible to produce rice economically without a well-planned weed control program. Weed control does not change the inherent production capacity of the plant or its biological potential for using nutrients, water and light. It helps to control the production environment, thereby allowing more of the inherent capacity of the plant to express itself in higher yields than otherwise would occur. Thus, the ability of weed control methods to increase output or to contribute to an increase is constrained by the upper level of the plant's potential output (Anon., 1975).

The weed problem persists because of the inability to cope with their great reproductive capacity and massive recycling potential. Another factor contributing to the weed problem is the shift in weed species as a consequence of the control measures applied. Because there are many kinds of weeds with varying germination periods and highly differing life cycles, weed management requires an integrated approach based on a thorough knowledge of the biology and the ecology of the species.

II. Yield Losses

Many factors cause loss of agricultural production and there is little doubt that weeds are of major significance. Herdt (1991) provides data on financial losses from the thirty most limiting biological factors (diseases, insect pests, soil problems, temperature and water-related problems, birds, rodents, and weeds) in rice production and concluded that weeds ranked first in terms of value of production foregone. (See Chapter 21 for a summary of country study evidence.)

To assess losses due to weeds is notoriously difficult because of how weeds interfere with human affairs is very subjective and the value of damage varies greatly from year to year. Smith et al. (1977) highlighted the need to consider not only direct yield and quality losses due to weeds in rice but also indirect losses due to weed control efforts such as the cost of chemicals, mechanical weeding, and labour.

Yield losses due to weeds vary with many factors, such as the type of rice culture, cultivar grown, plant spacing, amount of fertilizer applied, duration and time of weed infestation, weed species, amount of weed growth, cropping season, and ecological and climatic conditions. The cultivated field is maintained in favour of crop plants and a deliberate attempt is made to remove weeds from the field using different techniques. Thus, the nature and magnitude of the weed infestation is largely determined by man.

Since 1950, world rice production has increased remarkably. However, there has not been an accompanying decrease in crop losses due to pests, including weeds, despite an intensification in crop protection measures. Current technology does not utilize procedures for lasting solutions to specific weed problems except as they result fortuitously from the more economical, practical and immediate objectives. Farmers usually fight the same weed problems year after year (ARS National Research Program, 1976). *Echinochloa crus-galli* (L.) P. Beauv. still remains a serious weed in rice despite concerted efforts over the years using different techniques to control it.

The effect of weeds on rice yield is a function of the amount of weeds at the critical time during which the components of yield are determined with number of panicles being most sensitive to weed damage.

While there are numerous reports of yield losses due to uncontrolled weed growth in research station experiments, few attempts have been made to determine the nature and magnitude of yield losses due to weeds in farmers' fields. Losses in the absence of any weed control are unrealistic because most farmers do some weed control. Therefore, a comparison between weeded and unweeded plots overstates the additional benefits of weed control. A more realistic approach is to compare the added benefit from additional weeding compared to the farmers' weed control methods (Moody, 1983). The limited data available show that production losses can reach 30–40% for fields that are poorly weeded (Anon., 1988). In the Philippines, yield increases averaging 15.7% for irrigated transplanted rice, 40.8% for rainfed transplanted rice (Rao and Moody, unpublished), and 62.2% for upland rice (Elliot and Moody, 1987) due to additional weeding in farmers' fields have been reported. Thus, the way in which weed control methods are used by farmers is inadequate to prevent substantial yield losses.

Competition from weeds is greater when rice is seeded into dry soil than when it is wet seeded or transplanted. Weed control is more difficult, the cost is greater, the chances of success are less, the adoption of new technology is slower, risks are greater in dry-seeded and upland rice – rice cultures in which weed control is needed the most (Moody, 1983). In Laos, upland rice has a huge labour requirement for weeding. Families spend up to 200 days for weeding 1 ha of rice or 10–20 days to produce 100 kg of rice (Roder, personal communication).

III. Cost of Weed Control

Weed control for the majority of small-scale farmers is one of the most time-consuming activities. Work requires time (his or her own, the family's or hired), energy (the availability of which is determined by physical strength, endurance, health, and diet) and perhaps the financial or other sources to hire labour. Hence, opportunity costs must be added to those caused by crop loss and the costs of control. If it takes 10 days to weed 1 ha of rice, 30 million days would be required to weed the total rice area in the Philippines. It would take one person over 80,000 years to complete the task. Thus weeding of rice should be considered as a major occupation.

Hand weeding is generally not a very efficient method. Probably 10–20% or more of the plants with 10 cm or more growth are left in the field after weeding. The percentage of smaller weeds remaining in the field is much higher. On the average, the efficacy of this method is not more than 70%. The first weeding operation is done 3–4 weeks after transplanting and needs 25–34 labourers ha^{-1} depending on the weed density. The second weeding is generally done 15–30 days after the first and usually requires 12–25 labourers ha^{-1}. The second weeding operation is needed to pull out the weeds which escaped the first weeding (Negi, 1976).

The cost of direct methods of weed control as a percentage of total cost of production is 6.3% in transplanted rice in Japan (Chisaka and Noda, 1983), 11.1% in transplanted rice in Indonesia, 13% in dry-seeded rice in Bangladesh (Ahmed and Azizul Islam, 1983), and only 2% in rainfed transplanted rice in the Philippines (Rao and Moody, unpublished). In the Muda area of Peninsular Malaysia, the herbicide cost is 5.7% of the total cost of production (Wong, 1992). In the USA, weed control costs were 7% of the crop production value; most rice growers would regard this expenditure as a definite bargain in view of the loss they might sustain without weed control (Seaman, 1983).

The use of herbicides is often a more economical means of weed control than hand weeding. As a general rule, the higher the wage rate or the more severe the weed infestation, the more likely that herbicides will be economical. When the prices of rice and herbicides are considered, weed control with herbicides is economically attractive to farmers. In February 1993, in Iloilo province, Philippines, a farmer would spend $24.20 to control weeds in wet-seeded rice using pretilachlor at the recommended rate of 0.3 kg ai ha^{-1}. With a farm gate price of rice of $0.20 kg^{-1}, a yield increase of only 121 kg ha^{-1} would be needed to cover the cost of the herbicide. In contrast, hand weeding once would cost $128 (40 people × 2 days × $1.60 day^{-1}) – 5.3 times the cost of the herbicide (Moody, 1994).

A similar situation occurs in the Mekong Delta, Vietnam, where hand weeding is at least five times more expensive than herbicides for weed control in wet-seeded rice (Moody, 1992). In west Java, Indonesia, hand weeding of transplanted rice costs about $33 ha^{-1} which is equivalent to 280 kg rough rice (16% of the average yield). This is three to four times the cost of herbicide application (Burhan, 1993).

IV. Direct Seeded Rice

Although irrigation water, in most countries, is supplied at a highly subsidized rate, increased urban and industrial demand for water is likely to put an upward pressure on the price of water used for irrigating agricultural crops. This will lead to increased demand for methods which save irrigation. This means that rice must be grown with less water and less dependence has to be placed on water to suppress weeds. Wet-seeded rice and dry-seeded rice require less irrigation inputs and are economically more attractive to farmers than transplanted rice. In rapidly growing economies of Southeast Asia, the shift to wet seeding may be an intermediate phase in a transition to dry seeding as a method of rice crop establishment. As no puddling is required, dry seeding can substantially reduce the cost of irrigation. Thus, wherever agroclimatic conditions permit, economic incentives will lead ultimately to the adoption of dry seeding (Pandey, 1994). However, cost effective and sustainable integrated weed management systems need to be developed to meet this new demand.

To the extent that wet seeding is adopted in response to a cost-price squeeze, prices of labour, rice and herbicides are the major economic factors determining the adoption of wet seeding. An increase in the price of labour and decreases in prices of rice and herbicides are expected to lead to increased adoption of wet seeding (Pandey, 1994).

Weed problems are much greater in dry-seeded rice than in wet-seeded rice or transplanted rice. A heavy infestation of weeds, mostly grasses, is characteristic of dry-seeded rice (Moody and Mukhopadhyay, 1982).

A. Weed species

The recent changes from transplanting to direct seeding of rice in Asia have resulted in dramatic changes in the types and intensity of weeds and their distribution. Studies conducted in Malaysia clearly show that direct seeding techniques cause weed populations to shift from less competitive broadleaved weeds to more problematical grasses. Weed surveys in the Muda area revealed that in the late 1970s when direct seeding was in the incipient stage of development (less than 1% of the planted area), there were 21 weed species belonging to 13 families. The hierarchical order of dominance was *Monochoria vaginalis* (Burm. f.) Presl > *Ludwigia hyssopifolia* (G. Don) Exell > *Fimbristylis miliacea* (L.) Vahl > *Cyperus difformis* L. > *Limnocharis flava* (L.) Buch (Ho and Zuki, 1988). In the first season in 1989, when 82% of the area was direct seeded, 57 weed species belonging to 20 families were recorded. The order of severity was *E. crus-galli* > *Leptochloa chinensis* (L.) Nees > *F. miliacea* > *Marsilea minuta* L. > *M. vaginalis* (Ho and Itoh, 1991).

In Thailand, farmers are trying to reduce the severity of infestation of *E. crus-galli* in wet seeded rice by maintaining a longer dry period after planting but this results in the replacement of *E. crus-galli* by *L. chinensis*, *Cyperus iria* L., *F. miliacea*, *Echinochloa colona* (L.) Link and *Ischaemum rugosum* Salisb. (Vongsaroj, 1993).

Ali and Sankaran (1984) reported that the dominant weeds under puddled conditions were *E. crus-galli*, *C. difformis*, *Eclipta prostrata* (L.) L., *Ammannia baccifera* L. and *Marsilea quadrifolia* L. whereas *E. colona*, *C. iria* and *E. prostrata* were dominant under non-puddled conditions. Unchecked weed growth caused 53% reduction in grain yield in puddled conditions and 91% yield reduction in non-puddled conditions.

V. Herbicide-Resistant Weeds

One important development as a result of over-dependence on herbicides for weed control and continued use of the same herbicide or herbicides with the same mode of action is the evolution of weed species that have developed resistance to herbicides. In Malaysia, a resistant form of *F. miliacea* has been found in rice fields where 2, 4–D has been applied for 25 years; the weed was not controlled with six times the recommended rate of 2, 4–D (Watanabe *et al.*, 1994). Butachlor-resistant *E. crus-galli* has been reported in China (Huang and Lin, 1993). More resistance is observed where butachlor has been applied for 8–12 years and where two rice crops are grown per year.

As a result of the greater use of herbicides, resistance is expected to become a much more serious economic problem within the next five to ten years (LeBaron and McFarland, 1990). The problem can be avoided or reduced by exploiting a wide range of crop protection measures rather than over-relying on chemical inputs (Tan *et al.*, 1992) and by rotating herbicides having different modes of action or using tank mixtures of herbicides having different modes of action.

VI. Transgenic Rice Plants with Herbicide Resistance

Weeds closest in genetic terms to crops are generally the hardest to control. In these cases, breeding genetic herbicide resistance into the crop provides an attractive way of providing, or improving, herbicide selectivity (Bright, 1991). Transgenic rice plants resistant to bialaphos (Christou *et al.*, 1991; Toki *et al.*,

1992) and glufosinate-ammonium have been developed (Cao *et al.*, 1992; Datta *et al.*, 1992; Rathore *et al.*, 1993).

According to Toenniessen (1991), herbicide-resistant rice plants are likely to be one of the first practical applications of rice genetic engineering. However, engineering herbicide resistance in rice plants raises several issues:

1. Unnecessary high levels of resistance would permit the use of herbicides at extremely high rates.
2. Rice should not be made resistant to herbicides which have harmful environmental or toxicological effects.
3. A genetic barrier to outcrossing should be introduced into the herbicide-resistant crop to prevent the transferring of herbicide resistance to weed species, e.g. wild and weedy rice.

Risk of introgression between genetically-engineered crops and related weed species largely is unknown. However, hybridization between glufosinate-resistant canola (*Brassica napus* L.) and closely related weed species has been observed in the United States (Brammer *et al.*, 1995).

VII. Competitive Cultivars

A vigorous rice crop will suppress many weeds minimizing the need for weed control. The competing power of a crop against weeds is relative to and conditional upon such factors as the cultivar grown, cultural practices such as land preparation, row spacing, seeding rate and fertilizer rate and application time, time of planting, time of weed infestation, duration of weed control, and time of weeding. Thus several aspects of crop establishment can influence the effects of weeds.

Although plant breeders have successfully incorporated both insect and disease resistance into cultivars of many crops, genetic solutions to weed control are only now being explored. It may be necessary to sacrifice some feature of an optimal canopy for crop photosynthesis to achieve a canopy that is more competitive with weeds.

Determination of rice plant characters that contributed to competitiveness is difficult. The first step is to identify the competitive cultivar then determine the phenotypic as well as the genotypic characters responsible for competitive ability. Increased competitive ability has been attributed to early emergence, seedling vigour, increased rate of leaf expansion, rapid creation of a dense canopy, increased plant height, early root growth, and increased root size. Wheat cultivars have been shown to differ in their competitiveness with grass weeds. However, this could not be explained by differences in height, tillering, or dry matter accumulation between the cultivars (Martin and Gill, 1993).

Roots are the foundation of plants and yet they remain relatively unknown compared to the rest of the plant. A strong root system will better withstand occasional water shortages as well as competition with other cultivars, intercropped species and weeds. Cultivars differ as much in plant parts below the soil surface as in parts above the ground. For example, cultivar differences are known in root elongation, degree of branching, overall length of roots for a given soil volume and diameter of roots.

Rice cultivars differ in the main parameters associated with competitiveness; different strategies are used to out-compete weeds. There are also tradeoffs between the ability to tolerate weeds and the ability to compete for nutrients, light, and water. Using a competitive cultivar and enhancing its competitive ability through good crop husbandry can minimize the weed control inputs needed to achieve optimum yields (Harwood and Bantilan, 1974). Garrity et al. (1992) concluded that the choice of a relatively competitive cultivar may provide the practical equivalent of one or two hand weedings.

VIII. Allelopathy

Considerable evidence has been accumulated to implicate secondary plant metabolites as defensive agents in plant to plant relationships. Allelopathy, which is the direct or indirect effect of one plant on another through production and liberation of chemical compounds into the environment, is a widespread phenomenon among crop plants. The effect is direct if the chemical remains unaltered after release from the 'donor' plant, indirect if the chemical compound is altered in form or activity before affecting the 'recipient' plant.

Rice cultivars which produce toxic quantities of natural chemicals that can suppress and kill weeds in field environments have been found by screening germplasm collections for allelopathic types (Dilday et al., 1991; Fujii, 1992; Hassan et al., 1995). The allelochemicals are released into the soil either through root exudation or from decaying plant materials. Incorporation of allelopathic traits into commercial cultivars may enable crop plants to gain an advantage over other species through biochemical activity and subsequent competitiveness. The use of herbicides could be reduced, perhaps drastically, if genes for allelopathic effects can be transferred into commercial rice cultivars (Dilday et al., 1992) through conventional breeding or other genetic transfer techniques.

Development of rice cultivars that contain allelochemicals is a new control strategy that warrants additional research.

IX. Biological Control

Augmentive biological control of weeds refers to the utilization of endemic natural enemies against endemic weed species or exotic weed species which were introduced long ago and have become naturalized in the present habitat. In the bioherbicide approach, excesses of pathogen inoculum are applied to the entire population of an indigenous weed in the same manner as chemical herbicides, causing infection and death of the contacted host plants.

The bioherbicide approach should be used for weeds that are difficult to control by other (chemical, cultural, and mechanical) methods. Fungi would seem to have the greatest potential as bioherbicides because they offer a wide range of virulence, reproductive capacity, specificity, and stability. However, bioherbicides often underperform in the field compared to the laboratory and there is a need for more research on their formulation and application (Hislop, 1993). Also, in order for biological weed control agents to compete with chemical herbicides, they must be reasonably priced, effective, and reliable or have significant toxicological or environmental advantages (Jutsum, 1988).

Biological control must be considered as one component of a weed control system. It must be integrated with chemical, nonchemical, and crop management practices for control of the total weed biomass if the biological agent is to be effective and economical.

X. Priority Research Areas

In 1994, a questionnaire was sent to 148 scientists in South and Southeast Asia who were actively involved in rice weed control to determine research priorities; sixty-eight responded.

A. Research topics

From a listing of twenty-two research topics on various aspects of weed science, the respondents were asked to select six that they considered to be the most important. Over 40% of the respondents indicated that research should be conduced in the following areas:

1. Weed surveys – to develop research priorities and to assess effects of changes in cultural practices on weed populations. An understanding of the nature and magnitude of weed problems in rice is a prerequisite to the formation of weed control programs.

2. Development of integrated weed management systems including greater herbicide efficiency through improved crop husbandry practices. More than for other groups of pests, farmers have many options for the control of weeds. They include manual methods, mechanical methods, cultural techniques, herbicides and, in a few cases, biological control. Weed control is best accomplished when it is considered in the overall management program of crop production.

3. Ecological characteristics of weed species including preferred habitats, weed ecotypes, associated species, conditions required for a species to become a major weed and shifts in weed species as a result of changes in cultural practices (including tillage) or in response to prolonged herbicide use. Improved understanding of the biology and ecology of major weeds would allow prediction of weed appearance and give ideas on possible control measures. Trends in weed population shifts can be predicted with a certain level of probability from knowledge of the life history of the weed and cultural practices used. To be able to make this prediction, a complete understanding of the population dynamics of the weed species is necessary. By understanding the population dynamics of a weed it is possible to: (i) determine the factors that govern weed abundance; (ii) define the conditions and times most suitable for improving control measures; and (iii) predict the weed's response to various control measures and cropping patterns. It is important to have an accurate knowledge of the state of emergence and the progress of initial growth of weeds compared to that of rice in order to establish a rational weed control system (Yamane, 1976).

4. Monitoring of the fate of herbicides in the environment, including persistence, residues in surface water and ground water, effects on soil microbial activity and non-target organisms (other crops, fish, and other aquatic life), and effects on human health. Herbicides have played a dominant role in the control of weeds and will continue to do so in the near future. Reliance on this line of defence has introduced problems of herbicide resistance, environmental and human health problems and increasing regulatory constraints.

5. Assessment of the economic importance of weeds including determining yield losses at different crop stages (also development of predictive models for weed-related yield losses). Alternative weed management systems with minimal herbicide use require detailed understanding of crop–weed interactions and factors that influence the process of competition. Smith (1988) noted that research on crop–weed relations has focused primarily on the effects of the association rather than on the process itself. Detailed quantitative understanding of the underlying process when species are competing to capture the same growth resources is needed.

It is desirable to know how a weed infestation will respond to a particular management regime and what crop yield losses will result. Rational weed management systems require adequate models of weed–crop systems which show the effects of the weed on crop yield and which can predict the size of the weed population (Firbank and Watkinson, 1990). Farmers need timely yield loss information to make weed control decisions during crop growth and after harvest for purposes of assessing control efficacy and profits relative to the choices to be made on future management tactics.

Lowest priority (less than 10% of the responses) was given to the following topics:

1. Research on the greenhouse effect (e.g. high temperature, high CO_2) on weed species and the effect of weeds and weed control practices on gas emission (e.g. methane).

2. Monitoring herbicide resistance in weeds and development of strategies to address this problem.

B. Rice culture

The respondents (40% or greater) indicated that most emphasis should be placed on developing appropriate weed control technologies for wet-seeded rice (pregerminated seed sown, usually by broadcasting, on puddled soil) and dry-seeded, rainfed rice (dry seed sown into dry, poorly-drained sol which becomes flooded to a shallow depth (< 25 cm) at some time during crop growth). Less than 10% thought that emphasis should be placed on deepwater rice (usually dry seed sown into dry, poorly-drained soil which becomes flooded to a depth < 50 cm usually for a major portion of crop growth), and water-seeded rice (usually pregerminated seed sown into water).

Past and much current research on rice weed control has focused on transplanted rice. Because of the rapid expansion of direct seeding in tropical Asia, there is an urgent need for research on weed control under direct seeded conditions.

C. Information dissemination

The greatest needs in this area are the development of an information collection, retrieval and dissemination system (e.g. weed identification and control methods), and the publication of a manual that provides guidelines for extension workers on integrated weed management.

D. Training

The respondents indicated that there is a great need to train more people to develop strong weed research programs. People with good practical knowledge and skill are needed for research and extension. Short-term group training courses are needed for extensionists and short-term individual training courses on specific research topics are needed for researchers. Emphasis should also be placed on MSc degree training; researchers with BSc degrees in agronomy, soil science, botany or plant physiology have the necessary background for advanced degree training in weed science.

XI. Conclusions

The central issue for rice research is how to balance the need for ever-greater food production at prices affordable by the urban poor and the rural landless and profitable to farmers, against very real concerns about protecting natural resources and the environment for the generations to come. This will require scientific breakthroughs that will allow farmers to produce more rice not only with less land but also with less labour, less water and less harmful chemicals (Hossain, 1994).

Modern weed control technology has played a major role in the development of highly productive farming systems during the past three decades. The sustainabilty of farming systems with respect to soil, weed, disease and insect populations is now being questioned and will be a major research challenge for the 1990s and beyond. More than ever farmers and researchers will need to focus on integrated weed management.

The steady emergence of herbicides as a preferred technology for weed control in Asian rice systems follows a twenty-year period of widespread growth in insecticide use that is just beginning to subside. Asian farmers now realize that their dependence on insecticides has often been unnecessary, expensive and sometimes even dangerous. Although herbicides are much less toxic and persistent than the majority of insecticides used in Asian rice production, the inevitable question arises: 'Twenty years from now, will Asian societies regret having gone down the herbicide path?' (Naylor, 1994).

References

Ahmed, N.U. and Azizul Islam, A.J.M. (1983) Farmers' weed control technology for dry-seeded rice. In: *Weed Control in Rice*. International Rice Research Institute, Los Baños, Philippines, pp. 207–212.

Ali, A.M. and Sankaran, S. (1984) Crop-weed competition in direct seeded, flooded and rainfed bunded rice, *International Rice Research Newsletter* 9(2), 22.

Anon. (1975) *Pest Control: An Assessment of Present and Alternative Techniques. Vol. II. Corn/Soybeans Pest Control*. National Academy of Sciences, Washington, DC.

Anon. (1988) Tropical weeds: A growing menace. *Spore* 12, 4–6.

ARS National Research Program, (1976) Weed control technology for protecting crops, grazing lands, aquatic sites, and noncropland. NRP No. 20280 Weed Control, United States Department of Agriculture, Washington, DC.

Brammer, T., Thill, D.C., Brown, J., Brown, A. and Mallory-Smith, C.A. (1995) Risk assessment of pollen movement between herbicide-resistant transgenic canola and *Brassica* weed species, *Weed Abstracts* 35, 46.

Bright, S.W.J. (1991) Opportunities for introducing herbicide-resistant crops. In: Caseley, J.C., Cussans, G.W. and Atkins, R.K. (eds), *Herbicide Resistance in Weeds and Crops*. Butterworth-Heinemann, Oxford, pp. 365–374.

Burhan, H.A. (1993) Studies on weeds in low-land rice in the northern coastal plain of west Java, Indonesia, PLITS 1993/11 (5), Institut für Pflanzenproduktion in den Tropen und Subtropen, Universität Hohenheim, Stuttgart, Germany.

Cao, J., Duan, X.L., McElroy, D. and Wu, R. (1992) Regeneration of herbicide resistant transgenic rice plants following microprojectile-mediated transformation of suspension culture cells, *Plant Cell Reports* 11, 586–591.

Chisaka, H. and Noda, K. (1983) Farmers' weed control technology in mechanized rice systems in East Asia. In: *Weed Control in Rice*. International Rice Research Institute, Los Baños, Philippines, pp. 153–165.

Christou, P., Ford, T.L. and Kofron, M. (1991) Production of transgenic rice (*Ozyza sativa* L.) plants from agronomically important indica and japonica varieties via electric discharge particle acceleration of exogenous DNA into immature zygotic embryos. *Biotechnology*, 9, 957–962.

Datta, S.K., Datta, K. and Soltanifar, N., Donn, G. and Potrykus, I. (1992) Herbicide-resistant indica rice plants from IRRI breeding line IR72 after PEG-mediated transformation of protoplasts, *Plant Molecular Biology* 20, 619–629.

Dilday, R.H., Nastasi, P., Lin, J. and Smith, R.J. Jr (1991) Allelopathic activity of rice (*Oryza sativa* L.) against ducksalad [*Heteranthera limosa* (Sw.) Willd]. In: Hanson, J.D., Shaffer, M.J., Ball, D.A. and Cole, C.V. (eds) *Proceedings of the Symposium on Sustainable Agriculture for the Great Plains*, ARS–89. Department of Agriculture, Agricultural Research Service, Washington, DC, pp. 193–201.

Dilday, R.H., Frans, R.E., Semidey, N., Smith, R.J. and Oliver, L.R. (1992) Weed control with crop allelopathy. *Arkansas Farm Research* 41(4), 14–15.

Elliot, P.C. and Moody, K. (1987) Determining suitable weed control practices for upland rice (*Oryza sativa*) in Claveria, Misamis Oriental. *Philippines Journal of Weed Science*, 14, 52–61.

Firbank, L.G. and Watkinson, A.R. (1990) On the effects of competition: From monocultures to mixtures. In: Grance, J.B. and Tilman, D. (eds) *Perspectives on Plant Competition*. Academic Press, San Diego, pp. 165–192.

Fujii, Y. (1992) The potential biological control of paddyweeds with allelopathy – allelopathic effect of some rice varieties. In: *Biological Control and Integrated Management of Paddy and Aquatic Weeds in Asia*. National Agricultural Research Center, Tsukuba, Japan, pp. 305–320.

Garrity, D.P., Movillon, M. and Moody, K. (1992) Differential weed suppression ability in upland rice cultivars, *Agronomy Journal* 84, 586–591.

Harwood, R.R. and Bantilan, R.T. (1974) Integrated weed management. 2. Shifts in composition of the weed community intensive cropping systems, *Philippines Weed Science Bulletin*, 1(2), 37–59.

Hassan, S.M., Aidy, I.R. and Bastawisi, A.O. (1995) Allelopathic potential of rice varieties against major weeds in Egypt, *WSSA Abstract* 35, 60.

Herdt, R.W. (1991) Research priorities for rice biotechnology. In: Khush, G.S. and Toenniessen, G.H. (eds) *Rice Biotechnology. Biotechnology in Agriculture No. 6*. CAB International, Wallingford, UK, pp. 19–54.

Hislop, E.C. (1993) Application technology for crop protection: An introduction. In: Matthews, G.A. and Hislop, E.C. (eds) *Application Technology for Crop Protection*. CAB International, Wallingford, UK.

Ho, N.K. and Itoh, K. (1991) Changes in weed flora and their distribution in the Muda area. Paper presented at the 8th MADA/TARC Quarterly Meeting, November 3, 1991, Alor Setar, Malaysia.

Ho, N.K. and Zuki, I. Md. (1988) Weed population changes from transplanted to directed seeded rice in the Muda area. In: *Proceedings of the National Seminar and Workshop on Rice Field Weed Management*. Penang, Malaysia, pp. 55–78.

Hossain, M. (1994) Recent development in Asian rice economy: Challenges for rice research. Paper presented at the Workshop on Rice Research Prioritization, 21–22 February, 1994, International Rice Research Institute, Los Baños, Philippines.

Huang, B.Q. and Lin, S.X. (1993) Study on the resistance of barnyardgrass to butachlor in paddy fields in China (in Chinese, English abstract), *Journal of South China Agricultural University* 14, 103–108.

Jutsum, A.R. (1988) Commercial application of biological control: Status and prospects, *Philosophical Transactions of the Royal Society of London* Ser. B, 318, 357.

LeBaron, H.M. and McFarland, J. (1990) Overview and prognosis of herbicide resistance in weeds and crops. In: Green, M.B., LeBaron, H.M. and Moberg, W.M. (eds) *Managing Resistance to Agrochemicals: From Fundamental Research to Practical Strategies*. ACS Symposium Series 421, Washington DC, pp. 336–352.

Martin, R.J. and Gill, G.S. (1993) Weed competition and crop yield. In: Dodd, J., Martin, R.J. and Howes, K.M., (eds) *Management of Agricultural Weeds in Western Australia*, Bull. 4243. Department of Agriculture, Perth, Western Australia, pp. 81–84.

Moody, K. (1983) Weeds: definitions, costs, characteristics, classification and effects. In: Walter, H. (ed.) *Weed Management in the Philippines*, PLIT 1(1). Institut für Pflanzenproduktion in den Tropen und Subtropen, Universität Hohenheim, Stuttgart, Germany, pp. 15–26.

Moody, K. (1992) Weed management in wet-seeded rice in tropical Asia. In: *Biological Control and Integrated Management of Paddy and Aquatic Weeds in Asia*. National Agricultural Research Center, Tsukuba, Japan, pp. 1–20.

Moody, K. (1994) Economic and technological forces leading to increased herbicide use and some consequences. Paper presented at a conference on herbicide use in Asian rice production, 28–30 March, 1994, Stanford University, California.

Moody, K. and Mujkhopadhyay, S.K. (1982) Weed control in dry-seeded rice – problems, present status, future research directions. In: *Rice Research Strategies for the Future*. International Rice Research Institute, Los Baños, Philippines, pp. 147–158.

Naylor, R. (1994) Herbicides in Asian rice production: Perspectives from economics, ecology, and the agricultural sciences. Paper presented at a conference on herbicide use in Asian rice production, March 28–30, 1994, Stanford University, Stanford, California.

Negi, N.S. (1976) Weed control in rice, *Pesticide Information* 2(3), 94–102.

Pandey, S. (1994) Socio-economic research issues on wet seeding. Paper presented at the International Workshop on Constraints, Opportunities, and Innovations for Wet-seeded Rice, May 31–June 3, 1994, Bangkok, Thailand.

Rathore, K.S., Chowdhury, V.K. and Hodges, T.K. (1993) Use of *bar* as a selectable marker gene and for the production of herbicide-resistant rice plants from protoplasts, *Plant Molecular Biology*, 21, 871–884.

Seaman, D.E. (1983) Farmers' weed control technology in water-seeded rice in North America. In: *Weed Control in Rice*. International Rice Research Institute, Los Baños, Philippines, pp. 167–177.

Smith, R.J., Jr (1988) Weed thresholds in southern US rice (*Oryza sativa* L.). *Weed Technology* 2, 232–241.

Smith, R.J. Jr, Flinchum, W.T. and Seaman, D.E. (1977) Weed control in US rice production, *Agric. Handbook No. 497*, United States Department of Agriculture, Washington, DC.

Stoskopf, N.C. (1985) *Cereal Grain Crops*. Reston Publishing Co., Reston, Virginia.

Tan, S.H., Yeoh, H.F., Ramasamy, S. and Salleh, A.M. (1992) Pesticide use in Malaysia in the year 2000. In: Kadir, A.A.S.A. and Barlow, H.S. (eds) *Pest Management and the Environment in 2000*. CAB International, Wallingford, UK, pp. 165–174.

Toenniessen, G.H. (1991) Potentially useful genes for rice genetic engineering. In: Khush, G.S. and Toenniessen, G.H. (eds) *Rice Biotechnology. Biotechnology in Agriculture No. 6*. CAB International, Wallingford, UK, pp. 271–280.

Toki, S., Takamatsu, S., Nojiri, C., Ooba, S., Anzai, H., Iwata, M., Christensen, A.H., Quail, P.H. and Uchimiya, H. (1992) Expression of a maize ubiquitin gene promoter-*bar* chimeric gene in transgenic rice plants, *Plant Physiology* 100, 1503–1507.

Vogsaroj, P. (1993) Weeds in paddyfield and their control. In: Technical Report 11, *Report on Training Materials on Basic Course on Field Crops Production*. Bangkok, Thailand: FAO and Department of Agriculture, pp. 15–32.

Watanabe, H., Zuki, I.Md. and Ho, N.K. (1994) 2, 4–D resistance of *Fimbristylis miliacea* in direct seeded rice fields in the Muda area. In: Rajan, A. and Ibraim, Y. (eds) *Proceedings of the 4th International Conference on Plant Protection in the Tropics*. Kuala Lumpur, Malaysia, pp. 353–356.

Wong, H.S. (1992) Padi production and farm management first (off) season 1991. Report 2. Farm Management and Socio-economic Series, Muda Agricultural Development Authority, Alor Setar, Kedah, Malaysia.

Yamane, K. (1976) Ecology of weed emergence and their control in direct-seeded rice cultivation on upland field after flooding, *Hyogo Prefecture Agricultural Experiment Station Special Bulletin 51*.

18 Yield Loss Due to Drought, Cold and Submergence in Asia

M.M. Dey[1] and H.K. Upadhyaya[2]

[1]*International Center for Living Aquatic Resources Management (ICLARM), MC PO Box 2631, 0718 Makati, Metro Manila Philippines;* [2]*Center for Environmental and Agricultural Policy Research, Extension and Development (CEAPRED), PO Box 5752, Kathmandu, Nepal*

I. Introduction

A common objective of the country studies presented in Part II of this volume was to identify the various rice production constraints contributing to the present yield gap and prioritize them for research based on the corresponding yield losses. Accordingly, these country studies have generated a list of the more important rice research problem areas and have estimated the yield loss caused by each of them. The estimates of yield loss at the individual country level can be aggregated to generate a broad list of priority research problem areas for the Asian region.

This chapter presents data on yield loss caused by three abiotic factors – drought, cold and submergence. Unfortunately, data from only four countries – Bangladesh, China, India (eastern and southern) and Nepal – are included here because comparable methodology and data were not obtained for other countries. However, these four countries together account for 66% of Asia's rice area and 69% of Asia's rice production (IRRI, 1991), and so the data provide insights for a large fraction of the relevant area. Section II illustrates the concept and magnitudes of the yield gap observed in these countries. The third section illustrates the magnitudes of yield loss caused by drought, cold and submergence. The fourth section extrapolates the findings of five country studies and presents the annual production loss caused by each of these abiotic stresses in different agroecological zones and rice ecologies of Asia.

II. Rice Yield Gaps

As discussed in the introductory chapter in this volume, yield gap I is the difference between experiment station yield and potential farm yield, and is mainly the result of environmental differences and non-transferable technology, which cannot be managed or eliminated in farmers' fields (IRRI, 1977). Yield gap II is the difference between actual and potential farm yield, and is caused by various biological constraints and socioeconomic forces, which can be managed or eliminated in farmers' fields. The set of biological constraints have direct implications for research priorities and are called the technical constraints. Successful research on these constraints is hypothesized to be able to narrow yield gap II by as much as they presently contribute to this gap.

The central idea of the yield gap approach is to estimate the contribution of the technical constraints by measuring the yield loss caused by each and to evaluate their relative importance to help in setting priorities for research. The benefits from successful research on a constraint can be assumed to be equal or proportional to the size of production lost at present to that constraint. In this chapter, we focus on assessing the importance of drought, cold and submergence in eight important Asian rice-growing regions.

Table 18.1 presents estimates of yield gap II in different parts of Asia representing all the major agroecological zones of the region (Table 18.2), derived using the methodology discussed in the respective country chapters. Among the countries, China has the highest potential and average farm yields. China has one of the highest national average rice yields in the world, and is a pioneer in large-scale hybrid rice production, with nearly 20% of the area in hybrid rice and most of the rest of the rice area covered by modern varieties (see Chapter 10). Almost all rice production is under irrigated conditions, and the application of chemical fertilizers to rice is the highest in developing Asia. In 1990, the consumption of chemical fertilizers in China (25.3 million tons) accounted for almost half of the consumption in Asia (IRRI, 1991). In contrast, the adoption of modern varieties, the extent of irrigation and water control facilities, and the application of chemical fertilizers to rice production in

Table 18.1. Rice yield gap II in different agroecological zones and rice ecosystems in Asia (kg ha^{-1}).

Country/region	Agroecological zones	Potential farm yield	Average farm yield	Yield gap II
Southern India	1	4562	4012	550
Eastern India	2	3802	2041	1761
Bangladesh	3	3937	3055	882
Northeastern China	5	8654	5617	3037
Central China	6	9080	5297	3783
Nepal	6	3940	2267	1673
Northern China	7	8361	5257	3104
Western China	8	9207	5465	3742

Table 18.2. The agroecological zones (AEZ) of Asia and the study areas.

Agroecological zone	Country/region	Study areas	Represented rice ecologies
AEZ1: Warm arid and semi-arid tropics	Southern India	Southern India	Irrigated
AEZ2: Warm subhumid tropics	Part of eastern India, Kerala, Thailand, Myanmar	Eastern India, Kerala	Irrigated, rainfed lowland, flood-prone, upland
AEZ3: Warm humid tropics	Bangladesh, Indonesia, Malaysia, Philippines, Laos, Cambodia, Vietnam, Sri Lanka	Bangladesh, Indonesia, Philippines	Irrigated, rainfed lowland, flood-prone, upland
AEZ5: Warm arid and semi-arid subtropics	Rajasthan of India, northeastern China	Northeastern China	Irrigated
AEZ6: Warm subhumid subtropics	Uttar Pradesh, Haryana, Punjab of India, central China, Korea, Nepal	Central China, Nepal	Irrigated, rainfed lowland, upland
AEZ7: Warm, cool, humid subtropics	Southern China, Taiwan	Southern China	Irrigated
AEZ8: Cool subtropics	Northwestern corner of India, western China	Western China	Irrigated

Bangladesh, eastern India and Nepal are some of the lowest in Asia. As a result, the average farm yields are only between 2 and 3 tons ha^{-1}. The potential farm yield in these three countries is also only about 4 tons ha^{-1}, about half of the potential farm yield in China.

The difference between potential and actual farm yield, yield gap II, in these four countries ranged from less than 1 ton ha^{-1} in Bangladesh to more than 3 tons ha^{-1} in China. As a percentage of the actual yield, yield gap II ranged from 13% in southern India to 86% in eastern India. As discussed earlier, this gap can be interpreted as the extent to which the actual farm yield can be increased by eliminating all associated constraints, technical and socioeconomic, including management.

III. Yield Loss Due to Drought, Cold and Submergence

Technical constraints include both biotic and abiotic factors that limit rice yields in farmers' fields. Biotic constraints are, at least conceptually, solvable through research. Abiotic constraints can be affected but not solved by research. They include adverse climate and soils, insects and diseases, and others such as weeds, lodging, etc. The country studies show that, among the technical constraints, abiotic constraints are more prominent and yield-limiting than biotic constraints; and among abiotic constraints, droughts and floods are the most devastating to yields.

A. Drought

About 57% of Asia's rice area is irrigated, but the quality of irrigation varies considerably across regions. In many cases, the supply of irrigation water is unstable or inadequate to meet the crop water requirement at critical stages, subjecting the crop to the uncertainties of monsoon rains. The rest of Asia's rice area (43%) is completely dependent on rainfall and is largely prone to drought, irrespective of the production environments – lowland, upland and deepwater. Even if sufficient moisture is received over the growing season to support the physiological needs of the rice crop, the precipitation may not be evenly distributed enough to satisfy the crop water requirements at various stages of growth. The uneven distribution may result in heavy precipitation during certain periods leading to floods and to dry periods in between leading to drought conditions, such as are important in eastern India (see Chapter 8).

Drought affects rice plants at three critical stages of growth – seedling, vegetative and anthesis. Seedling stage drought is common in areas where broadcast seeding or dry seeding is practiced followed by erratic rainfall, such as in upland or deepwater rice areas. Vegetative stage drought occurs mostly in upland rice areas as a result of unpredictable monsoons. Drought during anthesis is the most serious and devastating to yields, as it adversely affects pollination and the flowers are left sterile. These three forms of drought, in general, and anthesis drought, in particular, are chronic causes of yield instability, especially in eastern India and Nepal, where access to quality irrigation and water control facilities throughout the rice-growing season is greatly limited. Almost every alternate year, drought appears as a major yield-limiting constraint in many parts of the South Asia region. Even in China, where virtually all rice is grown with irrigation, less than ideal control over irrigation water or actual shortages of water result in significant yield losses.

How much rice yield is lost to drought and other technical constraints, and what portion of observed yield gap II is explained by them in those countries? Table 18.2 presents the estimated yield loss per hectare due to all technical constraints (including drought) and drought alone. Table 18.4 shows yield loss due to drought as a proportion of all technical constraints, yield gap II and average yield in Asia. For simplicity and ease of comparison, yield-loss estimates reported by the respective country studies (Chapters 7–15) have been aggregated here for different rice-growing seasons and environments, using the proportion of rice area under each of them as weights. As shown in Table 18.3, the estimated average amount of yield loss due to all technical constraints varied from 468 kg ha^{-1} in southern India to 1515 kg ha^{-1} in central China. The average yield loss due to drought alone ranged between 17 kg ha^{-1} in

Table 18.3. Estimated yield loss due to various abiotic factors.

	Yield loss (kg ha^{-1})				
Country/region	All technical constraints	Abiotic factors	Drought	Cold	Submergence
Southern India	468	117	17	4	0
Eastern India	658	306	144	18	24
Bangladesh	635	284	93	10	140
Northeastern China	1350	1156	153	194	95
Central China	1515	1444	250	317	163
Nepal	1422	406	236	0	13
Southern China	1091	990	143	159	112
Northern China	1288	1033	169	160	79

southern India and 250 kg ha^{-1} in central China. In relation to the yield loss due to all technical constraints, the yield loss due to drought only is the lowest in southern India (4%) and the highest in eastern India (22%). Southern India is predominantly an irrigated area, while around 60% of the total cropped area in eastern India is rainfed.

Drought alone accounted for between 3% of the yield gap in southern India and 14% of the yield gap in Nepal. A large number of constraints (insects, diseases, adverse soils and others) are included as technical constraints; and considering the yield loss caused by each of them individually, the portion of yield gap accounted for by drought is quite significant, and in aggregate one of the largest of all the constraints. Though not important in the irrigated environment of southern India, yield loss due to drought is around 10% in Nepal, 7% in eastern India, 3% in Bangladesh and 3–5% of the average yield in China.

Table 18.4. Yield loss due to drought as a proportion of all technical constraints, Yield gap II and yield in different countries/regions of Asia.

	Yield loss due to drought (%)		
Country/region	All technical constraints	Yield gap II	Yield
Southern India	4	3	0
Eastern India	22	8	7
Bangladesh	14	11	3
Northeastern China	11	5	3
Central China	16	7	5
Nepal	17	14	10
Southern China	13	5	3
Northern China	13	5	3

B. Cold

Yield loss caused by cold in any region depends, among many other things, on temperature pattern, rice-growing seasons and on the type of variety farmers grow. Figures 18.1–18.5 show temperature patterns and rice-growing season(s) in different parts of Asia. Critical temperatures for rice vary depending on the stage of growth: the critical temperature at the seedling stage is 12–13°C, 9–16°C at the tillering stage, 15–20°C at the panicle initiation stage and 19–22°C at the anthesis stage (Yoshida, 1981). Thus, Fig. 18.1 shows that low temperature is not a limiting factor in the warm semi-arid tropics (southern India). In eastern India and Bangladesh, low temperature is a problem for the irrigated rabi (winter) crop commonly known as boro rice. For late-planted kharif (wet season) rice, cold is sometimes a problem particularly during the anthesis stage (Fig. 18.2). In northeastern China, representing warm arid and semi-arid subtropics, where only one rice crop is grown, cold is a problem both at early and later stages of the rice plant's life cycle (Fig. 18.3). In the more tropical areas of central and southern China, farmers follow two different rice-cropping patterns – a crop of single mid-season rice and a first crop plus second crop pattern (Fig. 18.4). Cold is a limiting factor during the early stage of the first crop, and during the late stage of single mid-season rice and second rice crops. Cold affects rice plants both at early and later stages of their life cycle in the cool subtropics (Fig. 18.5).

The per hectare yield loss due to cold is estimated to be very high in China – 317 kg ha^{-1} in central China, 194 kg ha^{-1} in northeastern China and 160 kg ha^{-1} in both southern and northern China (see Table 18.3). These

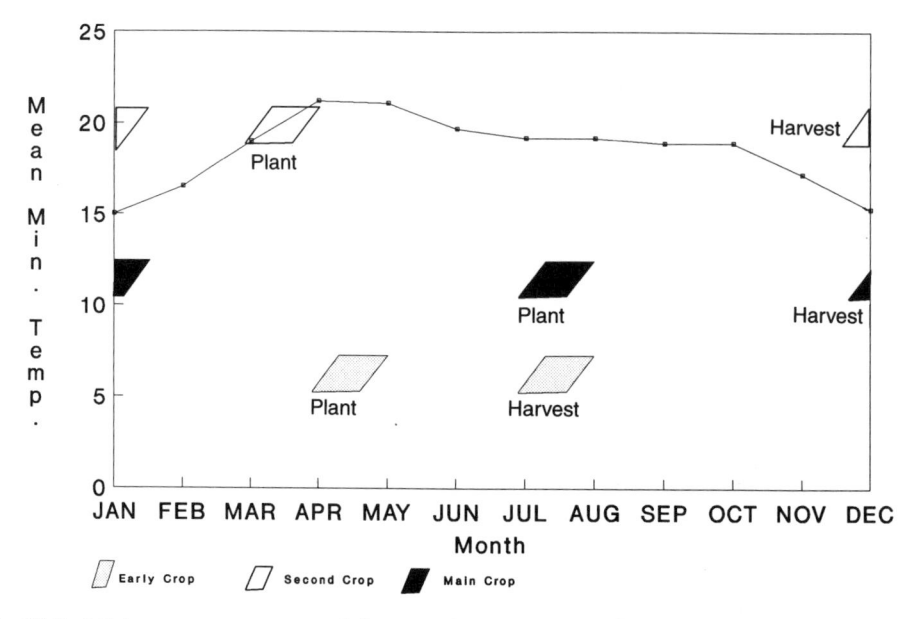

Fig. 18.1. Minimum temperature and rice-growing seasons: southern India, warm arid and semi-arid tropics (AEZ1).

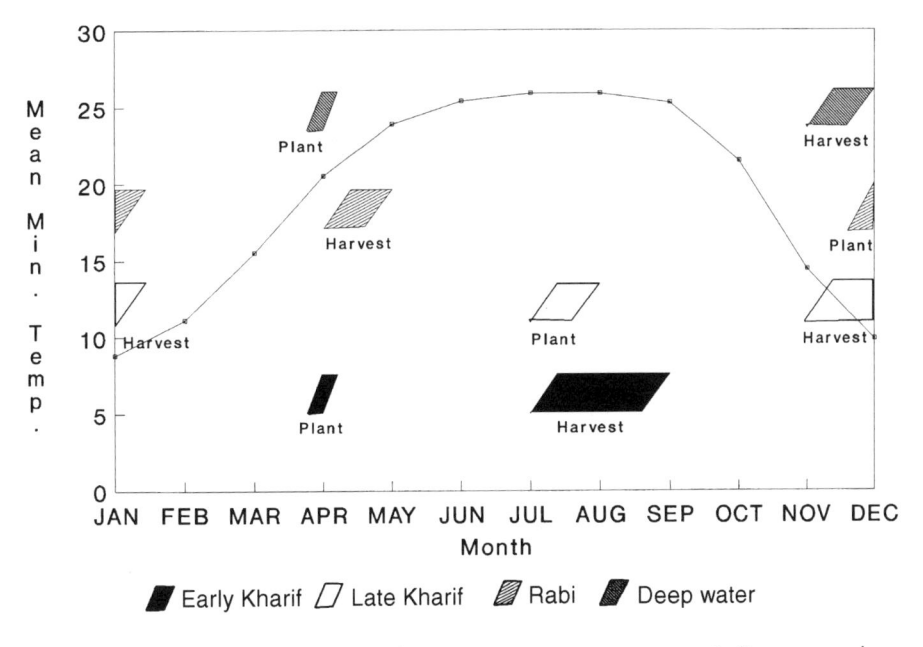

Yield Loss Due to Drought, Cold and Submergence 297

Fig. 18.2. Minimum temperature and rice-growing seasons: eastern India, warm sub-humid tropics (AEZ2).

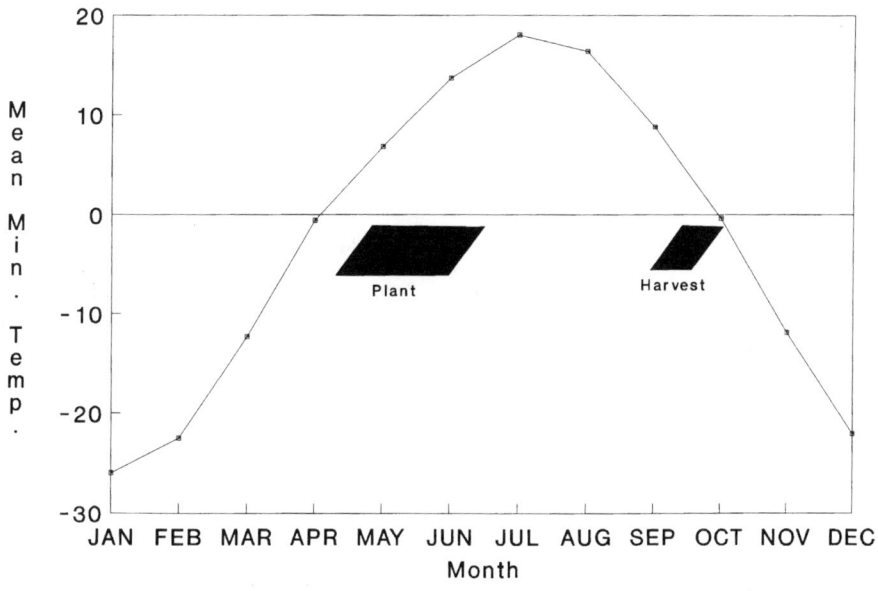

Fig. 18.3. Minimum temperature and rice-growing season: northeastern China, warm arid and semi-arid subtropics (AEZ5).

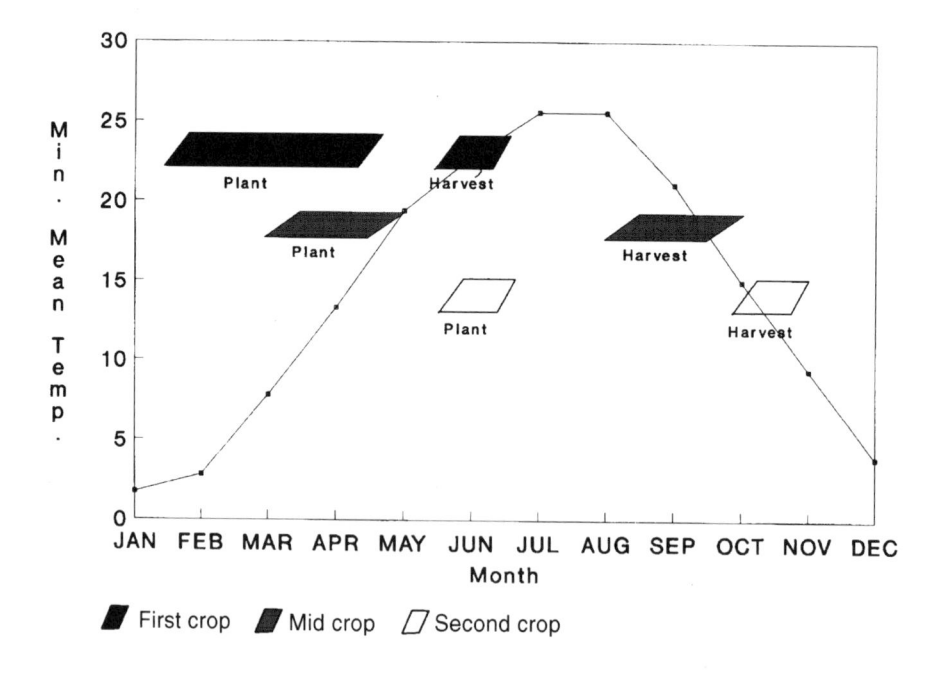

Fig. 18.4. Minimum temperature and rice-growing seasons: southern China (AEZ6/7).

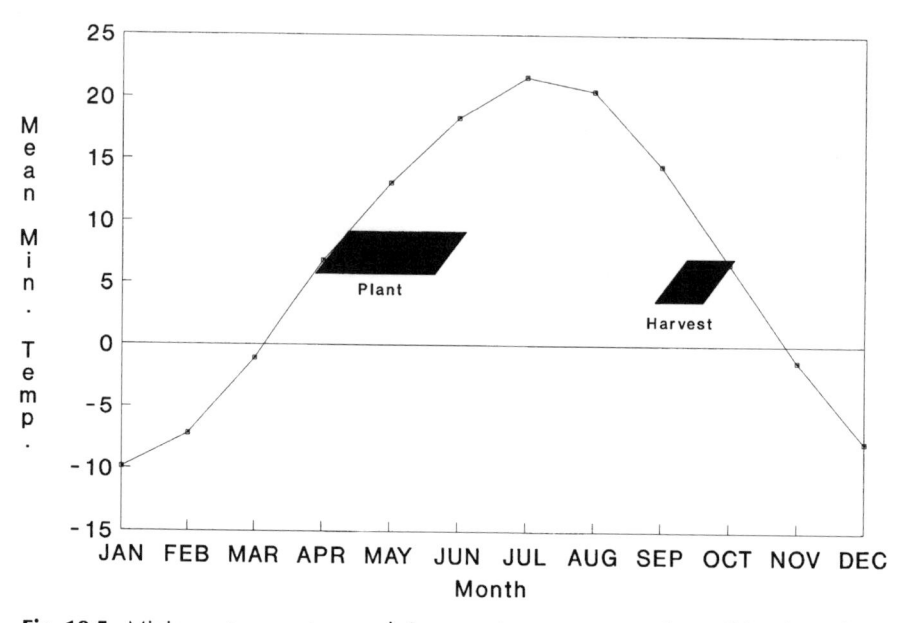

Fig. 18.5. Minimum temperature and rice-growing seasons: northern China (AEZ8).

Yield Loss Due to Drought, Cold and Submergence

Table 18.5. Yield loss due to cold in Asia.

Country/region	All technical constraints	Yield gap II	Yield
Southern India	1	1	0
Eastern India	3	1	1
Bangladesh	2	1	0
Northeastern China	14	6	3
Central China	20	8	6
Nepal	0	0	0
Southern China	15	5	3
Northern China	12	4	3

Column header: Yield loss due to cold (%)

losses represent 12–20% of the total loss caused by all technical factors and 4–8% of the total yield gap II (Table 18.5). They are equivalent to 6% of average yield in central China and 3% of average yield in the other parts of China.

Production loss due to cold is also substantial in Bangladesh and eastern India, but the country studies did not cover tropical high-altitude areas where cold temperature is most important. About 7 million ha of rice area are situated in the high altitudes of tropical countries like Nepal, India, the Philippines, Thailand and Indonesia and these areas are subject to very cold temperatures. Farmers currently do not grow high-yielding varieties in these areas, in part because of their lack of cold tolerance.

The negative effect of cold on the rice plant is not uniform across all growth stages (Table 18.6). Cold at the panicle initiation and anthesis stages of irrigated

Table 18.6. Relative intensity of yield loss due to cold at various growth stages in different cold-affected areas.

Country/region	Growing season	Rice ecology	Seedling	Vegetative	Anthesis
Eastern India (AEZ2)	Kharif	Irrigated	4	–	29
	Kharif	Rainfed lowland	7	–	21
	Rabi	Irrigated	1	–	–
Bangladesh (AEZ3)	Boro	Irrigated	10	19	16
Northeast China (AEZ5)	Single	Irrigated	43	29	150
Central and southern China (AEZ6)	Early	Irrigated	64	22	5
	Late	Irrigated	17	–	99
	Single	Irrigated	39	24	139
Northwest China (AEZ8)	Single	Irrigated	73	23	45

Column header: Yield loss due to cold (kg ha^{-1})

single season rice is reported to have a very detrimental effect on rice yield in northeastern China, resulting in an average yield loss of 150 kg ha^{-1}. In central and southern China, cold affects early rice at the seedling stage, and affects late and single season rice at the anthesis stages causing an estimated yield loss of 64, 99 and 139 kg ha^{-1}, respectively.

Cold at the anthesis stage of kharif rice in eastern India causes yield losses of 21–29 kg ha^{-1}. In the cool subtropical environment of northwest China, where rice is grown only in limited areas, cold affects rice plants at all growth stages and causes substantial yield loss.

The effect of cold on rice production is not static. The increasing importance of boro rice in Bangladesh and eastern India makes total rice production of those regions more vulnerable to cold. In south and central China, farmers are shifting from two rice crops to a single season high-yielding rice crop as the opportunity for non-farm work is increasing. Farmers are also increasingly growing japonica rice, which is more cold tolerant than the indica type, in late season, to meet the demand for high-quality rice. These changes make it likely that in the future, cold at the anthesis stage of single season rice will be a more important limiting factor to rice production in China.

C. Submergence

Submergence is a problem in rainfed lowland and deepwater environments of eastern India, Bangladesh, Myanmur, Thailand and Vietnam, and in irrigated environments of subtropical countries like China. Kharif season rice grown in Bangladesh, eastern India and Nepal, and single season and second rice crops in China, often get submerged during seedling and vegetative stages causing substantial yield losses. Winter season irrigated rice is also sometimes submerged during the panicle initiation and anthesis stages by flash floods. Average yield loss due to submergence ranges from negligible in southern India to 163 kg ha^{-1} in central China (Table 18.7). In Bangladesh, submergence is the most im-portant technical constraint; it accounts for 22% of all technical constraints and 16% of yield gap II and reduces average yield by 5% (Table 18.7).

IV. Extrapolation of Country Studies

Findings from the country studies, which give yield-loss estimates for different rice ecologies in different regions/countries, have been extrapolated to get an overall impression of yield loss due to various factors. We have classified rice growing into homogeneous agroecological zones (seven groups) and rice ecosystems (four groups) and have estimated the proportion of area covered by each group (W_i). We estimated the production loss per hectare for each of the groups i for a particular problem j, P_{ij}. This is estimated by taking the average estimates of losses for countries or states (India) and provinces (China) that represent the specific zone or ecosystem. The estimates of loss for Asia for a

Table 18.7. Yield loss due to submergence as a proportion of all technical constraints, yield gap II and yield in different countries/regions of Asia.

	Yield loss due to submergence (%)		
Country/region	All technical constraints	Yield gap II	Yield
Southern India	0	0	0
Eastern India	4	1	1
Bangladesh	22	16	5
Northeastern China	7	3	2
Central China	11	4	3
Nepal	1	1	1
Southern China	10	4	2
Northern China	6	2	1

specific problem area, P_j, was then obtained as the weighted average for the agroecological zones or ecosystems. Thus,

$$P_j = \sum_{i=1}^{n} W_{ij} P_{ij}. \qquad (18.1)$$

In the irrigated environments of Asia, drought and cold are the two most limiting factors to increased rice production (Table 18.8). Timely availability of irrigation water is a major problem in Asia and often causes drastic yield loss. As expected, intensity of drought is highest in upland environments; drought causes a yield loss of 190 kg ha^{-1} which is equivalent to almost 20% of average yield in this ecosystem. Submergence and drought are two constraints in lowland environments; uncontrolled water management regimes make rainfed environments vulnerable to both flood and drought. In Asia, as a whole, drought is the most limiting factor, followed by submergence and cold (Table 18.8). These three constraints together cause a loss of about 8% of total annual rice production in Asia, amounting to 36 million tons.

Table 18.8. Yield loss due to drought, cold and submergence in different agroecological zones and rice ecologies of Asia.

Zones/ecologies	Total rice area covered (%)	Yield loss (kg ha^{-1}) Drought	Yield loss (kg ha^{-1}) Cold	Yield loss (kg ha^{-1}) Submergence	Yield loss (% of production) Drought	Yield loss (% of production) Cold	Yield loss (% of production) Submergence
Agroecological zones							
Semi-arid tropics	11.2	97	11	14	3	0	0
Sub-humid tropics	26.0	112	13	23	5	1	1
Humid tropics	26.3	93	22	89	3	1	3
Semi-arid subtropics	3.1	153	194	95	4	6	3
Sub-humid subtropics	18.7	234	244	128	5	5	3
Humid subtropics	14.1	143	159	112	3	3	2
Temperate zone	0.6	243	135	107	5	3	2
Rice ecologies							
Irrigated	57.7	134	140	90	3	3	2
Rainfed lowland	24.7	160	18	80	7	1	3
Upland	9.7	190	1	23	17	0	2
Deepwater	8.9	11	0	11	1	0	1
Asia	100.0	130	71	79	4	2	2

Source: Rice statistics data base, IRRI and country studies in this volume.

References

International Rice Research Institute (IRRI) (1977) *Constraints to High Yields on Asian Rice Farms: an Interim Report.* IRRI, Los Baños, Philippines.

International Rice Research Institute (IRRI) (1991) *World Rice Statistics 1990.* IRRI, Los Baños, Philippines.

Yoshida, S. (1981) *Fundamental of Rice Crop Science.* International Rice Research Institute, Los Baños, Philippines.

19 Intercountry Comparison of Insect and Disease Losses

C. Ramasamy[1] and T. Jatileksono[2]
[1]Department of Agricultural Economics, Tamil Nadu Agricultural University, Coimbatore 641 003, India; [2]Department of Agricultural Economics, Gadja Mada University, Yogyakarta, Indonesia

I. Introduction

In this chapter, information on the nature and level of yield losses caused by major insect and disease pests across rice production environments in selected countries of Asia is considered. Knowledge of crop losses caused by specific pests is useful in formulating strategies to minimize losses and estimating the gains from research on specific pests.

II. Insects

Despite several decades of using insecticides and adopting other methods of controls, rice farmers in Asia still suffer considerable yield losses due to insects. Integrated pest management (IPM), the combination of improved crop management and careful varietal selection, integrated with judicious pest control, can reduce those losses. But this has happened in few countries and in few selected sites, and not in all places and in all countries, although the success in Indonesia is remarkable (see Chapter 15). Unequal access to technology for various farmer groups, capital inadequacy, knowledge gaps, supply constraints for modern inputs, failure of institutions to carry out their assigned roles and a host of other factors may be responsible for the continued importance of damage from insects in rice.

In the early phase of the use of modern rice varieties, severe pest problems emerged under farmers' conditions, particularly the planthopper and leafhopper and the virus diseases they vectored. Researchers responded to these challenges, both in the international research system and in national research institutions, by producing varieties resistant to insects. The incorporation of genetic resistance has minimized yield losses due to insects, but it is a time-consuming process. Since chemical control causes negative externalities on the environment and is often not available to poor credit-constrained farmers, incorporation of genetic host resistance through the biotechnology approach is one of the options. It has the advantage of being environmentally friendly, as well as being scale neutral.

However, modern photoperiod-insensitive varieties encouraged farmers and governments to develop intensive irrigation systems to realize year-round production potential. Multiple rice cropping over large areas also encouraged year-round pest development. Over the past two decades, the insect pest outbreaks that have occurred in different Asian countries may have been made more severe by the increased use of government-subsidized, broad-spectrum insecticides that kill the natural enemies of rice pests. In the absence of selective insecticides, insect-resistant varieties could not prevent such outbreaks (Litsinger, 1989).

Several studies have reported rice yield losses due to insects in Asia. Cramer (1967) reported that yield lost to all insects in tropical rice was 34%; Pathak and Dhaliwal (1981) reported 35–44%, and several other studies report losses of similar magnitude (Table 19.1).

The major rice insect pests which cause economic losses in Asia are: the stemborer, rice bugs, leaf folder, leaf and planthoppers, gall midge and rice hispa. Herdt and Riely (1987) assessed the relative intensity of problems in rice production based on personal interaction with a group of scientists experienced in Asian rice research. According to them, the stemborer, brown planthopper, gall midge and leafhopper were among the most important insects in Southeast Asia and China, and in South Asia the gall midge, brown planthopper and yellow stemborer were most important.

The insect pest fauna of rice in Asia is composed of species native to the broad reaches of Asia and those concentrated in selected regions of Asia. In addition, there are newcomers like rice water weevil which have invaded Japan, Korea and Taiwan in recent years (Kiritani, 1992). Thirty-four insects are identified as the principal insect pests of rice in Asia (Wilson and Claridge, 1991). Asia is reported to have a more complex fauna of insect pests than other regions, perhaps because Asia encompasses both tropics and temperate zones, while some of the pest species limit their habitat to either the tropics or the temperate zone. A comparison of rice pest fauna in temperate and tropical Asia

Table 19.1. Losses in rice production due to insects in Asian countries.

Study	Site/country	Loss (%)
Fernando, 1962	Sri Lanka	10–20
Way, 1976	India	35
Way, 1976	Philippines	16–30
1981	Tropical rice	35–44
Ahrens et al., 1983	East and southern Asia	23.7
Litsinger et al., 1987	Philippines	18.3
Kalode, 1987	India	28.8

shows that each niche in paddy fields is almost fully occupied by the same species or by the ecological homologues of different species (Kiritani, 1992). The major insect pests of Asia are briefly described below.

Yellow stemborer

It is widely viewed that yield loss due to stemborer injury is difficult to estimate and no precise method of assessing losses has been established, although it is an important insect-attacking rice in Asia. In India, borer activity is at its peak during the kharif season (October–November) resulting in 'white head'. In the rabi (summer) season, rice is subjected to borer attack both at the vegetative stage (February–March) causing 'dead heart' and at the heading stage (April–May) causing 'white head' (Chelliah et al., 1989). Estimates indicate that 2% white heads caused 4.4% yield loss in fields with a yield potential of 3 million tons ha^{-1}, and 6.4% yield loss in fields with a yield potential of 4 million tons ha^{-1}. Varieties with high silica content showed a negative reaction with dead heart. Higher amounts of silica in host plants adversely affect larval survival and dead heart formation. Stemborer is found to be an important insect in all countries for which yield losses have been estimated.

Rice bug

Species of the genus *Leptocorisa* are known by a number of popular names in different countries. The estimates of economic losses have been reported by various workers (Luh, 1982). Estimates of 10–100% yield losses are reported. An infestation of 100,000 per acre is stated to cause 25% yield loss. It seems that one rice bug reduces yield by 2.2 g per hill (Luh, 1982). In India, light to moderate occurrence is reported in Bihar, Gujarat, Haryana, Madhya Pradesh, Orissa, Tamil Nadu, Uttar Pradesh and West Bengal (Chelliah, 1989).

Leaf folder

This insect pest is widely distributed in rice-growing areas of Asia and Oceania, northeast Australia and Madagascar and its distribution varies widely between 48°N and 24°S latitude and 0°E to 172°W longitude (Khan et al., 1988). *Cnaphalocrocis medinalis* was a long distance migrant into temperate China, Japan and Taiwan (Sundarababu et al., 1993). Every year, the initial population migrated to these temperate countries from tropical regions. The insects migrated northward in the spring season and southward in the autumn. In the Philippines, this leaf folder remained the year-round in irrigated multiple rice crop systems but dispersed to a distance of 10 km in order to colonize the rainfed rice areas in the wet season (Sundarababu et al., 1993). *Marasmia patnalis* occurred in Indonesia, Malaysia, India and Sri Lanka. In the past 20 years several outbreaks have been reported in Bangladesh, China, Fiji, India, Japan, Korea, Malaysia, Nepal, Philippines, Sri Lanka and Vietnam.

Entomological studies report the damage due to this pest ranges from 18.3 to 58.4% in different countries. It has become increasingly abundant since the intensification of rice production (Sundarababu et al., 1993). It has assumed greater importance both in upland and lowland rice fields, particularly in areas where modern varieties are grown extensively. A look at the literature indicates that extended rice areas with assured irrigation systems, multiple rice cropping, reduced genetic variability due to the replacement of traditional varieties by high-yielding varieties and the application of higher doses of nitrogenous fertilizer have aggravated the problem of the leaf folder.

Brown planthopper

The brown planthopper (BPH) is one of the most important rice pests in Asian countries. It dies out over winter in temperate countries such as Japan, and annually migrates from overseas. BPH has been a major pest in India for the past two decades, causing great concern in certain intensive rice-growing tracts. In the kharif season of 1990, a severe outbreak took place in Bihar affecting more than 30,000 ha. In the past two decades, severe outbreaks have been reported from Taiwan. In Indonesia, the pest caused annual destructive infestations to rice fields during the 1970s, while it shifted around in the 1980s. A serious outbreak occurred in North Sumatra in 1982/83, Central and West Java in 1986/87 and in East Java and Lampung in 1987/88 (Sawada *et al.*, 1992).

Large-scale occurrence of BPH in Asia is favoured by mild winters followed by a hot summer with little rain during the May–June period.

Gall midge

The rice gall midge is a serious pest of rice in South and Southeast Asia. Studies indicate that for every 1% increase in rice gall midge infestation, yield loss was 0.502%, with losses varying among varieties and reaching as high as 2.5% in some of the varieties (Luh, 1982). Severe losses due to this pest are reported from India, Vietnam, Sri Lanka and Burma. In India it is most prevalent in Madhya Pradesh, Bihar, Orrisa, Andhra Pradesh and Maharashtra (Chelliah *et al.*, 1989). Evidence shows that the infestation is greater in dwarf varieties than in tall varieties.

Green leafhopper

The green leafhopper is the vector of tungro virus. Among green leafhoppers, the species *Nephotettix cincticeps* is restricted primarily to the temperate region, while two more species, *N. virescens* and *N. nigropictus* are distributed in wide areas extending from subtropical to tropical regions. Studies indicate the BPH and green leafhopper occupy similar niches in paddy fields (in Indonesia) implying that the BPH might be a possible competitor reducing the reproduction efficiency of green leafhopper (Widiarta *et al.*, 1993). It is an important pest as a vector of rice dwarf virus disease in southwestern Japan. In Okayama, western Japan, it lives in foxtail grass before invading transplanted rice (Widiarta *et al.*, 1993). It occurs widely in India but has assumed greater magnitudes in Andhra Pradesh, Madhya Pradesh, Orissa, Bihar, West Bengal, Tamil Nadu and Assam states, causing direct injuries to plants, besides being a vector of the rice tungro virus. Recently, rice tungro virus outbreaks occurred, especially in Andhra Pradesh, in more than 134,000 ha and in Orissa in 49,000 ha during the kharif season of 1990/91. Outbreaks of this pest are also reported from Bangladesh and Indonesia.

III. Estimates of Insect-Caused Yield Losses

Studies have been undertaken since the early part of this century to assess the magnitude of yield losses due to insects in various subregions and countries. The estimates of losses show significant variations by regions, as well as substantial differences across varieties, seasons, climatic regimes, rice production intensifications and production environments. Better estimates of the magnitude of losses will provide a clearer picture in order to plan research activities.

Table 19.2 shows the estimates of losses due to insects in recent periods as compiled in the country studies. Although different methodologies were used to derive the estimates, as explained in the individual country study chapters, with one exception the losses are of similar magnitude. The one exception is Indonesia, where the stemborer is reported as having caused a very high level of damage in the irrigated ecosystem of the 32 survey villages in 1990/91 (see Table 15.3). Even in those villages, however, loss from the stemborer was reported at less than 1% (i.e. below 20 kg ha^{-1}) in the rainfed and upland situations in the same year.

Leaving aside the stemborer, damage reported in all locations range from 1 to 110 kg ha^{-1} across the study locations. Rice leaf folder losses ranged from 9 to 44 kg ha^{-1}, and was reported in six of the seven locations. Brown planthopper losses were also reported in six of the seven locations, with damage ranging from 7 to 34 kg ha^{-1}. Ear head bug losses were reported in five of the seven locations, ranging from 3 to 40 kg ha^{-1}. Less frequently reported were losses from the green leafhopper, gall midge and rice hispa.

Table 19.2. Estimated average rice production losses (kg ha^{-1}) caused by main insects identified as constraints in selected rice ecologies.

Region	Stem-borer	Rice leaf folder	Brown plant-hopper	Green leaf-hopper	Ear head bug	Gall midge	Rice hispa	Other
East India[a]	35	9	7	15	3	8	na	27
West Bengal[b]	1	15	25	na	na	na	1	na
Southern India[c]	32	44	23	19	35	25	na	18
Bangladesh[d]	38	11	na	na	40	na	41	5
Indonesia[e]	346	na	25	na	28	na	na	na
Thailand[f]	1	9	21	12	na	na	na	40
Nepal[g]	110	42	34	41	20	na	89	96+

na, not applicable.
[a]Rainfed lowland, Table 7.6.
[b]Rainfed lowland aman, Table 8.6.
[c]All ecosystems, Table 9.2.
[d]Rainfed lowland aman, Table 11.7.
[e]Assumed 3.5 tons ha^{-1} yield, irrigated ecosystem, Table 15.3.
[f]All ecosystems (10 million ha), Table 13.9.
[g]Extension workers' views, Table 12.8.

IV. Rice Virus Diseases

It has been reported that the broad spread of high-yield varieties of rice in Asia during the green revolution has been associated with outbreaks of several virus diseases of rice (Saito, 1986). Many research activities have been devoted to overcome these virus diseases on rice, and developments in biotechnology may provide new opportunities to address the rice production constraints caused by virus diseases, though results are likely to be realized only in the long term.

Some misunderstanding or confusion had occurred in the aetiology of virus diseases due to the difficulty in studying them, as most of the rice virus diseases are transmitted by insects and not by mechanical means (Saito, 1986). This may have led to the identification of the same virus disease under different names in different countries, or in some cases, one name for different virus diseases in different countries (Table 19.3).

Investigation of rice diseases in Nepal was initiated in 1963–1964. So far, of 21 rice diseases identified in Nepal, two caused by viruses (rice dwarf virus and rice tungro virus) have been studied and the causal virus identified (Amatya and

Table 19.3. Estimated rice production loss caused by virus diseases in selected Asian countries.

Country	Type of rice	Viruses	Production loss (tons)	Yield loss (kg ha^{-1})	Rank of virus Biotic factors	Rank of virus All factors
Southern India	All rice	Tungro	131,580	16.98	13	21
Eastern India	All rice	Tungro	5,000	0.19	28	43
Bangladesh	Rainfed rice	Tungro	4,884	2.97	10	19
	Irrigated rice	Tungro	764	1.98	11	19
China	Early rice	Stripe	1,305	0.14	24	45
		Dwarf	93	0.01	36	58
		Yellow stunt	870	0.09	31	50
	Late rice	Dwarf	484	0.05	35	54
		Stripe	193	0.02	37	56
Indonesia	Lowland rice	Tungro	7,794	0.96	8	–
		Grassy stunt	3,261	0.40	14	–
		Yellow stunt	1,411	0.17	17	–
		Ragged stunt	445	0.05	20	–
	Upland rice	Tungro	20	0.02	18	–
		Grassy stunt	16	0.01	19	–
		Yellow stunt	10	0.01	21	–

Source: Ramasamy et al. (1994) for southern India; Widawsky and O'Toole (1990) for eastern India; Dey et al. (1994) for Bangladesh; Lin (1991) and Lin and Shen (1994) for China; and Jatileksono (1994) for Indonesia.

Manandhar, 1986). However, virus diseases were not among the top 26 problems either reported by rice farmers or listed by rice researchers and extension specialists in Nepal (see Chapter 12).

In India, rice is affected by four virus diseases: tungro, grassy stunt, ragged stunt and necrotic mosaic. Of these, tungro is widespread and occurs in epidemic form in several states of India. Grassy stunt is next in importance and is confined to southern India. Ragged stunt and necrotic mosaic are of minor importance in India and are reported only on experimental farms (Anjaneyulu, 1986).

Ramasamy et al. (1994) reported that rice tungro virus caused serious damage in southern India, creating total production losses of as much as 131,580 tons or average losses of 17 kg ha^{-1}. This loss is very high compared to other Asian countries (see Table 19.3). Within the country, however, rice yield loss due to tungro virus ranked only 13th among biotic factors and ranked 21st among all constraints in rice production. The top three disease constraints are blast, brown spot and sheath blight, causing rice yield losses of 30, 18.5 and 18.1 kg ha^{-1}, respectively.

In eastern India, the largest rice area of India, tungro virus is reported as a minor problem, creating paddy yield losses of only 0.19 kg ha^{-1} as compared to that caused by weeds (76 kg ha^{-1}), yellow stemborer (26 kg ha^{-1}), bacterial leaf blight (26 kg ha^{-1}), blast (22 kg ha^{-1}) and brown spot (11 kg ha^{-1}) (Widawsky and O'Toole, 1990).

Tungro virus has been reported to attack aus season rice in Bangladesh, but it is a minor problem (Dey et al., 1994). It caused 3 kg ha^{-1} paddy yield loss on average in rainfed areas, which is much lower than, for example, bacterial leaf blight (25 kg ha^{-1}), rice hispa (20 kg ha^{-1}), rice bug (18 kg ha^{-1}), stemborer (18 kg ha^{-1}) and blast (17 kg ha)$^{-1}$. In the irrigated area of Bangladesh, tungro caused 2 kg ha^{-1} yield loss, very little compared to that caused by stemborer (135 kg ha^{-1}), bacterial leaf blight (41 kg ha^{-1}), bacterial leaf streak (34 kg ha^{-1}), rice bug (26 kg ha^{-1}) and blast (24 kg ha^{-1}).

In Malaysia, four virus diseases – tungro, ragged stunt, grassy stunt and gall dwarf – are recognized in local rice crops. Of these viruses, tungro seems to be the most important and has been reported to have affected rice areas of 12,734 ha in 1981, 17,507 ha in 1982, 12,892 ha in 1983 and 1,964 ha in 1984 (Abubakar and Hashim, 1986).

Five virus diseases of rice are reported in Thailand, namely yellow orange leaf, ragged stunt, gall dwarf, transitory yellowing and grassy stunt. But they are kept under control by the use of suitable insecticides and resistant varieties (Chandrasrikul et al., 1986).

In the Philippines, the outbreak of tungro in the 1940s caused 30% annual yield losses, and in 1971–1981 tungro resulted in an average annual loss of 36% at one site in the country, whereas at another site, tungro appeared only in 3 years out of 10 (Teng and Revilla, 1994).

A total of 11 virus and virus-like diseases of rice have been observed in China, namely yellow dwarf, dwarf, transitory yellowing, black-streaked dwarf, stripe, grassy stunt, bunchy stunt, ragged stunt, orange leaf, tungro and gall dwarf. As far as the distribution and the damage caused by these diseases are concerned, transitory yellowing and dwarf are the most important (Hui, 1986). However, Lin (1991) reported that all these viruses were still minor problems because each of them resulted in less than 1 kg ha^{-1} of yield loss. The top three biotic constraints in China are sheath blight, weeds and blast, and they caused rice yield losses of 147, 48 and 29 kg ha^{-1} for early rice and 65, 43, and 38 kg ha^{-1} for late rice, respectively.

Eight viruses attack rice plants in Japan, i.e. dwarf, stripe, black-streaked dwarf, grassy stunt, transitory yellowing, ragged stunt, Waika and necrosis mosaic viruses (Shikata, 1986). Rice stripe virus causes the most severe damage to rice throughout Japan, and its outbreak highly depends upon the infective population of *Laodelphax striatellus*. The control measures of rice virus diseases consist of: (i) eradication of the vector populations to decrease the number of active transmitters; (ii) use of resistant rice cultivars; (iii) cultural practices adapted to local conditions, cultivars and vector dynamics; (iv) elimination of the virus source; and (v) exclusion of weeds and crops that increase vector density.

In Indonesia, tungro, grassy stunt and ragged stunt virus diseases have been a potential problem to rice production (Tantera, 1986). Research on these virus diseases in recent years has resulted in an improvement of the understanding of their biology and ecology, which has enabled the development of effective methods to control these diseases. An integrated control technique combining the use of resistant varieties, cultural practices and insect control has brought about a considerable decrease in virus infection in Indonesian rice. A varietal rotation scheme has been successfully implemented for tungro virus in South Sulawesi. Varietal resistance is used extensively for grassy stunt virus, in combination with cultural practices and insecticidal spray. The figures in Table 19.3 show that each virus disease in Indonesia caused rice yield losses of less than 1 kg ha^{-1}.

V. Other Diseases of Rice

Some idea of the importance of four other rice diseases for one important rice season or ecology in the study countries can be obtained from the data in Table 19.4. More information is available in the individual country study chapters, but bacterial blight was reported as a relatively important constraint in most countries except China and Thailand. It was clearly more important in the higher-yielding wet season (kharif season) than in the rainfed systems in eastern India (see Table 7.6).

Among the country studies, blast was most important in Nepal, as would be expected because the temperature regime is more favourable for blast in Nepal than in the other regions represented. Consistent with general knowledge, blast was much more important in the upland ecology of eastern India (the aus crop) than in the other ecologies (see Table 7.6), and was one of two diseases falling in the top 20 constraints for early season rice in China (see Table 10.11).

Sheath blight was reported as a significant disease in five of the studies. It is the most important disease causing losses in China averaging 30–40 kg ha^{-1} in

Table 19.4. Estimated average rice production losses (kg ha^{-1}) caused by bacterial and fungal diseases of rice in selected rice ecologies.

Region	Bacterial leaf blight	Blast	Sheath rot	Sheath blight	Rice ecology or season
China[a]	na	19	na	37	Early season
Eastern India[b]	26	27	15	7	Rainfed lowland
West Bengal[c]	63	9	na	28	Rainfed lowland (aman)
Southern India[d]	19	30	18	19	All ecosystems
Bangladesh[e]	57	26	na	na	Rainfed lowland (aman)
Indonesia[f]	25	na	na	na	Irrigated
Thailand[g]	na	10	na	na	All ecosystems
Nepal[h]	73	167	4	26	All ecosystems

na, not applicable.
[a] Early season rice, Table 10.11.
[b] Rainfed lowland rice, Table 7.6.
[c] Rainfed lowland (aman), Table 8.6.
[d] All ecosystems, Table 9.2.
[e] Rainfed lowland aman, Table 11.7.
[f] Table 15.3, assumed 3.5 tons ha^{-1} average yield, irrigated ecosystem.
[g] Table 13.9, assuming 10 million ha rice area.
[h] Extension workers' views, Table 12.8.

all three seasons. In eastern India, it was reported to cause nearly twice the yield loss in irrigated rice as in rainfed rice (see Table 7.6), and substantially more per hectare in Kerala than in the rest of southern India (see Table 9.2).

VI. Conclusions

Insect-caused damage is quite apparent to farmers, and in many cases they tend to associate more damage with the insects than scientific studies can substantiate. For example, in the Nepal country study, farmers attribute 739,000 tons of production losses to insects, while researchers attribute 572,000 tons (see Table 12.8). At the same time, farmers report damage from only five insects, while researchers report that eight are important. Farmers also report much higher damage (529,000 tons) from storage pests than researchers do (174,000 tons).

There is no doubt that heavy insect infestations can cause yield loses reaching 100%, but the information available on average national losses, based on the substantial efforts undertaken in the country studies reported in this volume suggest that even losses of 10% on a provincial or national level would be exceptionally high. Indeed, with one exception, losses attributed to a single insect ranged from 100 kg ha^{-1} and below, with most below 50 kg ha^{-1}. The most important insects are the stemborer, leaf folder, brown planthopper, green leafhopper, ear head bug, gall midge and rice hispa. Alone among countries, China reported extremely low levels of insect-caused losses (see Table 10.11).

Tungro and other virus diseases appear to be much less important problems in Asian rice production than reported earlier, and other diseases seem far more serious. It has been reported that bacterial leaf blight is more serious than tungro in eastern India, Bangladesh, China and Indonesia; rice blast is more serious than tungro virus in eastern and southern India, Bangladesh and China; and sheath blight is much more damaging than any other disease in China. The difference between earlier observations and those reported here may have been caused by the broad adoption of varieties incorporating a good level of tungro resistance, which only were available in the 1980s.

The experience of technological change led by varietal improvement in Asia has contributed significantly to the growth of rice production and to improved grain quality. Varietal improvement complemented by government policies of subsidized fertilizers and irrigation have contributed to remarkably increased rice yield with limited production loss, and farmers have approached potential yields in most irrigated areas. This suggests that future research on rice should give higher priority to increasing the potential yield rather than to yield-loss prevention (see Chapter 22).

References

Abubakar, A.K.B. and Hashim, H.B. (1986) Virus diseases of rice and leguminous crops in Malaysia. In: *Proceedings of a Symposium on Tropical Agriculture Research, Tsukuba, Japan, 1–5 October, 1985.*

Ahrens, C., Cramer, H., Megk, M. and Peschel, H. (1983) Economic impact of crop losses. In: *10th International Congress of Plant Protection.* Vol. I proceedings, Brighton, UK, 20–25 November.

Amatya, P. and Manandhar, H.K. (1986) Virus diseases of rice and legume crops in Nepal: status and future strategies. In: *Proceedings of a Symposium on Tropical Agriculture Research, Tsukuba, Japan, 1–5 October, 1985.*

Anjaneyulu, A. (1986) Virus diseases of rice in India. *Proceedings of a Symposium on Tropical Agriculture Research, Tsukuba, Japan, 1–5 October, 1985.*

Barker, R. and R. Herdt. 1985. *The Rice Economy of Asia.* Washington, D.C.: Resources for the Future, Inc., p. 2.

Chandrasrikul, A., Disthaporn, S. and Kittipakorn, K. (1986) Virus diseases of rice and leguminous crops in Thailand. *Proceedings of a Symposium on Tropical Agriculture Research, Tsukuba, Japan, 1–5 October, 1985.*

Chelliah, S., Bentur, J.S. and Prakasa, P.S. (1989) Approaches in rice pest management – achievements and opportunities. *Oryza* 26, 12–26.

Cramer, H.H. (1967) *Plant Protection and World Crop Production.* Pflanzenschutz Nachrichten Bayer 20.

Dey, M.M., Miah, M.N.I., Mustafi, B.A.A., Ahmed, R., Islam, A.S.M.N. and Alam, M.S. (1994) Rice production constraints in Bangladesh: implications for future research priorities. Paper presented at the Workshop on Rice Prioritization in Asia organized by the Rockefeller Foundation and the International Rice Research Institute at Los Baños, Philippines, 15–22 February, 1994.

Fernando, H.E. (1962) *Entomology Division: Administrative Report of the Director of Agriculture for 1960.* Colombo, Ceylon, pp. 204–208, 252, 267.

Herdt, R.W. and Riely, F.Z. (1987) International Rice Research Priorities: Implications for Biotechnology Initiatives. Rockefeller Foundation Workshop on Allocating Resources for Developing Country Agricultural Research. Bellazio, Italy, 6–10 July, 1987.

Hui, X.L. (1986) Research on rice virus diseases in China. In: *Proceeding of a Symposium on Tropical Agriculture Research, Tsukuba, Japan, 1–5 October, 1985.*

Jatileksono, T. (1994) Current state of rice production and research activities in Indonesia. Paper presented to the Workshop on Rice Prioritization in Asia organized by the Rockefeller Foundation and the International Rice Research Institute at Los Baños, Philippines, 15–22 February, 1994.

Kalode, M.F. (1987) Insect pests of rice and their management. In: Veerabhadra Rao, M. and Sithanantham (eds) *Plant Protection in Field Crops.* Directorate of Rice Research, Hyderabad, India, pp. 61–74.

Khan, Z.R., Barrion, A.T., Litsinger, J.A., Castilla, N.P. and Joshi, R.C. (1988) A bibliography of rice leaf folders. *Insect Science Applications* 9(2), 129–174.

Kiritani, K. (1992) Prospects for integrated pest management in rice cultivation, *Journal of Agriculture Research Quarterly* 26, 81–87.

Lin, J.Y. (1991) Rice production constraints in Zhejiang Province: a pilot study for agricultural research priority in China. Paper presented to the Workshop on Rice Prioritization in Asia organized by the Rockefeller Foundation and the International Rice Research Institute at Los Baños, Philippines, 15–22 February, 1994.

Lin, J.Y. and Shen, M. (1994) Rice production constraints in China: implications for biotechnology initiative. Paper presented to the Workshop on Rice Prioritization in Asia organized by the Rockefeller Foundation and the International Rice Research Institute at Los Baños, Philippines, 15–22 February, 1994.

Litsinger, J.A. (1989) Second generation insect pest problems on high yielding rices. *Tropical Pest Management* 35(3), 235–242.

Litsinger, J.A., Canapi, B.L., Bandong, J.P., Dela Cruz, C.G., Apostol, R.F., Pantua, P.C., Lumaban, M.D., Alviola III, A.L., Raymundo, F., Libecario, E.M., Loevinsohn, M.E. and Joshi, R.C. (1987) Rice crop loss from insect pests in wetland and dryland environments of Asia with emphasis on the Philippines. *Insect Science and its Application* 8, 677–692.

Luh, B.S. (1982) *Rice Production and Utilization.* Department of Food Science and Technology, University of California, Davis.

Ramasamy, C., Shanmugam, T.R. and Kumar, D.S. (1994) Constraints to higher rice yields in different rice production environments and prioritization of rice research in southern India. Paper presented to the Workshop on Rice Prioritization in Asia organized by the Rockefeller Foundation and the International Rice Research Institute at Los Baños, Philippines, 15–22 February, 1994.

Saito, Y. (1986) Progress and trend of research activities in the Asian region in relation to virus diseases in rice. In: *Proceedings of a Symposium on Tropical Agriculture Research, Tsukuba, Japan, 1–5 October, 1985.*

Sawada, H., Subroto, S.W.G., Suwardiwijaya, E., Mustaghfirin, and Kusmayadi, A. (1992) Population dynamics of the brown planthopper in the coastal lowland of West Java, Indonesia. *Journal of Agriculture Research Quarterly* 26, 88–97.

Shikata, E. (1986) Virus diseases of rice and leguminous crops in Japan. In: *Proceedings of a Symposium on Tropical Agriculture Research, Tsukuba, Japan, 1–5 October, 1985.*

Sundarababu, P.C., Balasubramaniam, G., Subramaniam, A. and Gopalan, M. (1993) *Integrated Management of Leaffolder in Rice.* Department of Agriculture and Entomology, Tamil Nadu Agricultural University, Coimbatore (mimeographed).

Tantera, D.M. (1986) Present status of rice and legume virus diseases in Indonesia. In: *Proceedings of a Symposium on Tropical Agriculture Research, Tsukuba, Japan, 1–5 October, 1985.*

Teng, P.S. and Revilla, I. (1994) Crop losses due to diseases and insects. Discussion notes presented to the Workshop on Rice Prioritization in Asia organized by the Rockefeller Foundation and the International Rice Research Institute at Los Baños, Philippines, 15–22 February, 1994.

Way, M.J. (1976) Entomology and the world food situation. *Bulletin of the Entomological Society of America* 22, 125–129.

Widiarta, N., Suzuki, Y., Saki, K.F. and Nakasuji, F. (1993) Comparative population dynamics of green leaf hoppers in paddy fields of the tropics and temperate regions. *Journal of Agriculture Research Quarterly* 26, 115–123.

Widawsky, D. and O'Toole, J.C. (1990) *Prioritizing the Rice Biotechnology Research Agenda for Eastern India.* The Rockefeller Foundation, New York.

Wilson, M.R. and Claridge, M.F. (1991) *Handbook for the Identification of Leafhoppers and Planthoppers of Rice.* CAB International, Wallingford, UK.

20 The Economic Impact of Rice Blast Disease in China

Minggao Shen[1] and Justin Yifu Lin[2]*

[1]Development Institute Research Center for Rural Development, 5 Liuligiao Beili, Fengtaiqu, Beijing 100055, People's Republic of China; [2]China Center for Economic Research, Peking University, Beijing 100871, People's Republic of China

I. Introduction

Rice is the most important grain crop in China, with the highest yield of all grain crops. The annual growth rate of rice output from 1950 to 1990 was 3.13%. The main source of increase in rice output has been yield increase, which was made possible through increased uses of modern varieties and modern inputs. As China's agriculture changed from a traditional to a modern one, a number of yield constraints have been ameliorated, including unfavourable climatic factors (such as temperature and light), water, soil conditions, insects, pests and diseases, but these remain the focus of China's rice research program (see Chapter 10). Of the diseases affecting rice production in China today, blast is one of the most serious. Its persistence continues to be a potential threat to the nation's ability to further increase and stabilize rice output.

This chapter discusses the occurrence of rice blast disease in China and examines the extent of its economic impact. The remaining parts of the chapter are organized as follows. Section II describes the occurrences of rice blast disease in China, focusing on its changes over time. Section III describes the losses that arise from rice blast disease and section V is the conclusion.

II. Factors Contributing to the Occurrence of Rice Blast Disease

Prior to the 1960s, China's agriculture was traditional with little use of modern inputs. Since then, the modernization of Chinese agriculture has been rapid. In addition to extending a range of new varieties and the use of modern inputs, a number of other technological improvements in the system of cultivation and

---

* We are indebted to Paul Teng for very helpful comments on an early draft.

plant management have been introduced. These changes, in part, have served to create an environment in which rice blast disease has evolved from a regional disease to one capable of creating a nationwide disaster.

A. Spread of modern varieties

China's agricultural research was fairly weak until the end of the 1950s. The main varieties of rice were bred by farmers themselves and were typically used for several generations. Although these varieties were well adapted to their region of origin, their yields were low. Under the conditions in the 1950s and 1960s, rice blast remained a regional disease occurring mainly in mountainous areas with low temperature and high humidity (GSPP, 1988). Since the 1960s, with the development of agricultural research in China, modern varieties with high-yield potential have been replacing traditional farm-bred varieties. The release of modern varieties was much faster than was possible with traditional methods. During this period, the Chinese rice economy experienced two major breakthroughs: one was the substitution of semidwarf fertilizer-responsive varieties for long-stalked varieties in the 1960s; the other was the diffusion of hybrid varieties in the 1970s. In the 1980s, the proportion of total area sown to improved varieties was more than 80% of the total sown area (Lin and Shen, 1991).

The expansion of area sown to improved varieties significantly increased rice yield, but simultaneously decreased the number of varieties. This increase in relative genetic homogeneity has resulted in some incidence of widespread disease occurrence. Also, since blast physiological races change in composition with varieties in different rice regions, the substitution of modern varieties has caused a differentiation of race groups. This differentiation process has included a loss of varietal resistance, facilitating the spread of rice blast disease.

B. Changes in cropping systems

Chinese producers have switched from single season rice to two season rice to increase grain output since the 1950s. In the southern Yangzi River areas, farmers have moved from rotations which included other crops to continuous cropping systems. North of the Yangzi River and south of the Huai River, winter wheat and rapeseed areas were expanded, and rice–wheat systems were popularized. In the 1970s, in the middle and lower reaches of the Yangzi River and in the southern rice regions, a three-harvest system (including two rice crops) was developed. Production rotations such as rice–rice–green manure, rice–rice–wheat and rice–rice–rapeseed spread rapidly (Lin and Shen, 1991). With the rise of the cropping index, single season rice and double-harvest rice coexisted. Early season rice overlapped with single season rice and single season rice overlapped with late season rice in the same locations in many parts of China. This cropping pattern led to a serious incidence of cross-infection. Atavism of rice blast disease was common (GSPP, 1988). The change of cropping systems also altered the growing season for rice, which increased the possibility of weather-induced blast disease. Early season rice and single season rice varieties have been especially vulnerable since a large part of the crop's growth then occurred during the rainy season which is characterized by long spells of fog and less sunlight.

C. Changes in cultural practices

Research has shown that the occurrence of rice blast disease is closely related to environmental factors, among which temperature and relative humidity are the most important. Also, high rates of nitrogen fertilizer make rice stems more succulent and the plant produces more leafy material making infection more probable (Huang and Miao, 1993). With the spread of modern varieties, rice cultivation practice in China radically changed. The shortening of the growing seasons, the douse planting, the increased application of fertilizers and the expansion of irrigated area has made rice blast disease more likely.

Due to these changes, rice blast disease changed from a regional disease in the 1950s and 1960s to one which threatened China with nationwide disaster in the 1970s. By the late 1970s and the early 1980s, rice blast disease epidemics began occurring nationwide. Figure 20.1 shows the proportion of rice area in China affected by leaf blast and neck blast between 1980 and 1990, based on data from an intensive survey of all of the 176 rice-producing

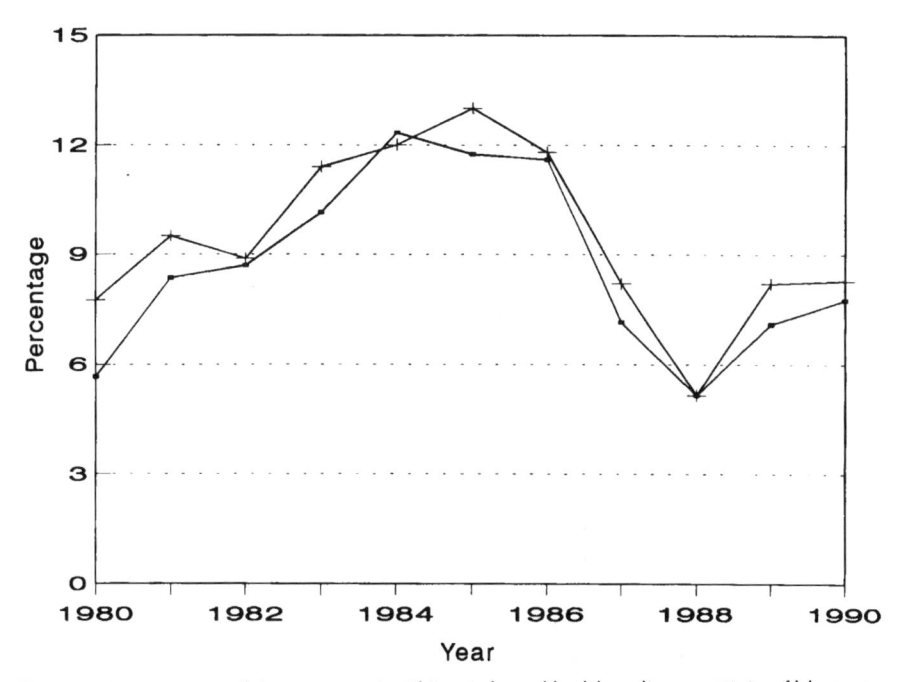

Fig. 20.1. Percentage of rice acreage in China infected by blast disease. ■, Leaf blast; +, neck blast.

prefectures in China carried out by the authors in 1991. In 1985, leaf blast affected 12% of national rice acreage and neck blast affected another 12%. From the figure we find that the incidence of leaf blast disease and neck blast disease are highly correlated. The incidence of blast disease fluctuated from year to year. Table 20.1 shows that in 1984 and 1985 blast disease was a significant disease (20% or more of rice infested) in prefectures with about 40% of China's rice area. In 1988, however, blast disease was a serious problem in prefectures that only covered 7% of China's rice area, and in almost half of the rice-producing prefectures in China blast disease did not occur at all. In 1989 and 1990, blast was again more widespread. Table 20.2 lists the occurrence separately for early, late and single season rice. The information shows that early season rice and single season rice, which are grown during periods of low temperature and continuous rain, are more likely to be affected by blast disease than late season rice.

Table 20.1. The occurrence of rice blast disease in China (%)

(1)	Rank 5 (2)	Rank 4 (3)	Rank 3 (4)	Rank 2 (5)	Rank 1 (6)
1980	54	19	12	6	9
1981	41	23	13	7	16
1982	30	26	14	9	21
1983	29	24	12	10	25
1984	23	14	12	14	37
1985	22	19	13	8	38
1986	28	19	10	10	32
1987	36	23	17	8	16
1988	50	25	11	8	7
1989	32	21	14	14	18
1990	31	26	18	10	16

Rank 5: Blast disease occurred in 0–5% of the rice area in a prefecture's total rice area.
Rank 4: Blast disease occurred in 5.1–10% of the rice area in a prefecture's total rice area.
Rank 3: Blast disease occurred in 10.1–15% of the rice area in a prefecture's total rice area.
Rank 2: Blast disease occurred in 15.1–20% of the rice area in a prefecture's total rice area.
Rank 1: Blast disease occurred in 20% of the rice area in a prefecture's total rice area.
Numbers in columns 2 to 6 indicate the percentage of the nation's rice acreage in the prefectures falling in the indicated rank infested with blast.
Source: The data reported in the table are taken from an intensive survey of all 176 rice-producing prefectures in China by the authors in 1991 for the project 'Agricultural Research Priorities in China'.

Table 20.2. The occurrence of blast disease for early, late and single season rice in China (%).

	Early season		Late season		Single season	
	Rank 2	Rank 1	Rank 2	Rank 1	Rank 2	Rank 1
(1)	(2)	(3)	(4)	(5)	(6)	(7)
1980	98	2	85	15	89	11
1981	79	21	80	20	94	6
1982	64	36	84	16	90	10
1983	67	33	69	31	86	14
1984	44	56	82	18	62	38
1985	63	37	71	29	54	46
1986	51	49	83	17	69	31
1987	82	18	87	13	83	17
1988	96	4	93	7	91	9
1989	80	20	88	12	79	21
1990	75	25	94	6	84	16

Rank 2: Blast disease occurred in 0–20% of the rice area in a prefecture's total rice area.
Rank 1: Blast disease occurred in 20% of the rice area in a prefecture's total rice area.
Numbers in columns 2 to 7 indicate the percentage of the respective rank's rice acreage in the nation's total rice acreage.
Source: See Table 20.1.

III. Yield Loss Caused by Rice Blast Disease and Prospects for its Control

Based on our prefecture-level survey in 1991, we found that the output loss caused by rice blast disease has been greatly reduced by prevention and control measures. The average loss of early season rice caused by blast disease in 1981–1990 was only 27.2 kg ha^{-1}, about 0.53% of average yields in that 10-year period. Late season rice had an even smaller loss, only about 0.23% of the average yields in 1981–1990. Single season rice loss was about 0.45% of the average yield.

Table 20.3 shows that the early half of the 1980s experienced heavier blast disease losses, especially during the two large outbreaks in 1984 and 1985. In those 2 years the output loss was around 0.7–0.8% of the yield. In 1985, the blast disease caused the single season rice to suffer a loss of 130 kg ha^{-1} (about 2.2% of the average yield in China). In 1988, the loss was lightest, less than 0.1% of the yield. A correlation analysis, based on the yearly observations of the prefecture-level data in 1980–1990 in our national survey, shows that the percentage of yield loss in a prefecture is negatively correlated with the yield level in that prefecture but that the correlation is not significant statistically. It suggests that rice blast disease influences the yields, but its impact is limited.

Overall, rice blast is the number two rice disease in China, second only to banded sclerotia blight, while higher than bacterial leaf blight (see Table 20.3).

Table 20.3. Actual losses caused by rice diseases in China (kg ha^{-1})

Year	Total	Sclerotia blight	Blast	Bacterial leaf blight
1980	55.8	22.0	23.6	8.7
1981	58.7	22.3	28.7	7.0
1982	65.4	25.2	21.3	12.9
1983	66.8	31.6	26.9	5.9
1984	85.5	36.6	40.6	7.0
1985	112.1	64.7	40.9	5.9
1986	56.2	28.4	19.0	3.7
1987	68.2	46.1	11.5	5.9
1988	50.4	35.6	7.0	2.8
1989	72.4	40.6	18.3	8.5
1990	71.4	38.0	15.3	13.6
Average	69.4	35.5	23.0	7.4

Source: See Table 20.1.

For early season rice in the 1980s, the loss due to blast was 39.4% of the total losses from diseases (also second to banded sclerotia blight). For late season rice, the loss was only 19.8%, behind both banded sclerotia blight and bacterial leaf blight. For single season rice, the loss due to rice blast disease was 30.5%, much heavier than that due to bacterial leaf blight. Among all the rice diseases, the loss caused by banded sclerotia blight, blast disease and bacterial leaf blight totalled between 84 and 97% of total losses from diseases, while the losses caused by all the other diseases were negligible.

The success of controlling rice blast disease in China has been achieved primarily by setting up a comprehensive plant protection system as well as the ability to mobilize farmers. With higher farmer incomes, however, some of these successes may be weakened or even disappear. The proportion of agricultural output to the national economy is decreasing, and the structure of vested interests and investment is changing. First, the regional governments at different levels, previously called 'agricultural governments', have in many places changed their focus to relatively more profitable sectors, such as industry, business and the provision of services. The proportion of government investment in agricultural infrastructure to total investment declined from 17.6% in 1963–1965 to 3.3% in 1986–1990 (State Statistic Bureau of China, 1991). The infrastructure accumulated in the 1960s and 1970s may have declined. Second, the downturn of government investment in agriculture and the dissolution of the collective economy have weakened the agricultural research and extension system, which has public goods characteristics (Lin, 1991). Third, the rural areas have diversified after the rural reforms and the proportion of income from grain production in the farmers' overall income has fallen. Moreover, the government often suppresses grain markets to lower the grain prices. Grain production is less profitable than other activities. Farmers' incentives to invest in grain production are inadequate. With the shift of labour from the agricultural sector to industries, the quality of the remaining farming population in terms of education decreases. Therefore, it is difficult to effectively implement prevention and protection measures (such as blast disease management) over a large area as had been done previously.

Resistance tests made by the Jiangxi Forecasting Co-ordination Group (1989) in the late 1980s found that the proportions of strong, moderate, weak and non-resistant hybrid rice varieties among total tested varieties were 37.5, 57.1, 7.2 and 0.0%, respectively, and for conventional varieties, they were respectively 19.2, 41.4, 19.2 and 19.2%. One reason for the hybrids' high resistance may arise from the fact that hybrid seeds are replaced freshly each year. The proportion of China's rice area sown to hybrid rice increased from 14% in 1980 to 26% in 1985 and then to 50% in 1990. The large-scale spread of hybrid rice after the mid-1980s may in part be responsible for stopping the outbreak of rice blast disease. Because currently the quality of hybrid rice is inferior to conventional rice and is therefore less profitable, the sown area of hybrid rice has been decreasing since 1990. This trend will probably not be reversed until hybrid varieties with better quality rice are bred and extended. The expansion of area sown to conventional rice varieties threatens to lower the resistance of China's rice crops to blast; meanwhile, seeds of the conventional varieties with blast resistance are in short supply. This may also lead to greater outbreaks of rice blast disease in the future.

The burden of controlling rice blast disease in the future falls on rice breeders. Since the blast races are continually changing, one kind of resistant variety may lose its resistance in a period as short as several years. Formerly popular conventional varieties – Nongken 58, Jingying 59 and Zhenshan series – are now not grown since they have lost their resistance to rice blast disease. The loss of resistance by Xianyou 2 was the main cause of the blast outbreaks in 1984 and 1985. Since breeding a new variety requires at least 5–7 years, a strong research effort is needed now more than ever.

IV. Conclusions

Intensive cultivation of rice is a characteristic of Chinese agriculture. Given the constant population pressure, the Chinese government and the farming community have put much effort into variety-improvement research and control of diseases and insect pests. As a result, China has one of the world's highest rice yields. Nevertheless, for the time being, the incentive to increase rice output has been weakened by a number of factors. The ability to increase output by traditional cultivation methods has been exhausted, while the rice sector is losing ground in its competition with other farm and non-farm sectors. Rice production needs support and subsidies from the government. As a result, it must continue to rely on the public sector's research and support system for new advances in science and technology.

With economic development, the requirements of agricultural technology are also changing. The prevention and control of rice blast disease in China has been a great success. But new, appropriate technology may be needed to preserve China's achievement. The technological choices of Chinese farmers are more and more determined by economic considerations (Lin and Shen, 1991). In order to develop and strengthen the prevention and control technology of rice blast disease, it is very important to consider the cost of agricultural technology innovation, the cost of learning and acquiring these technologies and the economic benefits that can be derived after the application of these new technologies.

References

General Station of Plant Protection (GSPP), Ministry of Agriculture (1988) Thirty-six years in prevention and control of rice diseases and pests in China. In: Zhen, Z. (ed.) *The Development of Prevention and Control of Rice Diseases and Pests in China*. Zhejiang Science and Technology Press, Zhejiang, pp. 1–8.

Huang, Q. and Miao, F. (1993) *The Disaster Reducing Agriculture*. Chinese Science and Technology Press, Beijing.

Jiangxi Forecasting Co-ordination Group of Rice Blast Disease Race Group and Resistance of Varieties (1989) The appraising results of rice blast disease resistance and its implications. In: National Plant Protection Station, Ministry of Agriculture (ed.) *Disease and Pest Forecast*, No. 5 (suppl). Agricultural Press, Beijing, pp. 37–37.

Lin, J.Y. (1991) The reforming of family responsibility system and the application of hybrid rice in China. *Journal of Development Economics* 36(1).

Lin, J.Y. and Shen, M. (1991) *Policy-making and the Technology Change in Chinese Agriculture*. Work Paper, Development Research Center.

State Statistic Bureau of China (1991) *Statistical Yearbook of China 1991*. The Statistical Press of China, Beijing.

IV Priority-Setting Applications

21 An Application of Priority-Setting Methods to the Rice Biotechnology Program

R.E. Evenson
Economic Growth Center, Yale University, 27 Hillhouse Avenue, New Haven, CT 06520, USA

I. Introduction

In this and the following chapter we report two applications of priority-setting methods to rice research. This chapter reports an application to research that has a high degree of uncertainty but lends itself to the use of *ex post* evidence for all but one component of priority setting, the 'time to achievement' (W_t) component. For this component, subjective probability estimates (SPEs) are utilized. In the next chapter, priority setting is applied to a broader range of rice research problem areas (RPAs) which are subject to less uncertainty, but where SPE methods are applied to two components, the 'benefits per unit' (b/u) component of priority setting, as well as the W_t component. In both applications, the units component is based on *ex post* or prior evidence.

In section II of this chapter we review the objectives and RPAs of the Rockefeller Foundation-funded International Program on Rice Biotechnology. Section III discusses the application of priority-setting methods (discussed more generally in Chapter 6) to the rice biotechnology program. Section IV reviews the *ex post* evidence used to establish the b/u components in the exercise. Section V reports the time-to-achievement estimates obtained by SPE methods. Section VI discusses the priority-setting and economic implications.

II. The Rice Biotechnology Program

As of early 1994 the rice biotechnology program had supported more than 130 projects conducted in 26 countries (including projects in several International Agricultural Research Centers, IARCs). Sixty-nine projects were located in developing countries. The research objectives of the rice biotechnology program are characterized as follows (note some projects had more than one objective):

Biotechnology tool development: 68 projects.
Developing yield-enhancement technologies: 75 projects.
Developing disease-resistance technologies: 38 projects.
Developing insect-resistance technologies: 19 projects.
Developing stress-tolerance technology: 11 projects.
Developing grain-quality technologies: 12 projects.
Other objectives: 8 projects.

These research objectives guided the RPA definitions used in the priority-setting exercise. They also reflect the priority traits strategy built into the design of the program. In particular, the disease-resistance, insect-resistance, stress-tolerance and grain-quality objectives are all related to traits controlled by one (or few) genes. These traits, in general, have formed a significant part of conventional rice-breeding strategy over the past two decades. This close parallel between the Rockefeller Foundation program design and conventional breeding and rice research makes it possible to utilize *ex post* evidence for the benefits per unit component of the priority-setting assessment. It should be noted here, however, that while the priority traits strategy emphasizes single gene traits, it does not preclude yield-enhancement technology strategies. Indeed, a larger number of projects in the program are seeking yield-enhancement objectives than are seeking specific resistance or tolerance objectives.

III. Methods: a Review (see Chapter 6)

The general expression for the present value of the benefits stream from a given research program is:

$$\begin{aligned} PVB_0 &= \sum_{t=0}^{n} B_t/(1+r)^t \\ &= \sum_{t=0}^{n} (b/u)_t U_t (1+r)^t \\ &= \sum_{t=0}^{n} (b/u)_t/(1+r)^t. \end{aligned} \qquad (21.1)$$

Deriving a quantitative value of present value of benefits, PVB, requires quantitative values for the three components in this expression: benefits per unit affected after t periods $= (b/u)^*$; units affected after t periods $= U^*$; and time to achievement weights $= (W_t)$. The time to achievement weights indicate the proportion of benefits achieved in each period.

Rigorously derived quantitative values for all three components would be ideal, but are impossible. Realistically, the priority setter has several options for estimating this benefits stream; but all involve obtaining SPEs of all or some of the information needed to calculate PVB_0.

Option 1: Obtain SPEs of B_t by period.
Option 2: Obtain SPEs for $(b/u)_t$ and U_t.

Option 3: U^* could be estimated from other evidence (e.g. past experience or secondary data) and SPEs obtained for $(b/u)_t$.
Option 4: Both U^* and the $(b/u)^*$ terms could be estimated from other evidence and SPEs obtained for W_t.
Option 5: U^* and W_t can be estimated from other evidence and SPEs for $(b/u)^*$ obtained.

Options 1–3 require those providing the SPEs to incorporate spatial and time dimensions in a single SPE. If SPEs can be confined to only one dimension, better estimates may be expected when uncertainty is high, as in this case. In many cases it is feasible to estimate U^*, the number of units potentially affected by an RPA (e.g. the number of hectares of rice affected by drought stress), from sources other than the individuals who provide SPEs. Thus, in practice, options 3, 4 and 5 are better than 1 and 2. And options 4 and 5 are better than 3 if one can use past experience to estimate both U^* and either $(b/u)^*$ or W_t.

A critical part of priority setting is the definition of the RPAs. If these can be related to past experience, this facilitates SPE estimation. In the case of rice biotechnology RPAs, the matter of continuity or consistency with past (pre-biotechnology) RPAs is relevant. If meaningful biotechnology RPAs are consistent with pre-biotechnology RPAs, this offers the possibility that option 4 (where U^* and $(b/u)^*$ are based on past experience and SPEs are obtained from W_t) can be pursued. (See Chapter 6 for a fuller discussion.)

The biotechnology objectives in the program, except for tool development, are all consistent with pre-biotechnology objectives. Several national priority-setting studies have recently provided estimates of units potentially affected by ecosystem regions (U^*) for traits objectives (see Chapters 7–19). Hedonic trait valuation studies (see Chapter 1) have provided estimates of $(b/u)^*$. On these grounds it was decided that it was most appropriate to obtain SPEs for time to achievement (W_t) for RPAs related to the rice biotechnology program's objectives.

SPEs must be provided by scientists with technical and scientific skills and objectivity. They cannot reasonably be obtained from non-scientists (whereas non-scientists may estimate U^* and $(b/u)^*$); but scientists are subject to possible bias. They would not be working on particular problems unless they have a relatively favourable assessment of likely outcomes. What one hopes is that the bias and uncertainty in the estimates can be 'judged' to some extent by two procedures. The first is to obtain independent SPEs from a number of informants so that one can examine the distribution of SPEs. The second is to specifically allow for uncertainty of estimates by obtaining from each respondent, not a single point estimate, but a subjective probability distribution (SPD) estimate.

The SPD dimension of the study was achieved by asking respondents for two estimates for time to achievement. All respondents were asked to presume that current international and national research programs would continue with a constant level of support. Researchers were, however, expected to change research programs and research objectives in response to scientific and technological developments. One of the SPD estimates was an optimistic estimate described as the earliest date when there was a 25% probability that the research objective would be achieved (or, alternatively, the date by which 25% of a program's multiple objectives would be achieved). Then a second, more conservative, date with a 75% probability of achievement was obtained. These two dates allowed each respondent to express the range of uncertainty (the SPD) for their estimates.

IV. Hedonic *Ex Post* Evidence of Trait Values

A number of *ex post* studies have used indexes of modern variety use (e.g. the percent of area planted to modern varieties) to reflect technology in statistical analyses. It is important to note, however, that the varieties composing the class of modern varieties has not been static through time. The original modern varieties of rice made available to farmers in the late 1960s, IR8, IR20 and related varieties) was largely replaced during the 1970s by a second generation of modern varieties that incorporated new traits, especially brown planthopper resistance and tungro resistance. That generation of modern varieties, in turn, has now been replaced by further generations of modern varieties, each with added traits for resistance to insect pests and diseases, for tolerance to abiotic stresses (heat, cold, drought, floods) and for agronomic traits (especially grain quality).

These traits have become important features of research policy and design. Rice breeders seek both quantitative genetic objectives and specialized trait-based genetic objectives. The IRRI plant type, as exemplified by IR8, represented a major advance in quantitative genetic traits which are complex and controlled by many genes (the IRRI has more recently started work on a second new plant type). The incorporation of specialized traits, which are controlled by a single (or few) genes, has been the objective of most rice-breeding work since the development of the IRRI plant type. The genetic evaluation and utilization program at the IRRI, for example, was directed toward incorporating a number of specialized traits into rice varieties (see Chapter 4). Specialized traits are also the objectives of rice biotechnology research. The tools of biotechnology allow breeders to use genes from alien sources, that is species other than rice, hence broadening the possible traits that might be the objective of research.

Hedonic regression methods can be used to infer trait values. These methods require a measure of value of the rice cultivars in which the traits are embedded. Traits have two means by which they contribute values. First, they may result in higher rice yields because of reduced losses from pests and disease. Second, they may also contribute value if they enable the high-yielding plant type to be produced in rice ecologies or environments where it was previously unsuited.

A. Hedonic estimates of trait values in rice

The hedonic specification is characterized as follows:

$$V_{ij} = F(T_{1ij}, T_{2ij}, ..., T_{nij}, Z_{ij}) \tag{20.2}$$

where: V_{ij} is a measure of economic value of a variety i in location j; T_{1ij}, T_{2ij}, etc. are trait content indexes for the variety i; and Z_{ij} is a vector of economic and ecological conditions that influence economic value and trait adoption.

Measures of economic value, V_{ij}, have included yields, total factor productivity, crop losses and pesticide use.

Trait content variables include:

1. Insect-resistance traits.
2. Disease-resistance traits.
3. Ecological stress (flood, drought, etc.) tolerance traits.
4. Agronomic (grain) quality traits.

Plant breeders have rated varieties in India and Indonesia according to the presence or absence of these traits.

The Z_{ij} variables include variables measuring climate, soils and other factors. Ideally a variable measuring the natural incidence of insect and disease pressures should be included in the specification. The absence of such data has biased most trait value estimates utilizing crop-loss or crop-yield data (see Evenson, 1994).

In Chapter 1 we reported a summary of hedonic estimates from two studies, one from India and one from Indonesia. These studies differed in approach and data but each contained a reasonably complete set of Z_{ij} variables – including price, weather, extension and infrastructure variables. In addition, a research variable capturing non-genetic contributions was included. Genetic variables included generational variables and specific host plant-resistance variables.

The Indian study was based on two sets of data on specific rice varieties. The first varietal dataset was compiled by the Indian Council of Agricultural Research (ICAR) for selected districts and years. ICAR reports yields for the three highest yielding varieties in farmers' yield trials in each district–year combination for irrigated and unirrigated kharif and rabi season rice. Fertilizer use was measured and yields reported by variety for a sample of farms in each district. Each variety was classified as to trait characteristics and hence yields can be related to these characteristics. This dataset was available for the years 1977–1989 and some 45 districts were covered.

The second Indian varietal dataset was based on state-level data reported by the State Departments of Agriculture of Punjab, Haryana, Andhra Pradesh, Tamil Nadu and Karnataka for different years. For each state–year combination, yield and area planted were reported from farmers' yield crop-cutting estimates for all important varieties planted. From these data, one can use the yields of other varieties in the state and year as a reference group, and for a given year the yield of varieties with trait x can be compared with the yields of all varieties in the state. Weather, insect and disease problems, etc. are thus assumed to have impacted on all varieties equally.

The Indonesian study utilized crop-loss data to examine the value of rice traits. There are at least three conceptual problems with crop-loss estimates. The first is the distinction between *actual* losses by farmers given that certain loss-reduction practices (pesticide application, crop rotation, etc.) were used and the *potential* losses if such practices were not used. This distinction is important because the incorporation of resistance typically will produce both lower actual losses and reduced pesticide use. The reduction in actual losses will not be a full measure of trait values. Potential losses may be a better measure, although pesticide use data should ideally be used.

The second conceptual problem is that crop losses may not be 'additive'. That is, total losses from two or more insect pests may be greater or less than the losses attributed individually to each pest (and similarly for diseases). Crop-loss data are typically attributed to individual pests.

The third conceptual issue is that the incorporation of insect or disease resistance into modern cultivars may result in the adoption of the higher-yielding cultivar in locations where the pests are endemic. Crop losses may not actually be reduced but yields will have been increased in this case.

A related issue that requires consideration in actual estimates is that the natural incidence of pest and disease pressure varies by location and over time. Variation by location creates a left-out variable problem. Locations with high natural incidence may have high losses even though resistance to pests and diseases is quite valuable. Variation over time means that pest and disease resistance may be of little or no value in some periods and of high value in others.

In Indonesia, losses by type (insect and disease) were measured by province and year. A matching set of trait indexes was developed from data on varieties planted and trait ratings by variety.

In Chapter 1 (see Table 1.3) the estimated coefficients for the two studies are summarized. Table 21.1 provides a summary of the economic implications of these studies. The column 'realized to date' is based on the estimated coefficients and varietal adoption rates to date. The column 'full resistance' indicates how losses would fall (or yields increase – losses are expressed relative to yields) if all rice cultivars had full resistance. The column 'potential with conventional breeding' is the author's judgement as to the degree to which conventional breeding programs can, over the next 20 years or so, incorporate more of these traits into rice varieties. These estimates are treated as *ex post* estimates of the benefits (b/u) from incorporating these traits in rice via biotechnology techniques.

B. Traits, rice yields and supply in India

Two additional studies based on Indian district data have been reported. One of these studies (Gollin and Evenson, 1990) was the first to use trait evidence to infer the value of genetic resources. The second, Evenson (1991), is the most recent study of this type.

An Application of Priority-Setting Methods

Table 21.1. Estimated yield and crop-loss reductions from genetic resistance to rice insects and diseases from conventional plant breeding.

| | Yield effects (or cost reduction) (%) ||||||
| | Insect resistance ||| Disease resistance |||
	Realized to date	Potential with conventional breeding	Full resistance	Realized to date	Potential with conventional breeding	Full resistance
Indonesia						
Reduced crop losses	1–2	3–5	10–12	1–3	1–2	5–10
Reduced pesticide	1	2	–	1	2	–
Higher crop TFP	11	25	50	3	5–7	10–15
India						
Higher district yields	3–5	6–10	44	2	7–10	50
Higher state yields	7–10	15–30	84	4.5	10–15	25

TFP, total factor productivity.

For the first study, Gollin and Evenson (1990), utilized different characteristics of varietal content and related them to rice yields at the district level. The study found that when varieties incorporating abiotic stress tolerance and agronomic characteristics were made available to farmers, yields were higher (This was not the case for disease and insect resistance.)

Gollin and Evenson also found strong positive impacts when the number of landraces from both national and international origins incorporated in varieties were associated with higher yields. This was evidence for genetic resource value, as Gollin and Evenson argued that the size of the germplasm collection enabled more of these rare traits materials to be built into modern rice varieties.

The second study, Evenson (1991), utilized similar genetic variables in a model that specifically recognized that modern variety adoption was itself a function of the genetic content of modern varieties. The model was a full rice supply model, treating the area planted to rice, the percent planted to modern varieties and rice yields as endogenous variables. Exogenous variables included price and infrastructure variables, non-varietal research and extension variables. The results showed that the availability of increased genetic content in modern rice varieties clearly expanded the area planted to modern varieties. This was true for the number of landraces as well.

The implications of these estimates are that the genetic content variables increased the area planted with modern varieties from 35–40% to roughly 70% by the late 1980s. The effects of genetic content on yields over and above the modern variety adoption effect were not large. The genetic content variables had a substantial effect on production because they caused farmers to plant the semidwarf plant type in more areas. These estimates indicate an additional 15–

20% contribution from second-generation genetic content with perhaps 10% from third-generation breeding which is highly dependent on rare traits. This is roughly double the effect originally estimated by Gollin and Evenson (1990) for yield effects only.

V. Assessing Time-to-Achievement for Rice Biotechnology Research

Assessing the expected timing of biotechnology achievements is complex and, to date, few formal assessments have been made. In 1991 an information request sent to the International Program on Rice Biotechnology participants included a question:

> Please provide a list of the *farmer usable technologies* your research is aimed at with a brief description of each and your best estimate of the time it will take to put each technology into farm level field tests.

Most of the respondents to the 1991 information request did provide descriptions of farmer-usable technology and many did provide time estimates. After examination of the responses, it was clear that they could not be easily organized into categories with clear economic meaning. Most of the technologies were very specific to the research projects. Furthermore the time estimates were often quite imprecise (e.g. 1–10).

After an analysis of the 1991 data, an alternative approach was developed. First, a specific set of RPAs was defined. These included four 'biotechnology tools' RPAs that would be useful to future researchers, five yield-enhancement RPAs, four 'disease–resistance' RPAs, four 'insect-resistance' RPAs and five 'stress-tolerance' RPAs. These RPAs were defined so as to be conformable with traits and achievements sought in conventional rice-breeding programs and with the priority trait strategy. Second, time to accomplishment subjective probabilities were obtained by asking for two estimates: an optimistic estimate which had only a 25% probability of achievement and a conservative estimate which had a 75% probability of achievement (Evenson, 1994).

A. Changes in subjective probability estimates (1993–1994)

The time-assessment SPEs reported in Tables 21.2–21.6 are the combined distributions from two groups of scientists participating in the International Program on Rice Biotechnology. The first group were 60 respondents to a postal survey sent out in May 1993. The second group was approximately 70 respondents from participants at the Seventh Rice Biotechnology Conference in Bali, Indonesia in May 1994. Fifteen respondents were common to both groups and they provide some evidence for changes in SPEs over time.

For the respondents who responded to both the surveys, a comparison of their 1993 and 1994 estimates was obtained. This comparison is of interest for two reasons. First, even though the sample is small, it may tell us something

Table 21.2. Time assessment: biotechnology tool development response distribution.

Biological tool	1994	1995	1996	1997	1998	1999	2000	2001	2002	2003	2004	2005	2006	2007	2008	2009	2010	Range	Median
I. Double haploid																			
25% prob.	13	11	22	8	3	2	4											1 1995–1997	1996
75% prob.	3	4	2	3	18	7	13	2		1		4		1				1 1997–2000	1998
II. Molecular markers																			
25% prob.	13	14	16	8	16	4	4		1		1							1 1995–1998	1996
75% prob.	5	2	5	2	9	9	18		1	3	2	1			1			1 1996–2000	2000
III. Pathogen markers																			
25% prob.	10	18	15	11	16	2	9			1		3						2 1995–2000	1997
75% prob.	1	3	5	7	8	8	19			5	0	14	1		1	1		10 1997–2005	2000
IV. Indica transformation																			
25% prob.	8	10	19	12	13	5	15	1	1	1		2					1	4 1996–2000	1998
75% prob.	3	2	11	7	9	4	20	3	8	3	1	6	2	1	1	1	1	16 1998–2005	2000
Respondent group (1994) IARC scientists																			
25% prob.	1	2	8	4	3	6	2		1				6			1	1		
75% prob.		1		3	1	5	5	2	2				5			6	6		
Less developed countries scientists																			
25% prob.	5	7	26	12	13	3	11		1				2						
75% prob.			3	3	12	8	24		5			1	14			6			
Developed countries scientists																			
25% prob.	12	9	21	7	10	1	12						1				6		
75% prob.	9	2	5	4	13	7	15		1				12				16		

Table 21.3. Time assessment: insect resistance response distribution.

Biological tool	1994	1995	1996	1997	1998	1999	2000	2001	2002	2003	2004	2005	2006	2007	2008	2009	2010	2015	2020	2030	2050	Range	Median
I. Sucking insects																							
25% prob.	2	5	10	2	5	2	4			2	2	2		1			1	1		1		1996–2000	1998
75% prob.		2	2	2	5	6	9	2	4	3	3	2		2			4			1		1998–2003	2000
II. Leaf folder																							
25% prob.	1	6	2	7	3	1		1	1	1							1	2				1995–1998	1996
75% prob.		2	4	5	1	8		1	2		1	1	1	1			2	1	3			1997–2000	2000
III. Stemborer																							
25% prob.	3	6	10	5	3	3	1	1				2	2				2	2				1996–2000	1996
75% prob.		2	5	2	4	8	8	5	4			2	2	1	3		2	2	1	2		1995–2003	2000
IV. Gall midge																							
25% prob.	2	4	2	2	6	4	4			2	1	2		2			2	1				1994–2003	1998
75% prob.			1	1	3	4	7	2		3	1	2			1	2	3	4	2			1999–2010	2000
Respondent group (1994) IARC scientists																							
25% prob.	2	6	4	6	7	1											2	2					
75% prob.		1	1	1	5	1	5	1	1			1				2	3	1					
Less developed countries scientists																							
25% prob.	1	2	12	4	4	3	1		1	1							3						
75% prob.			1		1	10	9		1	1	1		2		1			4					
Developed countries scientists																							
25% prob.	1	4	4	2	7	3	5			1	1	7	1	1				3					
75% prob.			1		3	4	1		4	4	4	2	2	2			4	5	3				

Table 21.4. Time assessment: disease resistance response distribution.

Biological tool	1994	1995	1996	1997	1998	1999	2000	2001	2002	2003	2004	2005	2006	2007	2008	2009	2010	2015	2020	2030	2050	Range	Median
I. Blast																							
25% prob.	1	8	8	2	21	4	9	3		2	2	3			1		3	3				1995–2005	1998
75% prob.		1	1	6	8	1	9		5	3		7			1	2	9	6				1995–2010	2000
II. Bacterial leaf blight																							
25% prob.	2	8	12	4	16	3	14			1		2			1		4	1				1996–2000	1998
75% prob.		1	1	4	9	5	11	2		2		7		1	1		8	4				1998–2010	2000
III. RSS virus																							
25% prob.	4	12		8	7	3	4					2				1	5					1996–2000	1997
75% prob.		1	2	2	8	2	5	2	2	1		3			1	1				2		1998–2010	2002
IV. Tungro virus																							
25% prob.	1	13	6	1	18	4	3	2			1	2					1	1				1995–1999	1998
75% prob.	2	3		2	11	3	7	1	5	1	1	4	1	1	2	1	5	3				1998–2002	2000
Respondent group (1994) IARC scientists																							
25% prob.		1	7	1	12	5	1																
75% prob.					6	0	4	2	4			4					4						
Less developed countries scientists																							
25% prob.	1	4	12	4	11	3	6	2									1						
75% prob.			1	1	11	6	7	3	6						2	2	2						
Developed countries scientists																							
25% prob.	5	12		1	13	1	8					6					2	2					
75% prob.		1			11	2	7	2	2			7		1	2		9	5					

Table 21.5. Time assessment: abiotic stress tolerance response distribution.

Biological tool	1994	1995	1996	1997	1998	1999	2000	2001	2002	2003	2004	2005	2006	2007	2008	2009	2010	2015	2020	2030	2050	Range	Median
I. Drought																							
25% prob.	1	3		1	5	6	20	1	1	2	1	6					6	1				1998–2005	2000
75% prob.				1		1	4		3	5	1	9	3	2			9	7			1	2000–2015	2005
II. Flood																							
25% prob.	2		1		6	6	13			1	1	2		1			3	1				1998–2000	2000
75% prob.							3	1		4	1	10	1	1			6	6	3		1	2003–2015	2005
III. Cold																							
25% prob.	1	4		6	6	10	5		2	1		2				1	7		1	1		1997–2005	1999
75% prob.				3	2	3	4		1	7	1	7					7	5			1	2000–2015	2005
IV. Salt																						
25% prob.	1	2	5	7	2	7	10	1	4		1	5		2			1	11				1997–2005	2000
75% prob.				4	1		5	1	3	4		6				1	15	4		1	1	2000–2010	2005
V. Nutrient deficiency																							
25% prob.		3			6	4	6	1	2			2					3					1998–2005	2000
75% prob.							5	1	3	6		3					5	34	2		1	2000–2015	2010
Respondent group (1994) IARC scientists																							
25% prob.	4	2	2	2	10	3			1			5		1			4	1					
75% prob.		1	1		1							8				1	7	4	1				
Less developed countries scientists																							
25% prob.	1	3		5	8	7	1	1	1		7	3	6	3	3	1	1	2					
75% prob.		1				1				2		3	2	6			5	4					
Developed countries scientists																							
25% prob.	1	4		1	7	3	17	1	2		4	1		1		10	7						
75% prob.							5		1	4	10		1	1		10	10	8					

Table 21.6. Time assessment: general yield enhancement response distribution.

Biological tool	1994	1995	1996	1997	1998	1999	2000	2001	2002	2003	2004	2005	2006	2007	2008	2009	2010	2015	2020	2030	2050	Range	Median
I. Male sterility																							
25% prob.		7	9	12	4	4	3	11			1	1	1	1			12					1995–2000	1996
75% prob.		3	3	3	6	6	1	6	2	1	3	5		5		1	8	2				2 1997–2010	2005
II. Starch metabolism																							
25% prob.	1		1	4	5	14	1	9	2	1		2					5					1996–2000	1998
75% prob.				1		4		11	3		3	7	2				5	3	1	1		1 2000–2006	2002
III. Photosynthetic effect																							
25% prob.	2		1	2	7	5		5		1	2	3					2	2				1997–1999	1999
75% prob.			1	2	12	2	1	2	2	2		5	2		2	1	3	9	1			2005–2015	2005
IV. Apomixis																							
25% prob.			1	4	2	4	1	10	1	1	2	3					5	1	1			2000–2010	2000
75% prob.				2	2	1		7				3			1	3	7	7	1	1		2000–2015	2010
V. Nitrogen fixation																							
25% prob.				2		1		4		1		4			1		8	12	1	5		1 2000–2020	2015
75% prob.						1		5		2		3				1	7	12	1	10	3	6 2010–2050	2020
Respondent group (1994) IARC scientists																							
25% prob.				2	1	3	3	4	1	1	0	4	0		0	0	5	3	1	1			
75% prob.				1	1	1	1	2			1	4	1			4	3	4	2	2			
Less developed countries scientists																							
25% prob.	3		0	7	6	6	3	4	1	1	3	3	1	0		1	2	1	1				
75% prob.	1		0	2	0	4	0	4	2	2	2	4	2	0		2	8	2	2				
Developed countries scientists																							
25% prob.	1		0	6	1	4	1	14				2	2	1			9	4	1	1	0		
75% prob.	1		0	1	1	2	0	6			2	5					9	11	2	2	2		

about the 'moving target' problem (i.e. if the SPE targets moved forward from 1993 to 1994, this would suggest unreliability). Second, this can indicate whether it is appropriate to pool the two sets of responses to obtain larger distributions.

An analysis of the data showed that there was a difference between the tool-development RPAs (Table 21.2) and the traits and yield-enhancement RPAs (Tables 21.3–21.6). For the biotechnology tool RPAs, the average 25% probability date moved later by 0.56 years between 1993 and 1994, and the 75% probability date moved later by 0.14 years. In contrast, for the traits and yield-enhancement RPAs both the 25% probability and the 75% probability dates moved earlier by 1.15 years on average.

Thus, while the sample of common respondents was small (30 responses for tools, 40 for others; not all respondents answered all questions), this does not suggest a serious moving target problem. If anything, it provides some evidence for consistency. It also supports the pooling or combining of the responses for detailed analysis.

B. Subjective probability estimates by research priority area

The SPE distributions by RPA are organized in five tables (Tables 21.2–21.6) reflecting the groups of RPAs. For each, the distribution of responses by both the 1993 and 1994 respondents for the 25 and 75% probability levels are reported. For each set of RPAs the 1994 responses by respondent category (IARC scientists, less developed countries and developed countries scientists) are summarized. The 'quality' of the SPEs can be assessed in several ways:

1. The dispersion can be assessed and consistency between the 25 and 75% probability estimates assessed. Good estimates should tend to have tight dispersions from both estimates.
2. The interval between the 25 and 75% levels can be assessed. Good estimates should show unimodal dispersion of the interval, and it is expected that the interval would be greater the later the 25% estimates are.
3. The unimodality of the estimates can be assessed. Good estimates should show unimodal distributions with relatively few long-term estimates.

By these criteria most SPEs meet reasonable standards, though this does vary by RPA. (There are also differences by respondent groups – see Table 21.7.)

Biotechnology tool development

Table 21.2 summarizes the data from four tool-development cases. In spite of considerable room for interpreting the questions, especially by those further from the applied end of the research spectrum, there is good agreement on the whole on tool development. Most of the relevant tools are in hand already and the uncertainty has to do with routine use. The ranges (covering 75% of the estimates) are tight for the 25% estimate and reasonable for the 75% estimate.

Table 21.7. Respondent group summary.

Group	Percent of respondents	Percent of 25% SPEs 1995 or earlier	2001 or later	Percent of 75% SPEs 2005–2015	2015 or later
IARC scientists	14	5	25	26	17
Less developed countries scientists	43	39	25	30	17
Developed countries scientists	43	56	50	40	66

Insect-resistance traits

Table 21.3 reports the evidence for transgenic insect-resistance-trait achievement to the field trial level (generally considered to be the routine breeding level). These estimates are generally of high quality (though poorest for gall midge resistance where substantial bimodalism is observed), the ranges are quite narrow and median estimates of time to achievement are close – 1996–1998.

Disease-resistance traits

Table 21.4 reports estimates for transgenic disease-resistance traits. These estimates are of lower quality than for insect-resistance traits. Bimodalism is observed in each case although it is mostly in the 75% estimates – hence the fairly broad ranges for these estimates (most bimodalism is from developed country scientists, see below). These might be regarded to be medium to good quality estimates.

Abiotic stress tolerance

Table 21.5 reports estimate for transgenic abiotic stress-tolerance traits. These estimates are of lower quality than those for insect and disease resistance. Ranges are broader, median dates are later and there is more biomodalism of the 75% estimates. But the estimates are still of medium quality.

Yield enhancement

Table 21.6 reports the estimates for general yield-enhancement RPAs. Most of these show a broad range of estimates with some, especially for nitrogen fixation, quite far in the future, but this does not necessarily mean that these are meaningless estimates.

Subjective probability estimates by respondent groups

Tables 21.2–21.6 each report distributions by respondent group. Table 21.7 summarizes these estimates by group. It is clear that the scientists from less developed countries have the most consistent unimodal distributions; IARC scientists have fewer very early estimates but otherwise have similar SPE distributions. Developed country scientists, on the other hand, have disproportionately more early and late estimates and hence more dispersion and bimodalism than scientists who work in developing countries. This appears to reflect both more scepticism and a lack of real world experience of applied problems.

VI. Expected Potential Benefits to the Program on Rice Biotechnology

To calculate expected (estimated) outcomes, we require estimates of the percentage change in rice production from achieving each of the above objectives and an estimate of the diffusion time and ultimate coverage. The trait value evidence summarized in Table 21.1 was utilized to calculate expected outcomes by RPA for the rice biotechnology program. Table 21.8 reports a summary of the time-to-achievement estimates used in the calculations.

Table 21.9 summarizes the estimates of expected pay-off to five separate components of the program. The table reports optimistic and conservative time assessments where time to production includes 4–6 years of farmer adoption time. For both the optimistic and conservative time-to-yield estimates, net present values of rice biotechnology benefits are computed. Both 5 and 10% discount rates are used. The most reasonable time assessment is probably the conservative estimate to farmers' fields. The optimistic estimates are reported to provide a perspective on the importance of timing. We consider the 5% discount rate the most appropriate rate for discounting to compare the net present value of the various components to the costs of achieving them. The 10% discount calculations provide a perspective on the importance of discounting.

For disease and insect resistance, the area figure reflects the judgement that the rice biotechnology program will achieve one-half of the difference between conventional breeding potential and full resistance. Yield estimates are conservative.

Table 21.8. Time-to-achievement summary (years).

Research priority area	Optimistic to field trial	Conservative to field trial	Conservative to farmers' fields
Tool development	3	6	–
Insect resistance	6	8–10	12–15
Disease resistance	6	10–15	12–18
Stress tolerance	8	13–18	15–20
Yield enhancement			
Hybrid rice	3	6	8
Other	12	20	25

Table 21.9. Estimates of the effects of the Rockefeller Foundation International Program on Rice Biotechnology.

	Time to field trials (years)		Time to production (years)		Annual effect after realization				Present value of benefits (billion US$)			
									5% Discount		10% Discount	
	Optimistic	Conservative	Optimistic	Conservative	Area (million ha)	Yield (%)	Quantity (million tons)	Value (1990) (billion US$)	Optimistic	Conservative	Optimistic	Conservative
1. Multiple insect resistance	8	15	12	21	37.5	30	41	8	91	50	20	10.5
2. Multiple disease resistance	10	18	15	22	50	15	27	5.4	25	19	5.4	4
3. Hybrid rice enhancement	5	10	9	16	30	15	16	3.2	41	28	14	7
4. Stress tolerance	13	18	17	22	50	15	27	5.4	23	19	5.4	4
5. General yield enhancement	20	25	25	40	100	20	70	14	82	39	13	3
Total with yield enhancement							181	36	260	164	63.8	28.5
Total without yield enhancement							111	22	178	125	50.3	25.5

Time to field trials based on time assessment (section II).
Time to production based on 4–6 years diffusion.
Area estimates based on incremental areas to conventional breeding.
Yield estimates based on India–Indonesia studies.
Program costs discounted at 10% = US$ 1 billions; benefits/costs ratio ranges from 14 to 39.
Program costs discounted at 5% = 2 billion; benefits/costs ratio ranges from 55 to 130.

Hybrid use enhancement is expected to be realized on only 30 million ha, reflecting the fact that substantial acreage is already planted to hybrid rice. Total rice area is expected to remain at approximately 150 million ha over the next three decades. At that level, expected yield increases will be roughly equal to expected demand increases.

Stress tolerance gains are expected to affect only one-third of the world's rice area and to be modest in effect. Since there is some concern that biotechnology might not produce this effect we have calculated net present values with and without this component.

Perhaps the point to make first regarding these calculations is that the most conservative estimates produce PVBs that are large relative to the costs of attaining them. At a 5% discount rate the conservative estimate produces a PVB of US$125 billion without yield enhancement. At 10% this falls to US$25.5 billion. If the cost of the program is fixed at a present value of US$1.8 billion at a 10% discount rate or US$3.2 billion at a 5% discount rate, we are looking at benefits/costs ratios ranging from 14 to 39. When yield-enhancement estimates are included, PVBs rise to US$164 billion at 5% and US$28.5 billion at 10% and benefits/costs ratios rise to 55–130.

For these calculations we assume that the present values of the Rockefeller program costs plus the biotechnology costs incurred by, or associated with, biotechnology programs are US$50 million per year for 3 years, US$ 100 million per year for the next 3 years and US$150 million per year for the next 25 years. This is more than is presently being expended and represents an effort to be both realistic and conservative in these calculations.

Are these estimates out of line with historical experience? The implied growth rates in productivity are that conventional breeding programs will produce gains over the next 29 years at only slightly lower rates than over the past 20 years. Then rice biotechnology gains will increase these growth rates, but to slightly less than the growth rates experienced during the 1965–1975 green revolution period.

Have conventional breeding programs been properly considered? Will they run out of steam if biotechnology does not take over? We would argue that the contributions of these programs have been properly considered and that they will run out of genetic steam, as it were, without biotechnology. This does not imply that other contributions of research will not continue. The structure of this analysis attributes only about half of total factor productivity gains in rice to research, and of those, only half are due to genetic importance.

One further calculation is associated with an important point, and that is that there are large externalities in the world of biotechnology. Programs such as the International Program on Rice Biotechnology produce information and technological tools of great value to related researchers and research institutions. This means that the rice biotechnology program produces benefits to other programs, particularly to researchers in the IARCS. Conversely, it also means that by delaying the biotechnology program the Rockefeller Foundation can benefit from the research of others and hence raise the returns to its investment. We have not attempted to estimate the value of the rice biotechnology program to other research institutions and no benefits of this sort are included in the calculations presented here. But we can calculate whether it made sense to start the program when the Rockefeller Foundation did.

The issue for the Rockefeller Foundation, is whether by delaying the program 1 year, the costs saved (say US$100 million) would be greater than the lost PVB from delaying the onset of the benefit streams. Obviously, if delaying the research program does not delay the onset of benefits in the future it pays to delay. If benefits at the conservative level discounted at 5% are delayed by 1 full year the loss in PVB will be US$8 billion (US$164 billion—US$156 billion). At 10% discount, the loss in PVB is US$1.5 billion (US$28.5—US$27 billion). At the 5% discount level it would not have paid to delay the program unless the benefit onset delay were less than one-tenth of a year (US$8 billion/US$70 million – the cost saved by delay). Even at 10% discount, the delay onset would have to be less than one-twentyfifth per year. And this does not count the benefits the program extends to other biotechnology programs.

It should be stressed here that the learning possibilities for IARC units and most National Agriculture Research programs in developing countries from this program are likely to be large and important. The IARCs have become increasingly conservative over the years and have been downright 'timid' in policies toward biotechnology. It will not only be the successes of this program that will be valuable; the failures will be too, because they will identify unpromising avenues of work.

The calculations in Table 21.9 can be made for each of the RPAs that are grouped together in Table 21.9. The benefits/costs ratios from this exercise can be considered to be a priority-setting exercise for these RPAs. It should be noted, however, that these calculations are for a set of RPAs where little experience has been obtained. For more conventional RPAs much tighter distributions can be obtained.

References

Evenson, R.E. (1991) Assessing economic value. In: *Valuing Environmental Benefits in Developing Countries*. Special Report 29. Michigan State University, Agricultural Experiment Station, September, 1991.

Gollin, D. and Evenson, R.E. (1990) Genetic resources and rice varietal improvement in India. Economic Growth Center, Yale University, New Haven (unpublished manuscript).

22 Rice Research Priorities: An Application

R.E. Evenson[1], M.M. Dey[2] and M. Hossain[3]
[1]*Economic Growth Center, Yale University, 27 Hillhouse Avenue, New Haven, CT 06520, USA;* [2]*International Center for Living Aquatic Resources Management (ICLARM), MC PO Box 2631, 0718 Makati, Metro, Manila, Philippines;* [3]*Social Sciences Division, International Rice Research Institute, PO Box 933, 1099 Manila, Philippines*

I. Introduction

In Chapter 21 a specialized application of priority-setting methods to rice biotechnology research was reported. In this chapter we report a broader priority-setting exercise for rice research. Priority setting requires measures of economic 'units', (u), benefits per unit (b/u) and time to achievement (W_t). (See below for a formal development of the methodology.) The exercise reported in this chapter utilizes the rice-loss data reported in the country studies (Chapters 7–15) as measures of units. In contrast to the previous chapter, it does not use *ex post* estimates for the benefits per unit component. Both benefits per unit and time-to-achievement components are estimated utilizing subjective probability estimate (SPE) techniques. This priority-setting exercise is accordingly a more general (and probably more useful) application than the exercise for rice biotechnology reported in Chapter 21.

The application reported in this chapter is broad in scope. It is directed to the National Agricultural Research (NAR) programs in Asia. The exercise is carried through to the development of research portfolios for different ecosystems in Asia. This chapter is intended to be a demonstration chapter in that methods and procedures are discussed so that further priority-setting applications (for regions or specific national programs) may be guided by this effort.

Section II of this chapter provides a review of the methods required to develop a 'priority portfolio' for research problem areas (RPAs) and for research techniques (RTs) for rice research. Section III describes the development of RPA–RT categories. Section IV discusses the measurement of units (u). Section V describes the process by which SPEs for the benefits per unit (b/u) and time-to-achievement (W_t) components were obtained. This process entailed utilizing the distinction between achievement to date and potential achievement, to obtain estimates of remaining potential. The procedure implicitly recognizes the diminishing returns element of research discussed in Chapter 5 of this volume. Section VI discusses the scientists' ratings and section VII reports the estimated priority portfolios by RPA–RT activities for each country data-set.

II. Methods for Obtaining the Research Priority Portfolio

The steps required to develop the research priority portfolio are:

1. RPA–RT categories must be developed. It is vital that the RPA–RT matrix be both comprehensive and flexible as regards research options.
2. For each RPA category, an estimate of units (u) over which research impacts can be expected must be established (in this chapter, we utilize crop loss and related data for these units).
3. For each RPA–RT combination, an estimate of the expected impact or benefits per unit (b/u) must be obtained (in this chapter, by SPE methods).
4. For each RPA–RT combination, an estimate of the time-to-achievement profile (W_t) must be made. This W_t estimate should be consistent with the principle of diminishing returns (see Chapter 5) and with the (b/u) estimates (we use SPE methods in this chapter).
5. The three components: units (u), benefits per unit (b/u), and time-to-achievement (W_t) weights can then be incorporated into the present value of benefits (PVB) relationships:

$$PVB_0 = \sum_{t=0}^{NB} W_t(b/u)(u)D_B \tag{22.1}$$

$$D_B = \sum_{t=0}^{NB} 1/(1+r)^t \tag{22.2}$$

This expression, which can be computed for each RPA–RT combination, is the present *value* at time $t = 0$ of the expected benefits stream associated with the research in the RPA–RT category. D_B is the discount term, where r is the discount rate. NB is the period of benefits.

We can also compute the present value of costs or expenditures for the same RPA–RT combination:

$$PVC_0 = \sum_{t=0}^{NC} EXP \cdot D_C \tag{22.3}$$

$$D_C = \sum_{t=0}^{NC} 1/(1+r)^t \tag{22.4}$$

where NC is the period of research expenditure. Suppose that *EXP* is a constant level of expenditure for NC years. We could set *EXP* at some level, r at some level and generate PVB_0 and PVC_0 (present value of costs). From these we could compute:

$$NPV_0 = \text{net present value} = PVB_0 - PVC_0 \qquad (22.5)$$

or

$$B/C = \text{benefit/cost ratio} = PVB_0/PVC_0. \qquad (22.6)$$

These financial calculations are often used in priority setting or resource allocation calculations. But it is critical to note that the term, b/u, is not independent of EXP, the implied costs or expenditures on research for the RPA–RT category. In the case of this study we obtained SPE estimates of b/u by specifying that the research programs, both in the IRRI and in NARs, would *continue* at approximately the level of the recent past (and that they would respond to research opportunities as they have in the recent past). Thus, we could set EXP (eqn 22.3) at levels of the recent past and compute NPV or B/C as shown in (eqns 22.5) and (22.6). We could utilize some discount rate, r, to do so. The size of NPV or B/C would then serve as a signal to increase or decrease expenditures on the RPA–RT profile item.

But we have another financial calculation option available to us which allows the *direct* calculation of the RPA–RT portfolio expenditure – and when computed for all portfolio items – the full portfolio. This option is to interpret the SPE estimate of b/u as being made under the presumption of a constant (and historically consistent) internal rate of return, IRR. At this IRR we can set:

$$PVB_0 = PVC_0$$

$$\sum_{t=0}^{NB} W_t(b/u)(u)(1+IRR)^t = \sum_{t=0}^{NC} EXP/(1+IRR)^t. \qquad (22.7)$$

Then if we know W_t, the time weights, b/u, the benefits per unit terms, u, the units, NB, NC and IRR, we can solve (eqn 22.7) for EXP, the annual expenditures for the portfolio item (presumed constant for NC periods). Provided that we have diminishing returns we can validly generate a full portfolio of EXP for every RPA–RT category. We can also do this for regions within a RPA–RT category provided that we have diminishing returns (note that we impose such diminishing returns over RT categories in an SPE of W_t or time-to-achievement weights below).

In this chapter we adopt the practice of generating a full portfolio by using (eqn 22.7). As noted above, we do need to be careful first that we are maintaining consistency as regards diminishing returns and second that we maintain consistency between the IRR that we use and the b/u estimates obtained. We address the matter of consistency or diminishing returns by specifying the time-to-achievement estimates so as to be consistent with diminishing returns. We do this by asking an SPE respondent to provide estimates of time to partial and full achievement – thus imposing a cap of full achievement for each b/u portfolio item. We do have to be careful, however, that we not define RPA categories spatially (i.e. for small regions) so as to violate the diminishing returns condition. (For example, if we defined the separate RPA categories from two districts in India where the benefits were equally available (valuable) to farmers in the two districts we would violate this condition. It would not make sense to have two separate research programs for the two districts. Accordingly, this procedure is relevant for relatively large regions where heterogeneous soil and climate conditions are such that research results are not equally valuable in the two regions.)

We deal with the second consistency item by choosing IRR to be 25%, an internal rate of return broadly consistent with historical experience (actually it is lower than IRRs estimated in most *ex post* studies but is probably consistent with the expectations of our SPE respondents). We also set NC, the period of research expenditure, to be equal to the period when 75% of the potential gain (b/u) is expected to be attained. For NB, the period of benefits, we assume that for management research RPA–RT items, $NB = NC$. For other research techniques we assume $NB = \infty$.

III. Defining Research Priority Areas and Research Techniques

Two principles guide the definition of RPAs. The first is that a 'full portfolio of RPAs for the relevant research organization must be defined. This means that RPAs both for loss reduction and for biological efficiency improvements must be included in the assessment. The second, an innovation compared to Herdt and Riely's (1986) original analysis, is that alternative research techniques for achieving potential gains must be recognized. Both are important to achieving a balanced set of priorities.

We have attempted to achieve a full set of RPAs by including all relevant crop-loss RPAs and a set of bioefficiency RPAs. We then assume that each can be addressed (with varying degrees of success) by any of four research technique categories:

1. *Managerial* – this covers agronomy, soils, economic and related research to improve technical and allocative management by rice farmers.
2. *Conventional plant breeding* – this covers conventional plant breeding for host plant resistance and tolerance as well as for biological efficiency (e.g. the new plant type).
3. *Wide crossing (tissue culture) breeding* – this covers the breeding techniques enabling the use of wild species and wild relatives of rice in rice-breeding programs.
4. *Transgenic breeding* – this covers transgenic rice breeding and marker-aided breeding.

We argue that each technique is potentially usable in each RPA and that the techniques are not mutually exclusive. They generally complement each other. Each, however, is subject to diminishing returns. Hence, each RPA–RT category is governed by the rules set forth in Chapter 5. Accordingly, the portfolio will typically include all four RT categories for each important RPA.

The RPA categories are reported in the tables given later in this chapter; they include insect loss, disease loss, abiotic stress and pest control RPAs as well as biological potential RPAs.

IV. The Measurement of Units

For the insect loss, disease loss, abiotic stress loss and pest control loss RPA–RT categories, we use the loss estimates reported in the country studies (see Chapters 7–15). For each of these loss estimates, we conclude that the regions for which losses are reported are sufficiently large that research programs targeted to one region will have lower impacts on other regions. In this chapter we treat RPA units as specific to rice ecosystems and countries.

For many crop-loss categories, a judgement must be made as to the potential for a research solution to the problem. It is quite feasible that a disease or insect loss can be eliminated as a result of research achievements. For example, host plant resistance achieved through wide crossing techniques may effectively eliminate the loss. For other losses – notably losses due to drought or submergence from flooding – research achievements cannot be expected to eliminate the loss. Drought-resistance traits in rice can reduce but not eliminate drought losses. Thus the units for these loss-related RPA–RT categories are scaled to reflect the potential for research achievement.

Similarly, for those RPA–RT categories where some form of biological efficiency gain is sought (as opposed to a loss reduction) some judgement as to the scope for biological improvement is required. In the case of rice, this scope has varied by rice ecology in the past. Most progress has been made for rice in irrigated systems, least on rice produced in deepwater conditions. It is also not always clear that scientists, when responding to SPE elicitation efforts, have in mind all rice produced in a region or some part of the rice produced in the region. In the portfolios developed below we report portfolios under alternative assumptions to deal with this problem. One of the country studies (China, see Chapter 10) provides estimates of yield gap I losses that we have used to achieve consistency between the biological efficiency and other loss categories.

Table 22.1 reports a summary of the units data (expressed in kg ha^{-1}) used in this exercise for each of the country ecosystem datasets reported in Chapters 7–15. The units for insect, disease, abiotic stress and pest RPAs are taken directly from Chapters 7–15 where crop-loss estimates are reported. The units for the bioefficiency RPAs for irrigated ecosystems are based on the China study where yield gap I accounting estimates were reported. The bioefficiency units for rainfed, upland and deepwater ecosystems were 'judgementally' set to be 66, 56 and 33%, respectively, of the irrigated ecosystem units to reflect lower bio-efficiency potential.

The units data then represent potential gains from research programs conducted over the relatively short terms (see time estimates below); they implicitly set an upper limit on research potential over the short term.

Table 22.1. Rice priority units (kg ha^{-1}).

Research priority areas	East India Irri-gated	East India Up-land	East India Rain-fed	Deep-water	South India Irri-gated	Nepal Irri-gated	Nepal Up-land	Bangladesh Irri-gated	Bangladesh Rain-fed	China Irriga-ted	Indonesia Irri-gated	Indonesia Up-land
Insects												
Stemborer	114	6	35	82	32	65	29	39	28	23	142	7
Brown planthopper	23	4	7		23	12	9				33	129
White-backed planthopper	37	3	5									
Leaf folder	10	2	9		44	29	19				9	4
Hispa						34	50	68	30			
Green leafhopper	21	19	15		19							
Gall midge	33		8		25	20	10				33	1
Caseworm		2	13									
Armyworm	26	26	9			20	30				33	35
Mealy bug						19	11	7	4			
Rice bug	23	24	3			112	52	41	29		38	20
Diseases												
Blast	10	54	37	21	30	112	125	35	21	19	8	6
Leaf scald												
Green leaf spot												
Brown spot	10	32	8	12	19	41	34				7	2
Sheath rot	16	2	15	10	18							4
Sheath blight	14	8	7	8	19	16	13			34	0	0
Stem rot						34	22					
Bacterial blight	54	3	26	42	19	98	55	72	41		33	35
Bacterial leaf streak								28	17			
False smut						13	5				0	

Diseases										
Atom blight					17					
Tungro							7	4	9	0
Ragged stunt									1	
Ufra										
Abiotic Stress										
Drought	92	281	177	63	44	146	115	240	164	141	47	22
Submergence	20	8	45	63		9	7	170	83	106	47	22
Cold	15	3	28		4	21	20			90		
Heat										36		
Acidity	64	70	8		6					39		
Alkalinity	8	15	15	78	3							
Salinity	20	17	4		23							
Nutrient deficiency	40	35	41		38	96	39	35	23	447		
Iron toxicity	7	5	14		2							
Pests											
Weeds	107	183	86	70	25	71	56	79	39	26	95	44
Crabs								16	10			
Rodents	21	13	15	20	16	40	34	39	19	20	95	44
Birds	6	33	7	21	13	10	10	20	10	26	24	11

(continued overleaf)

Table 22.1. Rice priority units (kg ha^{-1}). (continued)

Research priority areas	East India Irri-gated	East India Up-land	East India Rain-fed	East India Deep-water	South India Irri-gated	Nepal Irri-gated	Nepal Up-land	Bangladesh Irri-gated	Bangladesh Rain-fed	China Irri-gated	Indonesia Irri-gated	Indonesia Up-land
Bio-efficiency												
Plant design	315	79	151	33	292	228	120	434	135	574	331	131
Photosynthetic efficiency	229	39	129	22	212	166	60	315	117	418	284	65
Growth duration	286	53	172	22	265	207	80	394	156	522	378	87
Grain quality	200	53	129	22	186	145	80	276	117	365	284	87
Total yield	2864	1316	2151	1097	2655	2070	1994	3942	1949	5221	4730	2180

V. Obtaining Subjective Probability Estimates of Potential Benefits (b/u) and Time to Achievement (W_r)

A formal questionnaire designed to elicit 'ratings' from scientists that could be used to obtain estimates of potential benefits, the b/u component, and the time weight, W_t, was developed. For the purposes of this application, formal ratings were obtained from 17 rice scientists (nine from the IRRI and eight from NARs). Scientists were asked to provide four numbers for each RPA–RT cell for which they considered themselves scientifically qualified to provide estimates. These four numbers were:

1. A rating of the potential (RP) for a research contribution to the RPA–RT problem area. Ratings were on a 1–5 scale.
2. A rating of the achievement to date (RA) by research on the PPA–RT problem area.
3. An assessment of the date (years from now) by which either a 25% achievement of the *remaining potential* (potential minus achievement to date) would be achieved or by which there was a 25% likelihood of achievement.
4. An assessment of the date by which either a 75% achievement of the remaining potential was expected or by which a 75% likelihood of achievement was expected.

The elicitation of these four numbers was based on the following principles:

1. Scientists are more comfortable with a rating scale (1–5) than with a specific estimate of a productivity level. Rating scales linked to achievement were provided to scientists; These were:

> *Scale 1* – less than 10% achievement of loss elimination (or increase in biological efficiency);
> *Scale 2* – 10–25% achievement of loss elimination (or increase in biological efficiency);
> *Scale 3* – 25–50% achievement of loss elimination (or increase in biological efficiency);
> *Scale 4* – 50–75% achievement of loss elimination (or increase in biological efficiency);
> *Scale 5* – 75+% achievement of loss elimination (or increase in biological efficiency).

The distribution of these ratings obtained from the sample of respondents was then quantified into a mean percentage achievement measure (the variance was also computed).

2. The distinction between RA and RP was needed to clarify what was meant by remaining RP. By specifying both RP and RA we attempted to capture more clearly the *incremental* potential for further gains. In many RPA–RT classes, respondents indicated that while substantial RP for problem solutions existed in the past, research programs had already achieved all or most of this potential – i.e. they had exhausted much of the potential (see Chapter 5).

For research priority setting, we base the future RP on the remaining potential, i.e. RP–RA.

Achievement to date ratings were based on research programming to date. Respondents were asked to visualize the continuation of current research programs with some strengthening and normal responsiveness to research opportunities in estimating RP and RA. Note that by utilizing these RA–RP concepts in this way we are attempting to rule out the possibility of specifying an arbitrary research program to obtain RP and time-to-achievement (W_t) estimates. Respondents have the experience to rate actual programs better than hypothetical programs.

3. Scientists need some scope for expressing the variance in their SPEs. This could be done by eliciting a variance on the b/u rating component or on the time-to-achievement component, W_t. The evaluation has to choose to ask for one estimate of the time to achievement (W_t) and two estimates of the potential (b/u) or one estimate of the potential and two estimates of the time to achievement. It is impractical to attempt two estimates of each. Our experience with the rice biotechnology study (see Chapter 21) and with scientists indicates that eliciting two dates on time to achievement was an effective way to obtain a 'distribution' reflecting the degree of uncertainty of scientists.

VI. Scientists' Ratings

Tables 22.2 through 22.6 summarize the scientists' responses to the ratings elicitation. It should be noted that not all respondents completed each block of RPA questions. They did, however, complete each RT question for the RPAs for which they responded. This was designed to achieve comparative consistency over RT, but one cannot be certain that it did (see below).

For each RPA in Tables 22.2–22.6, four numbers are reported for each research technique:

- Mean years to 25% achievement of remaining potential (Y25).
- Mean years to 75% achievement of remaining potential (Y75).
- Mean estimated achievement percent of remaining potential (RP–RA).
- Standard deviation of estimated deviations percent of remaining potential (SD).

Obviously, standard deviations of Y25 and Y75 could also have been computed. We believe, however, that for purposes of displaying variation in estimated impacts of research programs, variation in RP–RA, our estimate of b/u, is more relevant than variation in Y25 and Y75 which were designed to allow scientists to express their subjective variances. Thus the differences in Y25 and Y75 reflect the 'within scientists' subjective variation in estimates, while the standard deviations reported in Tables 22.2–22.6 reflected variations in estimates between scientists.

A. Insect loss

We turn first to the insect-loss RPAs summarized in Table 22.2. We note that there are differences in the RP–RA estimates both by RPA and by RT. Given the small scientist sample and the relatively high standard deviations across scientists, few of these differences are statistically significant. Most standard deviations are lower than the estimated RP–RA terms (note that scientists reported separate ratings for RP and RA). Most standard deviations for RP and RA separately were roughly one-third or so of the mean RP and RA estimates. The standard deviation of the differences, however, are relatively high. Should this be construed to mean that few differences across RPAs actually exist? If so, we can simply use 'congruence' rules to allocate resources over RPAs (see Chapter 5).

We would argue that the procedure of separately identifying the RP and RA components probably results in an upward bias in the standard deviations and that differences over RPAs are meaningful. We also consider differences over RTs to be meaningful. Here we note that the highest RP–RA estimates are for the transgenic breeding techniques in all but one or two cases. Wide crossing and tissue-culture techniques tend to be located between conventional breeding and transgenic techniques in these estimates.

Timing estimates also do not vary substantially by RPAs, but clearly do by research technique. The management research techniques are expected to yield results earlier than the genetic improvement techniques. Interestingly, transgenic techniques do not appear to have very different time estimates from conventional breeding or wide crossing techniques.

B. Disease loss

Ratings for disease-loss RPAs (Table 22.3) show similar patterns of variation over RPAs and RTs to those observed for insect-loss RPAs. As with insect-loss RPAs, there is more variation in the expected gains from working on the more important diseases; and as with insect-loss RPAs, transgenic techniques generally have the highest expected gains and the largest expected periods to achievement.

C. Abiotic stress loss

Abiotic stress-loss RPAs (Table 22.4) again show patterns similar to those for other losses. Management solutions generally have lower expected contributions, however, and tend to have longer expected time-to-achievement estimates.

Table 22.2. Scientists' ratings: insect-loss research priority areas (RPAs).

RPA	Management research			Conventional breeding				Wide crossing				Transgenic breeding			
	Y25	Y75	RP-RA	SD	Y25	Y75	RP-RA	SD	Y25	Y75	RP-RA	SD	Y25	Y75	RP-RA	SD
Yellow stemborer	5	10	0.24	0.20	8	13	0.16	0.12	9	15	0.22	0.18	7	13	0.54	0.32
Striped stemborer	4	11	0.32	0.22	9	12	0.15	0.10	9	12	0.20	0.20	7	10	0.52	0.46
Brown planthopper	5	8	0.16	0.22	9	12	0.16	0.15	9	12	0.20	0.26	10	12	0.31	0.28
White-backed/brown planthopper	4	10	0.23	0.18	7	11	0.16	0.17	10	13	0.27	0.22	9	14	0.20	0.24
Leaf folder	5	9	0.28	0.18	9	12	0.17	0.15	9	13	0.10	0.12	9	13	0.20	0.36
Hispa	6	13	0.12	0.10	11	15	0.20	0.10	8	14	0.20	0.16	10	12	0.33	0.42
Green leafhopper	5	10	0.17	0.22	7	12	0.20	0.10	10	16	0.30	0.26	9	13	0.30	0.20
Gall midge	5	10	0.30	0.21	7	12	0.30	0.21	9	15	0.28	0.26	9	15	0.32	0.30
Caseworm	6	11	0.30	0.21	8	17	0.16	0.09	11	19	0.15	0.18	10	15	0.36	0.40
Armyworm	6	11	0.30	0.17	9	15	0.16	0.09	11	16	0.15	0.18	10	15	0.36	0.40
Grasshopper	4	6	0.20	0.20	7	9	0.14	0.12	9	11	0.07	0.10	7	10	0.14	0.22
Mealy bug	4	7	0.20	0.10	8	12	0.14	0.12	9	11	0.10	0.10	7	10	0.30	0.42
Rice bug	4	7	0.20	0.14	8	12	0.20	0.16	9	11	0.07	0.10	7	10	0.14	0.22

Table 22.3. Scientists' ratings: disease loss research priority areas (RPAs).

RPA	Management research Y25	Y75	RP-RA	SD	Conventional breeding Y25	Y75	RP-RA	SD	Wide crossing Y25	Y75	RP-RA	SD	Transgenic breeding Y25	Y75	RP-RA	SD
Blast	6	14	0.30	0.20	5	12	0.20	0.26	6	13	0.22	0.20	8	13	0.40	0.28
Leaf scald	5	10	0.70	0.28	9	17	0.20	0.10	11	20	0.26	0.12	10	18	0.14	0.12
Gerlachia leaf spot	5	25	0.26	0.12	10	17	0.30	0.12	11	20	0.30	0.12	11	19	0.20	0.16
Brown spot	8	15	0.20	0.16	8	12	0.30	0.12	9	15	0.30	0.12	10	18	0.20	0.16
Sheath rot	10	17	0.28	0.30	10	17	0.15	0.10	11	19	0.28	0.10	10	17	0.10	0.14
Sheath blight	6	15	0.36	0.32	10	16	0.08	0.10	8	16	0.24	0.20	7	13	0.34	0.26
Stem rot	10	17	0.20	0.10	10	15	0.20	0.10	8	16	0.20	0.10	7	13	0.20	0.10
Bacterial blight	9	12	0.20	0.16	6	13	0.22	0.28	5	11	0.36	0.22	8	12	0.25	0.20
Bacterial leaf streak	5	10	0.20	0.10	8	13	0.16	0.16	5	10	0.20	0.20	7	11	0.26	0.12
False smut	7	12	0.05	0.05	7	12	0.05	0.05	7	13	0.20	0.10	7	12	0.20	0.10
Atom blight	7	12	0.05	0.05	7	12	0.05	0.05	7	13	0.20	0.10	7	12	0.20	0.10
Tungro	10	17	0.22	0.22	5	14	0.22	0.12	7	14	0.32	0.10	8	15	0.48	0.40
Ragged stunt	10	17	0.20	0.10	7	12	0.16	0.16	7	14	0.20	0.10	8	15	0.20	0.20
Ufra	10	17	0.20	0.10	7	12	0.16	0.16	7	14	0.20	0.10	8	15	0.20	0.20

Table 22.4. Scientists' ratings: abiotic stress-losses research priority areas (RPAs).

RPA	Management research			Conventional breeding				Wide crossing				Transgenic breeding			
	Y25	Y75	RP-RA	SD	Y25	Y75	RP-RA	SD	Y25	Y75	RP-RA	SD	Y25	Y75	RP-RA	SD
Drought	6	14	0.12	0.12	8	16	0.22	0.14	10	17	0.22	0.24	12	16	0.32	0.22
Submergence	7	15	0.12	0.12	11	18	0.26	0.20	13	19	0.20	0.12	13	16	0.32	0.26
Cold	7	12	0.08	0.10	10	17	0.22	0.12	13	19	0.26	0.24	12	16	0.36	0.26
Heat	7	14	0.08	0.10	12	19	0.12	0.10	14	19	0.20	0.14	10	18	0.32	0.28
Lodging																
Acidity	7	12	0.14	0.14	6	11	0.20	0.16	7	14	0.12	0.16	10	16	0.24	0.16
Alkalinity	9	18	0.24	0.16	8	15	0.10	0.18	10	18	0.16	0.16	10	16	0.20	0.14
Salinity	8	13	0.20	0.14	7	14	0.20	0.14	9	16	0.20	0.18	10	15	0.30	0.20
Nutrient deficiency	7	14	0.16	0.10	7	17	0.14	0.16	11	18	0.08	0.10	13	17	0.12	0.18
Iron toxicity	7	14	0.14	0.14	7	19	0.16	0.14	9	15	0.24	0.26	10	16	0.20	0.20

Table 22.5. Scientists' ratings: rice pests research priority areas (RPAs).

RPA	Cultural management				Mechanical control				Biological control				Biopesticides				Transgenic breeding			
	Y25	Y75	RP-RA	SD	Y25	Y75	RP-RA	SD	Y25	Y75	RP-RA	SD	Y25	Y75	RP-RA	SD	Y25	Y75	RP-RA	SD
Weeds	5	11	0.20	0.10	12	18	0.25	0.10	4	10	0.20	0.10	5	11	0.15	0.20	10	19	0.30	0.11
Crabs	5	9	0.26	0.11	14	17	0.10	0.14	5	12	0.13	0.23	4	6	0.13	0.11	15	17	0.10	0.14
Rodents	5	9	0.25	0.10	15	18	0.06	0.11	4	9	0.20	0.16	11	16	0.15	0.10	14	17	0.10	0.14
Birds	5	15	0.00	0.00	5	17	0.00	0.00	2	5	0.20	0.16	5	17	0.10	0.14	4	18	0.00	0.00

D. General pest loss

The RTs specified for the control of weeds and other pests (Table 22.5) differ from those for other crop-loss categories. Cultural and mechanical control options are expected to play the major role in weed control. Research has expected contributions to make in terms of biological control methods and biopesticides. Transgenic options for control also have some promise.

E. Biological efficiency

It is important that bioefficiency RPAs be included in priority setting. Since they do not have natural 'loss' units it is sometimes difficult to specify meaningful RPAs. Consultation with scientists indicates that the RPAs in Table 22.6 are meaningful, but the priority setter should be particularly aware that the RPAs are subject to change as new scientific and technological options become available.

The plant design work at the IRRI has been particularly important in recent years as the new plant type has been developed (see Chapter 4). Further gains are expected through the use of new RTs (to date conventional breeding techniques have been used). Photosynthetic efficiency gains are indicated to have relatively low potential and relatively long times to achievement. Growth duration (shorter growing season) options are more promising. Improvements in grain quality options appear to be quite good; this is particularly important given demand trends favouring higher-quality rice (see Chapter 2).

F. Scientists ratings: summary comments

As noted in the analytical chapter (Chapter 5) in this volume, differences in the underlying research probabilities (b/u, here treated as RP–RA) and time to achievement (TTA, here treated as W_t) are required to justify research resource allocation differing from the 'congruence' rule (where research is allocated in proportion to units). From a statistical perspective, the standard errors reported in Tables 22.2–22.6 appear to suggest that it is difficult to justify departures from congruence. This is to some degree a realistic conclusion. Few *ex ante* or priority-setting exercises show large difference in (b/u) estimates across RPAs (even when applied to multicommodity research).

Yet, we would argue that the differences that do emerge from exercises like this should be looked at carefully. Evidence for high expected pay-offs (or low expected pay-offs) call for follow-up attention. It should be evident that the information reported in these tables is subject to both subjective and comparative uncertainty, some of which is due to different knowledge, but most of it is real.

Table 22.6. Scientists' ratings: biological efficiency potentials research priority areas (RPAs).

RPA	Hybridization Y25 Y75 RP-RA SD	Conventional breeding Y25 Y75 RP-RA SD	Wide crossing Y25 Y75 RP-RA SD	Tissue culture Y25 Y75 RP-RA SD	Transgenic breeding Y25 Y75 RP-RA SD	Marker-Aided selection Y25 Y75 RP-RA SD
Plant design	7 13 0.23 0.14	4 11 0.22 0.12	8 15 0.23 0.14	5 11 0.17 0.14	9 17 0.33 0.24	10 17 0.32 0.19
Photosynthetic efficiency	8 17 0.10 0.11	9 16 0.20 0.13	8 18 0.17 0.23	8 15 0.10 0.11	10 17 0.36 0.27	11 19 0.36 0.08
Growth duration	7 15 0.23 0.15	6 10 0.23 0.15	8 15 0.26 0.20	4 9 0.20 0.25	7 17 0.24 0.26	8 14 0.36 0.22
Grain quality	7 14 0.40 0.23	5 11 0.30 0.15	8 13 0.23 0.23	6 13 0.26 0.16	8 14 0.36 0.36	9 15 0.54 0.22

VII. Research-Priority Portfolios

A. Procedures

The information in each RPA–RT cell in Tables 22.2–22.6 can be combined with the units data in Table 22.1 to produce research portfolios through eqn 21.7. (Recall that with information on W_t, b/u and u, and given that we fix NB, NC and IRR, we can generate EXP, the annual expenditure justified by these parameters to realize the IRR returns (set equal to 25% for this exercise).)

We applied the priority portfolio procedure to data for each of the four rice ecosystems and report the results in Tables 22.7–22.10. Comparable portfolios are computed for each country dataset in Appendix Tables 22.1–22.12. We recognize the inherent uncertainty in the SPEs by computing a range of priorities based on the standard deviations of the RP–PA estimate. We should note that the procedure is most relevant for broad geographic regions, and is probably not relevant for small regions. (For example, we would not be justified in generating a different portfolio for two Indian districts based on the units data in Table 22.1, because research targeted to one district would probably be highly relevant (i.e. not subject to diminishing returns) in a nearby district.)

To summarize our procedures:

1. RPA units (in kg ha^{-1}) were taken from Table 22.1.
2. A low estimate of benefits per unit (b/u) was obtained by taking the mean minus half the standard deviation of the RP–RA estimates reported in Tables 22.2–22.6 for each RPA–RT category. A high estimate was generated by taking the mean RP–RA estimates plus half the standard deviation.
3. An adjustment was made for some 'double counting' across RTs. The rating procedure required that the scientist enter report RP and RA (and Y25 and Y75) ratings for each RT. It did not impose that the percent achievement on the remaining potential (RP–RA) add up to one. That is, it did not impose the condition that more than 100% achievement could be obtained. A review of the RP–RA estimates indicates that most do add up to more than 100%. The wide crossing tissue-culture techniques are generally regarded as middle ground techniques between conventional breeding and transgenic breeding techniques. All three are complementary since conventional breeding is inherent in each. We decided to adjust the wide–crossing RP–RA estimates downward by half to bring the estimates more in line with the accounting consistency argument.
4. The Y25 and Y75 estimates were utilized as W_t weights in computing the D_B and D_C discount factors. For the benefits discount we added 4 years to the Y25 and Y75 estimates for diffusion to farmers' yields. Scientists were asked to estimate Y25 and Y75 on the basis of experimental field development. We added an additional 3 years to the Y25 and Y75 estimates for the bioefficiency RPAs to reflect the longer process of varietal development for these techniques.
5. Equation 22.7 was then used to generate portfolio elements expressed by kilogram per RPA–RT category divided by total yield. This is equivalent to an

Table 22.7. Priority portfolios: irrigated rice ecosystems.

Research problem area	Management research Low	High	Conventional breeding Low	High	Wide crossing Low	High	Transgenic breeding Low	High	All techniques Low	High
Insects										
Stemborer	0.020	0.048	0.010	0.023	0.005	0.012	0.048	0.088	0.082	0.170
Brown planthopper	0.002	0.009	0.002	0.005	0.001	0.004	0.003	0.009	0.008	0.027
White-backed planthopper	0.004	0.009	0.002	0.006	0.001	0.002	0.001	0.004	0.008	0.021
Leaf folder	0.006	0.011	0.002	0.005	0.000	0.002	0.000	0.008	0.009	0.026
Hispa	0.000	0.001	0.000	0.001	0.000	0.001	0.000	0.002	0.002	0.005
Green leafhopper	0.001	0.007	0.003	0.006	0.001	0.002	0.003	0.006	0.008	0.20
Gall midge	0.009	0.019	0.009	0.018	0.002	0.005	0.004	0.012	0.024	0.053
Caseworm	0.001	0.001	0.000	0.001	0.000	0.000	0.000	0.001	0.001	0.003
Armyworm	0.005	0.008	0.002	0.003	0.000	0.001	0.002	0.007	0.009	0.019
Mealy bug	0.000	0.000	0.000	0.000	0.000	0.000	0.000	0.001	0.000	0.001
Rice bug	0.005	0.011	0.002	0.007	0.000	0.001	0.002	0.009	0.010	0.029
Total	0.053	0.125	0.033	0.075	0.011	0.030	0.064	0.146	0.161	0.375
Diseases										
Blast	0.005	0.010	0.003	0.014	0.002	0.005	0.006	0.013	0.016	0.043
Leaf scald										
Gerlachia leaf spot										
Brown spot	0.002	0.004	0.001	0.001	0.001	0.002	0.001	0.002	0.005	0.009
Sheath rot	0.001	0.003	0.001	0.002	0.001	0.001	0.000	0.001	0.003	0.007
Sheath blight	0.003	0.008	0.000	0.001	0.001	0.002	0.004	0.008	0.008	0.019
Stem rot										
Bacterial blight	0.003	0.007	0.006	0.026	0.010	0.018	0.008	0.018	0.026	0.069
Bacterial leaf streak	0.000	0.001	0.000	0.001	0.000	0.001	0.001	0.001	0.001	0.003
False smut										
Atom blight										
Ufra	0.000	0.000	0.000	0.000	0.000	0.000	0.000	0.000	0.000	0.000
Total	0.014	0.034	0.013	0.048	0.016	0.031	0.022	0.050	0.066	0.163
Abiotic stresses										
Drought	0.006	0.018	0.012	0.023	0.003	0.010	0.008	0.017	0.029	0.068
Submergence	0.002	0.005	0.004	0.008	0.001	0.001	0.002	0.005	0.008	0.019
Cold	0.000	0.001	0.001	0.001	0.000	0.001	0.001	0.001	0.002	0.004
Heat										
Acidity	0.002	0.006	0.006	0.015	0.001	0.004	0.003	0.006	0.012	0.030
Alkalinity	0.000	0.001	0.000	0.001	0.000	0.000	0.000	0.001	0.001	0.003
Salinity	0.001	0.003	0.003	0.007	0.001	0.002	0.002	0.004	0.007	0.015
Nutrient deficiency	0.004	0.007	0.003	0.011	0.000	0.001	0.000	0.003	0.007	0.022
Iron toxicity	0.000	0.001	0.001	0.001	0.000	0.001	0.000	0.001	0.001	0.003
Total	0.015	0.041	0.029	0.066	0.006	0.020	0.017	0.037	0.067	0.164

Table 22.7. Priority portfolios: irrigated rice ecosystems. (*continued*)

Research problem area	Management research Low	Management research High	Conventional breeding Low	Conventional breeding High	Wide crossing Low	Wide crossing High	Transgenic breeding Low	Transgenic breeding High	All techniques Low	All techniques High
Other pests										
Weeds	0.020	0.033			0.015	0.039	0.012	0.017	0.047	0.089
Crabs	0.000	0.000			0.000	0.000	0.000	0.000	0.000	0.001
Rodents	0.010	0.015			0.006	0.013	0.001	0.003	0.016	0.030
Birds	0.000	0.000			0.002	0.007	0.000	0.000	0.002	0.007
Total	0.030	0.049			0.024	0.059	0.012	0.020	0.066	0.127
Bioefficiency										
Plant design			0.061	0.107	0.014	0.027	0.028	0.061	0.104	0.195
Photosynthetic efficiency			0.013	0.026	0.004	0.019	0.018	0.041	0.035	0.085
Growth duration			0.035	0.069	0.013	0.030	0.020	0.068	0.068	0.167
Grain quality			0.039	0.065	0.007	0.028	0.023	0.070	0.069	0.163
Total			0.148	0.267	0.039	0.104	0.090	0.239	0.277	0.610
Grand total	0.113	0.248	0.223	0.456	0.095	0.244	0.206	0.491	0.638	1.439

annual research intensity (research spending/crop value) for each RPA–RT category.

B. Priority portfolios for irrigated rice ecosystems

We have six sets of units estimates for irrigated rice. The priority portfolios for these six locations – eastern India, southern India, Nepal, Bangladesh, China and Indonesia – are reported in Appendices 22.1–22.6. Table 22.7 reports the aggregate portfolio for irrigated rice ecosystems. Portfolios are reported in intensity form and a low and high portfolio is reported for each RPA–RT category.

Thus for the aggregate irrigated ecosystem the low (i.e. conservative) estimates of research potential generate a portfolio calling for a total research expenditure of 0.638% of rice value. For the high estimates of research potential the portfolio is 1.439% of value. For the low estimates (this is approximately the case for the high estimates as well) the allocation by RPA class is approximately 25% to insect RPAs (0.161/0.638), 10% to disease RPAs, 11% to abiotic stress RPAs, 10% to other pest RPAs and 43% to bioefficiency RPAs. By research technique, the allocations are 18% to management-related research, 35% to conventional breeding, 15% to wide crossing and 32% to transgenic breeding techniques.

These allocations, it should be noted, do not incorporate time specificity in the sense that those techniques requiring a longer time to achievement may be phased into the research program more slowly. (Higher discount factors are considered in the calculations.) If this were done the priority intensities for conventional breeding would be higher and for transgenic breeding lower in current periods and would decline (rise) as time passed.

When we compare this portfolio with current investments we find that the low portfolio is higher than current portfolios (which are roughly 0.3 to 0.4%) and that the high portfolio is considerably higher (probably four times as high). But this disjuncture is not inconsistent with the *ex post* studies reviewed in Chapter 1. *Ex post* estimates of internal rates of return are generally higher than 25%. *Ex post* investment portfolios would have had to be above our low portfolios and near our high portfolios to have generated internal rates of return of 25%.

When we compare priority portfolios across locations (see Appendices 22.1–22.7) we find that the *low* priority intensities range from 0.476% for China to 1.112 for Nepal. These differences, it should be noted are due to differences in the units, i.e. in the size of the research opportunities as reflected in Table 22.1. They are not due to differences in the b/u and W_t components. Note further that the units (as proportions of yield) for the bioefficiency RPAs do not differ by location.

Differences in insect units account for the largest differences in portfolio intensities by country. Eastern India and Nepal have large scope for insect-loss reductions, reflected in portfolio intensities of 0.234 and 0.283); southern India, Bangladesh and Indonesia have moderate scope for insect-loss reduction (portfolio intensities of 0.107, 0.068 and 0.145, respectively); while China has very low scope for insect-loss reduction (portfolio intensity of 0.014).

Considerable differences in disease-loss RPAs exist as well. Nepal shows the highest potential from these RPAs, while China and India have relatively low potential for disease-loss reduction. For abiotic stresses, southern India and Indonesia have the lowest research scope. For other pests, China reports low scope for research gains.

C. Priority portfolios for rainfed rice

We have two locations for rainfed rice portfolios, eastern India and Bangladesh. The portfolios are reported in Appendices 22.7 and 22.8, respectively. The aggregate portfolio is reported in Table 22.8. We note first that these portfolios differ from the irrigated rice portfolios in each country. The rainfed portfolio for eastern India is lower than the irrigated portfolio (0.652 vs. 0.770). This is partly accounted for by the lower bioefficiency RPAs portfolios (recall that these are based on judgement), and the insect RPAs have lower portfolios in the rainfed locations. As expected, the abiotic stress RPA portfolios are higher.

For Bangladesh the total rainfed and irrigated portfolios are approximately the same (0.611 vs. 0.612). The bioefficiency portfolios are lower for rainfed RPAs, but the insect and disease RPA portfolios are higher. The major

Table 22.8. Priority portfolios: rainfed rice ecosystems.

Research problem area	Management research Low	Management research High	Conventional breeding Low	Conventional breeding High	Wide crossing Low	Wide crossing High	Transgenic breeding Low	Transgenic breeding High	All techniques Low	All techniques High
Insects										
Stemborer	0.012	0.028	0.006	0.014	0.002	0.007	0.028	0.052	0.047	0.100
Brown planthopper	0.001	0.003	0.001	0.002	0.000	0.001	0.001	0.003	0.003	0.009
White-backed planthopper	0.002	0.004	0.001	0.002	0.000	0.001	0.000	0.002	0.003	0.008
Leaf folder	0.003	0.006	0.001	0.002	0.000	0.001	0.000	0.004	0.004	0.012
Hispa	0.001	0.003	0.001	0.002	0.001	0.002	0.001	0.006	0.004	0.012
Green leafhopper	0.002	0.008	0.004	0.007	0.001	0.002	0.003	0.006	0.009	0.023
Gall midge	0.003	0.006	0.003	0.006	0.001	0.002	0.001	0.004	0.007	0.017
Caseworm	0.004	0.008	0.002	0.003	0.000	0.001	0.002	0.006	0.007	0.018
Armyworm	0.003	0.005	0.001	0.002	0.000	0.001	0.001	0.004	0.005	0.012
Mealy bug	0.001	0.001	0.000	0.000	0.000	0.000	0.000	0.002	0.001	0.003
Rice bug	0.004	0.008	0.002	0.006	0.000	0.001	0.002	0.007	0.007	0.022
Total	0.033	0.078	0.020	0.045	0.006	0.019	0.040	0.094	0.098	0.236
Diseases										
Blast	0.013	0.027	0.007	0.036	0.005	0.013	0.017	0.035	0.043	0.110
Leaf scald										
Gerlachia leaf spot										
Brown spot	0.001	0.003	0.001	0.001	0.001	0.001	0.001	0.002	0.003	0.007
Sheath rot	0.001	0.003	0.001	0.002	0.001	0.002	0.000	0.002	0.004	0.010
Sheath blight	0.002	0.006	0.000	0.001	0.001	0.001	0.002	0.005	0.006	0.013
Stem rot										
Bacterial blight	0.003	0.008	0.007	0.030	0.011	0.025	0.009	0.021	0.030	0.083
Bacterial leaf blight	0.002	0.003	0.001	0.002	0.001	0.006	0.002	0.004	0.005	0.015
False smut										
Atom blight										
Tungro	0.000	0.000	0.001	0.001	0.000	0.000	0.001	0.001	0.001	0.003
Ragged stunt										
Ufra										
Total	0.023	0.050	0.018	0.072	0.020	0.049	0.032	0.069	0.092	0.240
Abiotic stresses										
Drought	0.020	0.061	0.041	0.079	0.010	0.035	0.028	0.059	0.100	0.233
Submergence	0.005	0.016	0.010	0.023	0.002	0.004	0.006	0.015	0.024	0.058
Cold	0.001	0.004	0.003	0.006	0.001	0.002	0.004	0.008	0.008	0.020
Heat										
Acidity	0.001	0.002	0.002	0.005	0.000	0.001	0.001	0.002	0.004	0.010
Alkalinity	0.001	0.003	−0.001	0.004	0.000	0.001	0.002	0.003	0.003	0.011
Salinity	0.000	0.001	0.001	0.002	0.000	0.001	0.001	0.001	0.002	0.005
Nutrient deficiency	0.006	0.012	0.005	0.019	0.001	0.002	0.001	0.005	0.013	0.037
Iron toxicity	0.001	0.003	0.002	0.006	0.001	0.003	0.001	0.003	0.005	0.015
Total	0.036	0.102	0.064	0.142	0.015	0.049	0.043	0.095	0.159	0.388

Table 22.8. Priority portfolios: rainfed rice ecosystems. (*continued*)

Research problem area	Management research Low	Management research High	Conventional breeding Low	Conventional breeding High	Wide crossing Low	Wide crossing High	Transgenic breeding Low	Transgenic breeding High	All techniques Low	All techniques High
Other pests										
Weeds	0.030	0.050			0.023	0.056	0.018	0.026	0.071	0.131
Crabs	0.001	0.002			0.000	0.002	0.000	0.000	0.002	0.004
Rodents	0.007	0.010			0.004	0.009	0.000	0.002	0.011	0.022
Birds	0.000	0.000			0.002	0.006	0.000	0.000	0.002	0.006
Total	0.038	0.063			0.030	0.073	0.018	0.028	0.086	0.163
Bioefficiency										
Plant design			0.039	0.068	0.009	0.017	0.018	0.039	0.066	0.124
Photosynthetic efficiency			0.010	0.019	0.003	0.014	0.014	0.031	0.027	0.064
Growth duration			0.028	0.055	0.010	0.024	0.016	0.055	0.055	0.133
Grain quality			0.033	0.055	0.006	0.019	0.020	0.060	0.059	0.134
Total			0.110	0.198	0.029	0.074	0.068	0.184	0.207	0.456
Grand total	0.130	0.292	0.212	0.458	0.099	0.264	0.201	0.469	0.642	1.483

difference is in the abiotic stress RPAs as expected. For the aggregate portfolio, insect RPAs account for a smaller proportion of the portfolio than for irrigated rice (15% vs 25%), while abiotic stress RPAs account for a higher proportion (25% vs 11%).

D. Priority portfolios for upland rice

Upland rice priority portfolios for three locations – eastern India, Nepal and Indonesia – are reported in Appendices 22.9–22.11. The aggregate upland rice portfolio is reported in Table 22.9.

These portfolios differ considerably by country. The eastern India portfolio for upland rice is actually higher than the irrigated and rainfed portfolios. The bioefficiency RPAs have lower intensities for upland rice, but the upland abiotic stress RPA intensities (drought) are higher, as are other pest RPAs. For Nepal, the upland portfolio intensities are the same as the irrigated portfolio intensities. The bioefficiency RPAs are, of course, lower, but other components are lower as well. For Indonesia, a very low research portfolio is called for in upland rice. Insect RPAs are significant, but other components have low units.

Table 22.9. Priority portfolios: upland rice ecosystems.

Research problem area	Management research Low	High	Conventional breeding Low	High	Wide crossing Low	High	Transgenic breeding Low	High	All techniques Low	High
Insects										
Stemborer	0.003	0.008	0.002	0.003	0.001	0.002	0.008	0.014	0.013	0.027
Brown planthopper	0.004	0.022	0.004	0.013	0.002	0.009	0.007	0.020	0.018	0.063
White-backed planthopper	0.002	0.004	0.001	0.002	0.000	0.001	0.000	0.002	0.003	0.008
Leaf folder	0.001	0.003	0.001	0.001	0.000	0.000	0.000	0.002	0.002	0.007
Hispa										
Green leafhopper	0.003	0.016	0.008	0.014	0.002	0.005	0.006	0.013	0.019	0.048
Gall midge	0.000	0.000	0.000	0.000	0.000	0.000	0.000	0.000	0.000	0.001
Caseworm	0.001	0.002	0.000	0.001	0.000	0.000	0.000	0.002	0.002	0.004
Armyworm	0.017	0.030	0.006	0.011	0.001	0.004	0.007	0.024	0.031	0.069
Mealy bug										
Rice bug	0.013	0.027	0.005	0.018	0.001	0.003	0.006	0.024	0.025	0.073
Total	0.045	0.112	0.027	0.064	0.007	0.025	0.035	0.100	0.114	0.301
Diseases										
Blast	0.027	0.055	0.016	0.074	0.010	0.027	0.034	0.071	0.087	0.227
Leaf scald										
Gerlachia leaf spot										
Brown spot	0.008	0.019	0.004	0.005	0.007	0.010	0.005	0.011	0.023	0.045
Sheath rot	0.000	0.001	0.000	0.001	0.000	0.000	0.000	0.000	0.001	0.002
Sheath blight	0.004	0.010	0.000	0.001	0.001	0.003	0.004	0.010	0.010	0.024
Stem rot	0.000	0.000	0.000	0.000	0.000	0.000	0.000	0.000	0.000	0.000
Bacterial blight	0.001	0.003	0.003	0.012	0.005	0.008	0.003	0.008	0.012	0.032
Bacterial leaf streak										
False smut										
Atom blight										
Tungro										
Ragged stunt										
Ufra										
Total	0.041	0.088	0.023	0.093	0.023	0.049	0.047	0.101	0.134	0.330
Abiotic stresses										
Drought	0.040	0.120	0.080	0.155	0.020	0.068	0.056	0.115	0.196	0.458
Submergence	0.001	0.004	0.003	0.006	0.001	0.001	0.002	0.004	0.006	0.015
Cold	0.000	0.001	0.001	0.001	0.000	0.000	0.001	0.001	0.002	0.003
Heat										
Acidity	0.009	0.027	0.030	0.069	0.004	0.020	0.013	0.026	0.056	0.142
Alkalinity	0.002	0.005	−0.001	0.007	0.001	0.002	0.002	0.005	0.004	0.018
Salinity	0.002	0.005	0.006	0.013	0.002	0.004	0.004	0.008	0.014	0.030
Nutrient deficiency	0.007	0.014	0.006	0.022	0.001	0.003	0.001	0.005	0.014	0.044
Iron toxicity	0.001	0.002	0.001	0.003	0.000	0.002	0.001	0.002	0.003	0.008
Total	0.063	0.178	0.125	0.275	0.028	0.100	0.079	0.166	0.296	0.719

Table 22.9. Priority portfolios: upland rice ecosystems (*continued*).

Research problem area	Management research Low	Management research High	Conventional breeding Low	Conventional breeding High	Wide crossing Low	Wide crossing High	Transgenic breeding Low	Transgenic breeding High	All techniques Low	All techniques High
Other pests										
Weeds	0.094	0.157			0.073	0.174	0.055	0.080	0.222	0.411
Crabs										
Rodents	0.011	0.017			0.006	0.015	0.001	0.004	0.018	0.035
Birds	0.000	0.000			0.012	0.033	0.000	0.000	0.012	0.033
Total	0.105	0.174			0.091	0.221	0.055	0.083	0.252	0.478
Bioefficiency										
Plant design			0.033	0.059	0.008	0.015	0.016	0.033	0.057	0.107
Photosynthetic efficiency			0.005	0.010	0.001	0.007	0.007	0.015	0.013	0.032
Growth duration			0.014	0.027	0.005	0.012	0.008	0.027	0.027	0.067
Grain quality			0.022	0.037	0.004	0.013	0.013	0.040	0.040	0.089
Total			0.074	0.133	0.019	0.046	0.044	0.115	0.137	0.294
Grand total	0.255	0.552	0.249	0.565	0.168	0.441	0.260	0.565	0.932	2.123

E. Priority portfolios for deepwater rice

We have only one deepwater rice portfolio for eastern India, (Appendix 22.12 and Table 22.10). This portfolio has a low bioefficiency RPA component. This portfolio indicates high stemborer RPA potential and relatively high disease potential (blast and bacterial blight).

VII. An Overview of the Exercise

Priority setting can be applied in a number of settings. In particular, national research programs and state or district (subnational) programs must set priorities. In this chapter we have demonstrated the application of priority-setting methods to several country and regional programs. We re-emphasize that the application reported here is for a single-commodity program, not for a multiple-commodity program.

How useful has this exercise been? Is it easily replicable? The reader must ultimately answer these questions. However, we offer the following comments.

1. It seems clear that meaningful RPA definition is important. It is especially important that RPAs be specified to allow a broad range of research focus and that the RPA structure should not constrain and restrict research options. We believe that the inclusion of the bioefficiency RPAs at least partially addressed this need.

Table 22.10. Priority portfolios: deepwater rice ecosystems.

Research problem area	Management research Low	Management research High	Conventional breeding Low	Conventional breeding High	Wide crossing Low	Wide crossing High	Transgenic breeding Low	Transgenic breeding High	All techniques Low	All techniques High
Insects										
Stemborer	0.055	0.133	0.029	0.064	0.014	0.033	0.133	0.244	0.230	0.474
Brown planthopper										
White-backed planthopper										
Leaf folder										
Hispa										
Green leafhopper										
Gall midge										
Caseworm										
Armyworm										
Mealy bug										
Rice bug										
Total	0.055	0.133	0.029	0.064	0.014	0.033	0.133	0.244	0.230	0.474
Diseases										
Blast	0.016	0.033	0.009	0.044	0.006	0.016	0.021	0.042	0.052	0.135
Leaf scald										
Gerlachia leaf spot										
Brown spot	0.005	0.011	0.002	0.003	0.004	0.006	0.003	0.006	0.013	0.025
Sheath rot	0.002	0.006	0.002	0.004	0.002	0.003	0.001	0.004	0.006	0.017
Sheath blight	0.006	0.016	0.000	0.002	0.002	0.004	0.007	0.016	0.015	0.038
Stem rot										
Bacterial blight	0.009	0.022	0.018	0.079	0.030	0.055	0.023	0.055	0.080	0.212
Bacterial leaf streak										
False smut										
Atom blight										
Tungro										
Ragged stunt										
Ufra										
Total	0.038	0.088	0.032	0.132	0.043	0.084	0.055	0.123	0.167	0.427
Abiotic stresses										
Drought	0.014	0.042	0.028	0.055	0.007	0.024	0.020	0.041	0.069	0.161
Submergence	0.011	0.034	0.023	0.051	0.005	0.009	0.014	0.032	0.052	0.126
Cold										
Heat										
Acidity										
Alkalinity	0.019	0.037	−0.009	0.056	0.006	0.017	0.019	0.039	0.034	0.150
Salinity										
Nutrient deficiency										
Iron toxicity										
Total	0.044	0.114	0.041	0.162	0.018	0.051	0.052	0.112	0.155	0.438

Table 22.10. Priority portfolios: deepwater rice ecosystems (*continued*).

Research problem area	Management research Low	Management research High	Conventional breeding Low	Conventional breeding High	Wide crossing Low	Wide crossing High	Transgenic breeding Low	Transgenic breeding High	All techniques Low	All techniques High
Other pests										
Weeds	0.055	0.091			0.043	0.101	0.032	0.046	0.130	0.239
Crabs										
Rodents	0.016	0.025			0.009	0.022	0.001	0.005	0.027	0.051
Birds	0.000	0.000			0.011	0.031	0.000	0.000	0.011	0.031
Total	0.071	0.116			0.063	0.154	0.033	0.051	0.167	0.321
Bioefficiency										
Plant design			0.017	0.029	0.004	0.007	0.008	0.017	0.028	0.053
Photosynthetic efficiency			0.003	0.007	0.001	0.005	0.005	0.010	0.009	0.022
Growth duration			0.007	0.014	0.003	0.006	0.004	0.014	0.014	0.033
Grain quality			0.011	0.018	0.002	0.006	0.007	0.020	0.020	0.045
Total			0.038	0.068	0.010	0.024	0.023	0.060	0.071	0.153
Grand total	0.209	0.451	0.140	0.425	0.147	0.347	0.295	0.591	0.791	1.813

2. It is equally important that alternative RTs techniques be recognized in priority setting. There are two reasons for this. First, the definition of research programs should not be technique-constrained. Second, priorities between techniques are important and this requires a clear comparison between techniques. This exercise is, to our knowledge, the first to develop a RPA–RT matrix approach to priority setting. We believe that this has been relatively successful in addressing the problem.

3. SPEs are difficult to obtain. There is a strong tendency for scientists to report roughly similar SPEs for RPAs in many studies (Fishel, 1970). Scientists also tend not to distinguish between accomplishments to date and potential. Scientists also tend to be uncomfortable with expressions of range between estimates. In this exercise we address these issues:

(a) by specifying that the level of research resources would continue at present levels and that research conduct would be best technique or best practice research for the RPA–RT category;

(b) by asking for both an achievement to date rating and a research potential rating. We did not ask for percentage estimates – only for ratings which were then scaled into percent increases. We did not ask for a range on ratings;

(c) by asking for estimates of years to 25% and to 75% achievement. We expected that these two estimates would allow the respondent to express uncertainty, and hence a range. We believe that we were successful in obtaining estimates of b/u and W_t from this exercise, and that scientists did express some of their range of uncertainty in Y25 and Y75 – our procedure was workable. The alternative would have been to ask for a probability range on the ratings scores.

(d) in combining the estimates from several scientists we chose to compute a high–low range from the RPA–RT estimates by taking the mean ±0.5 standard deviation. We expected the mean estimates of Y25 and Y75 to be timing estimates. Obviously, we could have either asked scientist respondents for individual ranges of their SPEs and then attempted to combine them into priority parameters. This combination is not straightforward. We have within-scientists estimates of the standard deviation and between-scientist estimates. It was our judgement at this point that we would not have obtained more reasonable estimates of the ranges by asking scientists for them directly. We could have used the standard deviations of the Y25 and Y75 estimates in the portfolio process, but it is not clear that this would have added much to the ±0.5 standard deviation procedure;

(e) in the end, our procedures were workable. For rice research programs we consider this exercise to be a richer approach than the approach taken in Chapter 21 where we used *ex post* estimates for b/u ratings and concentrated on time to achievement. The ratings exercise was accepted by scientists as a meaningful exercise. We obtained results from a relatively small sample of knowledgeable scientists.

Thus we regard this chapter as a demonstration of a workable priority-setting exercise. It is certainly not a final refinement of priority setting, but we consider it to be a useful forward step. We want to re-emphasize that priority-setting exercises do not replace the need for continuous management review. Priority setting itself is not static and needs to be undertaken periodically.

Finally, we want to emphasize the important need to confirm and check priority setting against *ex post* evidence.

References

Fishel, W. (1970) Uncertainty in public research administration and subjective probability estimates about changing the state of knowledge. PhD thesis, North Carolina State University, Raleigh, North Carolina.

Appendix Table 22.1. Priority portfolios: irrigated rice, eastern India.

	Management research		Conventional breeding		Wide crossing		Transgenic breeding		All techniques	
Research problem area	Low	High	Low	High	Low	High	Low	High	Low	High
Insects										
Stemborer	0.029	0.071	0.015	0.034	0.007	0.018	0.071	0.130	0.122	0.253
Brown planthopper	0.002	0.010	0.002	0.006	0.001	0.004	0.004	0.009	0.009	0.030
White-backed planthopper	0.012	0.027	0.005	0.016	0.002	0.006	0.003	0.012	0.022	0.061
Leaf folder	0.003	0.006	0.001	0.003	0.000	0.001	0.000	0.004	0.005	0.014
Hispa										
Green leafhopper	0.002	0.011	0.005	0.009	0.001	0.003	0.005	0.009	0.014	0.032
Gall midge	0.012	0.025	0.011	0.023	0.002	0.007	0.006	0.016	0.032	0.070
Caseworm	0.002	0.004	0.001	0.002	0.000	0.000	0.001	0.003	0.004	0.009
Armyworm	0.008	0.014	0.003	0.005	0.000	0.002	0.003	0.011	0.015	0.033
Mealy bug										
Rice bug	0.007	0.014	0.002	0.009	0.000	0.002	0.003	0.012	0.012	0.036
Total	0.077	0.181	0.046	0.107	0.016	0.043	0.095	0.207	0.234	0.539
Diseases										
Blast	0.003	0.006	0.002	0.008	0.001	0.003	0.004	0.007	0.009	0.024
Leaf scald										
Gerlachia leaf spot										
Brown spot	0.001	0.003	0.001	0.001	0.001	0.002	0.001	0.002	0.004	0.008
Sheath rot	0.001	0.004	0.001	0.002	0.001	0.002	0.000	0.002	0.004	0.010
Sheath blight	0.004	0.011	0.000	0.001	0.001	0.003	0.005	0.010	0.010	0.024
Stem rot										
Bacterial blight	0.005	0.011	0.009	0.039	0.014	0.027	0.011	0.027	0.039	0.103
Bacterial leaf streak										
False smut										
Atom blight										
Tungro										
Ragged stunt										
Ufra										
Total	0.014	0.034	0.013	0.051	0.019	0.036	0.021	0.048	0.066	0.169
Abiotic stresses										
Drought	0.008	0.024	0.016	0.030	0.004	0.013	0.011	0.023	0.038	0.090
Submergence	0.001	0.004	0.003	0.006	0.001	0.001	0.002	0.004	0.006	0.015
Cold	0.000	0.002	0.002	0.003	0.000	0.001	0.002	0.002	0.005	0.008
Heat										
Acidity	0.005	0.016	0.016	0.039	0.002	0.011	0.007	0.015	0.031	0.080
Alkalinity	0.001	0.001	−0.000	0.002	0.000	0.001	0.001	0.002	0.001	0.006
Salinity	0.002	0.004	0.005	0.009	0.001	0.003	0.003	0.006	0.010	0.022
Nutrient deficiency	0.005	0.009	0.004	0.015	0.000	0.002	0.000	0.004	0.010	0.030

	Management research		Conventional breeding		Wide crossing		Transgenic breeding		All techniques	
Research problem area	Low	High	Low	High	Low	High	Low	High	Low	High
Iron toxicity	0.000	0.002	0.001	0.003	0.000	0.001	0.000	0.002	0.002	0.007
Total	0.023	0.061	0.046	0.107	0.009	0.033	0.027	0.055	0.105	0.258
Other pests										
Weeds	0.032	0.054			0.025	0.059	0.019	0.027	0.076	0.140
Crabs										
Rodents	0.007	0.010			0.004	0.008	0.000	0.002	0.011	0.020
Birds	0.000	0.000			0.001	0.003	0.000	0.000	0.001	0.003
Total	0.039	0.064			0.030	0.071	0.019	0.029	0.088	0.164
Bioefficiency										
Plant design			0.061	0.107	0.014	0.027	0.028	0.061	0.104	0.195
Photosynthetic efficiency			0.013	0.026	0.004	0.019	0.018	0.041	0.035	0.085
Growth duration			0.035	0.069	0.013	0.030	0.020	0.068	0.068	0.167
Grain quality			0.039	0.065	0.007	0.022	0.023	0.070	0.069	0.157
Total			0.148	0.267	0.039	0.097	0.090	0.239	0.277	0.604
Grand total	0.153	0.340	0.253	0.533	0.112	0.281	0.252	0.579	0.770	1.733

Appendix Table 22.2. Priority portfolios: irrigated rice, southern India.

	Management research		Conventional breeding		Wide crossing		Transgenic breeding		All techniques	
Research problem area	Low	High	Low	High	Low	High	Low	High	Low	High
Insects										
Stemborer	0.009	0.021	0.005	0.010	0.002	0.005	0.021	0.039	0.037	0.076
Brown planthopper	0.002	0.011	0.002	0.007	0.001	0.005	0.004	0.010	0.010	0.033
White-backed planthopper										
Leaf folder	0.015	0.029	0.005	0.013	0.001	0.004	0.001	0.019	0.022	0.065
Hispa										
Green leafhopper	0.002	0.011	0.005	0.009	0.001	0.003	0.005	0.009	0.013	0.032
Gall midge	0.010	0.020	0.009	0.019	0.002	0.006	0.005	0.013	0.026	0.058
Caseworm										
Armyworm										
Mealy bug										
Rice bug										
Total	0.038	0.092	0.027	0.059	0.008	0.023	0.035	0.090	0.107	0.264

	Management research		Conventional breeding		Wide crossing		Transgenic breeding		All techniques	
Research problem area	Low	High	Low	High	Low	High	Low	High	Low	High
Diseases										
Blast	0.010	0.019	0.005	0.026	0.004	0.010	0.012	0.025	0.031	0.080
Leaf scald										
Gerlachia leaf spot										
Brown spot	0.003	0.007	0.001	0.002	0.002	0.004	0.002	0.004	0.008	0.017
Sheath rot	0.001	0.005	0.002	0.003	0.001	0.002	0.000	0.002	0.005	0.012
Sheath blight	0.006	0.016	0.000	0.002	0.002	0.004	0.007	0.015	0.015	0.037
Stem rot										
Bacterial blight	0.002	0.004	0.003	0.015	0.006	0.010	0.005	0.010	0.015	0.039
Bacterial leaf streak										
False smut										
Atom blight										
Tungro	0.001	0.002	0.006	0.011	0.004	0.005	0.007	0.016	0.017	0.034
Ragged stunt										
Ufra										
Total	0.023	0.053	0.018	0.059	0.019	0.035	0.032	0.073	0.091	0.219
Abiotic stresses										
Drought	0.004	0.012	0.008	0.016	0.002	0.007	0.006	0.012	0.020	0.047
Submergence										
Cold	0.000	0.001	0.000	0.001	0.000	0.000	0.000	0.001	0.001	0.003
Heat										
Acidity	0.000	0.002	0.002	0.004	0.000	0.001	0.001	0.002	0.003	0.008
Alkalinity	0.000	0.000	0.000	0.001	0.000	0.000	0.000	0.001	0.001	0.002
Salinity	0.002	0.005	0.006	0.011	0.001	0.004	0.004	0.007	0.013	0.027
Nutrient deficiency	0.005	0.010	0.004	0.016	0.000	0.002	0.000	0.004	0.010	0.031
Iron toxicity	0.000	0.000	0.000	0.001	0.000	0.000	0.000	0.000	0.001	0.002
Total	0.013	0.030	0.021	0.049	0.005	0.015	0.011	0.027	0.049	0.120
Other pests										
Weeds	0.008	0.014			0.006	0.022	0.005	0.007	0.019	0.042
Crabs										
Rodents	0.005	0.008			0.003	0.007	0.000	0.002	0.009	0.017
Birds	0.000	0.000			0.003	0.008	0.000	0.000	0.003	0.008
Total	0.014	0.022			0.012	0.037	0.005	0.009	0.031	0.067
Bioefficiency										
Plant design			0.061	0.107	0.014	0.027	0.028	0.061	0.104	0.195
Photosynthetic efficiency			0.013	0.026	0.004	0.019	0.018	0.041	0.035	0.085
Growth duration			0.035	0.069	0.013	0.030	0.020	0.068	0.068	0.167
Grain quality			0.039	0.065	0.007	0.044	0.023	0.070	0.069	0.179
Total			0.148	0.267	0.039	0.120	0.090	0.239	0.277	0.626
Grand total	0.087	0.198	0.214	0.433	0.082	0.229	0.174	0.437	0.556	1.297

Appendix Table 22.3. Priority portfolios: irrigated rice, Nepal.

Research problem area	Management research Low	Management research High	Conventional breeding Low	Conventional breeding High	Wide crossing Low	Wide crossing High	Transgenic breeding Low	Transgenic breeding High	All techniques Low	All techniques High
Insects										
Stemborer	0.023	0.056	0.012	0.027	0.006	0.014	0.056	0.103	0.097	0.200
Brown planthopper	0.001	0.007	0.002	0.005	0.001	0.003	0.002	0.007	0.006	0.022
White-backed planthopper										
Leaf folder	0.012	0.024	0.005	0.011	0.001	0.003	0.001	0.016	0.018	0.055
Hispa	0.005	0.011	0.004	0.007	0.004	0.008	0.005	0.023	0.018	0.050
Green leafhopper										
Gall midge	0.010	0.021	0.009	0.020	0.002	0.006	0.005	0.013	0.026	0.059
Caseworm										
Armyworm	0.009	0.016	0.003	0.006	0.001	0.002	0.004	0.012	0.016	0.036
Mealy bug	0.009	0.014	0.003	0.007	0.001	0.002	0.005	0.027	0.017	0.051
Rice bug	0.044	0.092	0.018	0.061	0.003	0.011	0.020	0.079	0.085	0.243
Total	0.112	0.242	0.056	0.143	0.018	0.049	0.097	0.280	0.283	0.715
Diseases										
Blast	0.046	0.092	0.026	0.125	0.017	0.046	0.058	0.120	0.148	0.383
Leaf scald										
Gerlachia leaf spot										
Brown spot	0.008	0.020	0.004	0.006	0.007	0.010	0.005	0.011	0.024	0.047
Sheath rot										
Sheath blight	0.007	0.017	0.000	0.002	0.002	0.004	0.007	0.016	0.016	0.040
Stem rot	0.004	0.007	0.006	0.009	0.004	0.007	0.011	0.018	0.025	0.041
Bacterial blight	0.011	0.027	0.022	0.098	0.036	0.068	0.029	0.068	0.099	0.262
Bacterial leaf streak										
False smut	0.001	0.001	0.001	0.002	0.002	0.004	0.005	0.008	0.009	0.016
Atom blight										
Tungro										
Ragged stunt										
Ufra										
Total	0.077	0.164	0.059	0.243	0.069	0.140	0.115	0.242	0.320	0.789
Abiotic stresses										
Drought	0.017	0.052	0.035	0.067	0.009	0.030	0.024	0.050	0.085	0.198
Submergence	0.001	0.002	0.002	0.004	0.000	0.001	0.001	0.002	0.004	0.009
Cold	0.001	0.004	0.003	0.006	0.001	0.002	0.004	0.008	0.009	0.020
Heat										
Acidity										
Alkalinity										
Salinity										
Nutrient deficiency	0.017	0.032	0.014	0.050	0.001	0.006	0.002	0.012	0.033	0.100
Iron toxicity										
Total	0.036	0.091	0.053	0.127	0.011	0.039	0.030	0.072	0.131	0.328

	Management research		Conventional breeding		Wide crossing		Transgenic breeding		All techniques	
Research problem area	Low	High	Low	High	Low	High	Low	High	Low	High
Other pests										
Weeds	0.030	0.049			0.023	0.055	0.017	0.025	0.069	0.129
Crabs										
Rodents	0.018	0.026			0.010	0.023	0.001	0.005	0.028	0.054
Birds	0.000	0.000			0.003	0.008	0.000	0.000	0.003	0.008
Total	0.047	0.075			0.036	0.085	0.018	0.030	0.101	0.191
Bioefficiency										
Plant design			0.061	0.107	0.014	0.027	0.028	0.061	0.104	0.195
Photosynthetic efficiency			0.013	0.026	0.004	0.019	0.018	0.041	0.035	0.085
Growth duration			0.035	0.069	0.013	0.030	0.020	0.068	0.068	0.167
Grain quality			0.039	0.065	0.007	0.022	0.023	0.070	0.069	0.157
Total			0.148	0.267	0.039	0.097	0.090	0.239	0.277	0.604
Grand total	0.273	0.572	0.317	0.779	0.172	0.411	0.351	0.864	1.112	2.627

Appendix Table 22.4. Priority portfolios: irrigated rice, Bangladesh.

	Management research		Conventional breeding		Wide crossing		Transgenic breeding		All techniques	
Research problem area	Low	High	Low	High	Low	High	Low	High	Low	High
Insects										
Stemborer	0.007	0.018	0.004	0.009	0.002	0.005	0.018	0.033	0.031	0.064
Brown planthopper										
White-backed planthopper										
Leaf folder										
Hispa	0.005	0.012	0.004	0.007	0.004	0.009	0.005	0.025	0.018	0.052
Green leaf hopper										
Gall midge										
Caseworm										
Armyworm										
Mealy bug	0.002	0.003	0.000	0.002	0.000	0.000	0.001	0.005	0.003	0.010
Rice bug	0.009	0.018	0.003	0.012	0.001	0.002	0.004	0.015	0.016	0.047
Total	0.023	0.050	0.011	0.029	0.007	0.016	0.027	0.078	0.068	0.173

	Management research		Conventional breeding		Wide crossing		Transgenic breeding		All techniques	
Research problem area	Low	High	Low	High	Low	High	Low	High	Low	High
Diseases										
Blast	0.007	0.015	0.005	0.021	0.003	0.008	0.009	0.020	0.024	0.063
Leaf scald										
Gerlachia leaf spot										
Brown spot										
Sheath rot										
Sheath blight										
Stem rot										
Bacterial blight	0.005	0.011	0.008	0.038	0.014	0.026	0.011	0.026	0.038	0.101
Bacterial leaf streak	0.006	0.009	0.002	0.007	0.002	0.007	0.007	0.012	0.018	0.035
False smut										
Atom blight										
Tungro	0.000	0.001	0.002	0.003	0.001	0.001	0.002	0.004	0.005	0.009
Ragged stunt										
Ufra										
Total	0.018	0.036	0.017	0.068	0.020	0.042	0.030	0.062	0.084	0.208
Abiotic stresses										
Drought	0.015	0.045	0.030	0.058	0.007	0.025	0.021	0.043	0.073	0.171
Submergence	0.009	0.025	0.017	0.038	0.004	0.007	0.010	0.024	0.039	0.094
Cold										
Heat										
Acidity										
Alkalinity										
Salinity										
Nutrient deficiency	0.003	0.006	0.002	0.009	0.000	0.001	0.000	0.002	0.006	0.019
Iron toxicity										
Total	0.027	0.076	0.049	0.105	0.011	0.034	0.032	0.069	0.119	0.285
Other pests										
Weeds	0.017	0.029			0.013	0.032	0.010	0.015	0.041	0.075
Crabs	0.004	0.006			0.001	0.006	0.000	0.001	0.005	0.013
Rodents	0.009	0.014			0.005	0.012	0.000	0.003	0.015	0.028
Birds	0.000	0.000			0.003	0.008	0.000	0.000	0.003	0.008
Total	0.030	0.048			0.023	0.058	0.011	0.019	0.063	0.125
Bioefficiency										
Plant design			0.061	0.107	0.014	0.027	0.028	0.061	0.104	0.195
Photosynthetic efficiency			0.013	0.026	0.004	0.019	0.018	0.041	0.035	0.085
Growth duration			0.035	0.069	0.013	0.030	0.020	0.068	0.068	0.167
Grain quality			0.039	0.065	0.007	0.022	0.023	0.070	0.069	0.157
Total			0.148	0.267	0.039	0.097	0.090	0.239	0.277	0.604
Grand total	0.097	0.210	0.226	0.469	0.099	0.247	0.190	0.467	0.612	1.394

Appendix Table 22.5. Priority portfolios: irrigated rice, China.

Research problem area	Management research Low	High	Conventional breeding Low	High	Wide crossing Low	High	Transgenic breeding Low	High	All techniques Low	High
Insects										
Stemborer	0.003	0.008	0.002	0.004	0.001	0.002	0.008	0.014	0.014	0.028
Brown planthopper										
White-backed planthopper										
Leaf folder										
Hispa										
Green leafhopper										
Gall midge										
Caseworm										
Armyworm										
Mealy bug										
Rice bug										
Total	0.003	0.008	0.002	0.004	0.001	0.002	0.008	0.014	0.014	0.028
Diseases										
Blast	0.003	0.006	0.002	0.008	0.001	0.003	0.004	0.008	0.010	0.026
Leaf scald										
Gerlachia leaf spot										
Brown spot										
Sheath rot										
Sheath blight	0.006	0.015	0.000	0.002	0.001	0.004	0.006	0.014	0.014	0.034
Stem rot										
Bacterial blight										
Bacterial leaf streak										
False smut										
Atom blight										
Tungro										
Ragged stunt										
Ufra										
Total	0.009	0.021	0.002	0.010	0.003	0.007	0.010	0.022	0.024	0.060
Abiotic stresses										
Drought	0.007	0.020	0.013	0.026	0.003	0.011	0.009	0.019	0.032	0.076
Submergence	0.004	0.012	0.008	0.018	0.002	0.003	0.005	0.011	0.019	0.044
Cold	0.002	0.007	0.006	0.010	0.001	0.004	0.007	0.014	0.015	0.035
Heat	0.001	0.003	0.001	0.002	0.001	0.001	0.002	0.007	0.005	0.013
Acidity	0.002	0.005	0.005	0.013	0.001	0.004	0.002	0.005	0.010	0.027
Alkalinity										
Salinity										
Nutrient deficiency	0.031	0.059	0.025	0.093	0.003	0.011	0.003	0.022	0.062	0.185
Iron toxicity										
Total	0.046	0.107	0.058	0.161	0.011	0.035	0.029	0.078	0.144	0.381

	Management research		Conventional breeding		Wide crossing		Transgenic breeding		All techniques	
Research problem area	Low	High	Low	High	Low	High	Low	High	Low	High
Other pests										
Weeds	0.004	0.007			0.003	0.008	0.002	0.004	0.010	0.018
Crabs										
Rodents	0.003	0.005			0.002	0.005	0.000	0.001	0.005	0.011
Birds	0.000	0.000			0.003	0.008	0.000	0.000	0.003	0.008
Total	0.007	0.012			0.008	0.021	0.002	0.005	0.018	0.038
Bioefficiency										
Plant design			0.061	0.107	0.014	0.027	0.028	0.061	0.104	0.195
Photosynthetic efficiency			0.013	0.026	0.004	0.019	0.018	0.041	0.035	0.085
Growth duration			0.035	0.069	0.013	0.030	0.020	0.068	0.068	0.167
Grain quality			0.039	0.065	0.007	0.022	0.023	0.070	0.069	0.157
Total			0.148	0.267	0.039	0.097	0.090	0.239	0.277	0.604
Grand total	0.065	0.148	0.210	0.442	0.061	0.162	0.140	0.359	0.476	1.110

Appendix Table 22.6. Priority portfolios: irrigated rice, Indonesia.

	Management research		Conventional breeding		Wide crossing		Transgenic breeding		All techniques	
Research problem area	Low	High	Low	High	Low	High	Low	High	Low	High
Insects										
Stemborer	0.022	0.054	0.011	0.025	0.006	0.013	0.053	0.098	0.092	0.190
Brown planthopper	0.002	0.009	0.002	0.005	0.001	0.004	0.003	0.008	0.008	0.026
White-backed planthopper										
Leaf folder	0.002	0.003	0.001	0.002	0.000	0.000	0.000	0.002	0.003	0.008
Hispa										
Green leafhopper										
Gall midge	0.007	0.015	0.007	0.014	0.001	0.004	0.003	0.009	0.019	0.043
Caseworm										
Armyworm	0.006	0.011	0.002	0.004	0.000	0.002	0.002	0.009	0.011	0.026
Mealy bug										
Rice bug	0.007	0.014	0.002	0.009	0.000	0.002	0.003	0.011	0.012	0.036
Total	0.045	0.105	0.026	0.060	0.009	0.025	0.065	0.139	0.145	0.329

	Management research		Conventional breeding		Wide crossing		Transgenic breeding		All techniques	
Research problem area	Low	High	Low	High	Low	High	Low	High	Low	High
Diseases										
Blast	0.001	0.003	0.001	0.004	0.001	0.001	0.002	0.004	0.004	0.012
Leaf scald										
Gerlachia leaf spot										
Brown spot	0.001	0.002	0.000	0.000	0.001	0.001	0.000	0.001	0.002	0.004
Sheath rot										
Sheath blight	0.000	0.000	0.000	0.000	0.000	0.000	0.000	0.000	0.000	0.000
Stem rot										
Bacterial blight	0.002	0.004	0.003	0.014	0.005	0.010	0.005	0.010	0.015	0.039
Bacterial leaf streak										
False smut	0.000	0.000	0.000	0.000	0.000	0.000	0.000	0.000	0.000	0.000
Atom blight										
Tungro	0.000	0.001	0.002	0.003	0.001	0.002	0.002	0.005	0.005	0.010
Ragged stunt	0.000	0.000	0.000	0.000	0.000	0.000	0.000	0.000	0.000	0.000
Ufra										
Total	0.004	0.009	0.006	0.022	0.008	0.014	0.009	0.019	0.027	0.064
Abiotic stresses										
Drought	0.002	0.007	0.005	0.009	0.001	0.004	0.003	0.007	0.012	0.028
Submergence	0.002	0.006	0.004	0.009	0.001	0.002	0.002	0.006	0.009	0.022
Cold										
Heat										
Acidity										
Alkalinity										
Salinity										
Nutrient deficiency										
Iron toxicity										
Total	0.005	0.013	0.009	0.018	0.002	0.006	0.006	0.013	0.021	0.050
Other pests										
Weeds	0.017	0.029			0.013	0.032	0.010	0.015	0.041	0.075
Crabs										
Rodents	0.018	0.027			0.010	0.024	0.001	0.005	0.029	0.056
Birds	0.000	0.000			0.003	0.008	0.000	0.000	0.003	0.008
Total	0.035	0.056			0.027	0.064	0.011	0.020	0.073	0.139
Bioefficiency										
Plant design			0.061	0.107	0.014	0.027	0.028	0.061	0.104	0.195
Photosynthetic efficiency			0.013	0.026	0.004	0.019	0.018	0.041	0.035	0.085
Growth duration			0.035	0.069	0.013	0.030	0.020	0.068	0.068	0.167
Grain quality			0.039	0.065	0.007	0.022	0.023	0.070	0.069	0.157
Total			0.148	0.267	0.039	0.097	0.090	0.239	0.277	0.604
Grand total	0.089	0.184	0.190	0.367	0.084	0.206	0.181	0.430	0.544	1.186

Appendix Table 22.7. Priority portfolios: rainfed rice, eastern India.

Research problem area	Management research Low	Management research High	Conventional breeding Low	Conventional breeding High	Wide crossing Low	Wide crossing High	Transgenic breeding Low	Transgenic breeding High	All techniques Low	All techniques High
Insects										
Stemborer	0.012	0.029	0.006	0.014	0.002	0.007	0.029	0.053	0.048	0.104
Brown planthopper	0.001	0.004	0.001	0.002	0.000	0.002	0.002	0.004	0.004	0.013
White-backed planthopper	0.002	0.005	0.001	0.003	0.000	0.001	0.000	0.002	0.004	0.011
Leaf folder	0.004	0.007	0.001	0.003	0.000	0.001	0.000	0.005	0.006	0.017
Hispa										
Green leafhopper	0.002	0.010	0.005	0.009	0.001	0.003	0.004	0.009	0.012	0.031
Gall midge	0.004	0.008	0.004	0.008	0.001	0.002	0.002	0.005	0.010	0.023
Caseworm	0.005	0.010	0.002	0.004	0.000	0.001	0.002	0.008	0.010	0.023
Armyworm	0.004	0.007	0.001	0.002	0.000	0.001	0.002	0.005	0.007	0.015
Mealy bug										
Rice bug	0.001	0.002	0.000	0.002	0.000	0.000	0.000	0.002	0.002	0.006
Total	0.034	0.083	0.022	0.048	0.005	0.019	0.041	0.093	0.102	0.242
Diseases										
Blast	0.015	0.030	0.008	0.039	0.006	0.015	0.018	0.038	0.047	0.122
Leaf scald										
Gerlachia leaf spot										
Brown spot	0.002	0.004	0.001	0.001	0.001	0.002	0.001	0.002	0.005	0.009
Sheath rot	0.001	0.005	0.002	0.003	0.001	0.002	0.000	0.003	0.005	0.013
Sheath blight	0.003	0.007	0.000	0.001	0.001	0.002	0.003	0.007	0.007	0.017
Stem rot										
Bacterial blight	0.003	0.007	0.006	0.025	0.009	0.017	0.007	0.017	0.025	0.067
Bacterial leaf streak										
False smut										
Atom blight										
Tungro										
Ragged stunt										
Ufra										
Total	0.023	0.052	0.017	0.070	0.018	0.038	0.030	0.067	0.089	0.227
Abiotic stresses										
Drought	0.020	0.061	0.041	0.078	0.010	0.034	0.028	0.058	0.099	0.232
Submergence	0.004	0.012	0.008	0.018	0.002	0.003	0.005	0.011	0.019	0.046
Cold	0.001	0.006	0.004	0.007	0.001	0.003	0.005	0.011	0.011	0.027
Heat										
Acidity	0.001	0.002	0.003	0.007	0.000	0.002	0.001	0.002	0.005	0.013
Alkalinity	0.002	0.004	−0.001	0.005	0.001	0.002	0.002	0.004	0.003	0.014
Salinity	0.000	0.001	0.001	0.002	0.000	0.001	0.001	0.002	0.003	0.006
Nutrient deficiency	0.007	0.013	0.006	0.021	0.001	0.002	0.001	0.005	0.014	0.041
Iron toxicity	0.002	0.005	0.003	0.007	0.001	0.003	0.001	0.004	0.007	0.019
Total	0.037	0.104	0.065	0.146	0.016	0.051	0.044	0.097	0.162	0.398

	Management research		Conventional breeding		Wide crossing		Transgenic breeding		All techniques	
Research problem area	Low	High	Low	High	Low	High	Low	High	Low	High
Other pests										
Weeds	0.034	0.057			0.027	0.064	0.020	0.029	0.081	0.150
Crabs										
Rodents	0.006	0.009			0.003	0.008	0.000	0.002	0.010	0.020
Birds	0.000	0.000			0.002	0.005	0.000	0.000	0.002	0.005
Total	0.041	0.067			0.032	0.077	0.021	0.031	0.093	0.175
Bioefficiency										
Plant design			0.039	0.068	0.009	0.017	0.018	0.039	0.066	0.124
Photosynthetic efficiency			0.010	0.019	0.003	0.014	0.014	0.031	0.027	0.064
Growth duration			0.028	0.055	0.010	0.024	0.016	0.055	0.055	0.133
Grain quality			0.033	0.055	0.006	0.019	0.020	0.060	0.059	0.134
Total			0.110	0.198	0.029	0.074	0.068	0.184	0.207	0.456
Grand total	0.135	0.305	0.213	0.462	0.100	0.259	0.204	0.472	0.652	1.498

Appendix Table 22.8. Priority portfolios: rainfed rice, Bangladesh.

	Management research		Conventional breeding		Wide crossing		Transgenic breeding		All techniques	
Research problem area	Low	High	Low	High	Low	High	Low	High	Low	High
Insects										
Stemborer	0.011	0.025	0.005	0.012	0.003	0.006	0.025	0.047	0.044	0.091
Brown planthopper										
White-backed planthopper										
Leaf folder										
Hispa	0.005	0.011	0.004	0.007	0.003	0.008	0.005	0.022	0.017	0.048
Green leafhopper										
Gall midge										
Caseworm										
Armyworm										
Mealy bug	0.002	0.003	0.001	0.002	0.000	0.001	0.001	0.007	0.004	0.012
Rice bug	0.012	0.025	0.005	0.017	0.001	0.003	0.005	0.022	0.023	0.067
Total	0.030	0.065	0.015	0.038	0.007	0.018	0.037	0.097	0.088	0.218

	Management research		Conventional breeding		Wide crossing		Transgenic breeding		All techniques	
Research problem area	Low	High	Low	High	Low	High	Low	High	Low	High
Diseases										
Blast	0.009	0.018	0.005	0.025	0.003	0.009	0.011	0.024	0.029	0.076
Leaf scald										
Gerlachia leaf spot										
Brown spot										
Sheath rot										
Sheath blight										
Stem rot										
Bacterial blight	0.005	0.012	0.010	0.044	0.016	0.046	0.013	0.030	0.045	0.133
Bacterial leaf streak	0.007	0.011	0.003	0.008	0.003	0.026	0.009	0.014	0.021	0.059
False smut										
Atom blight										
Tungro	0.000	0.001	0.002	0.003	0.001	0.002	0.002	0.005	0.006	0.011
Ragged stunt										
Ufra										
Total	0.021	0.043	0.020	0.080	0.024	0.083	0.036	0.073	0.101	0.279
Abiotic stresses										
Drought	0.021	0.062	0.041	0.080	0.010	0.035	0.029	0.059	0.102	0.237
Submergence	0.009	0.025	0.017	0.038	0.004	0.007	0.010	0.024	0.039	0.094
Cold										
Heat										
Acidity										
Alkalinity										
Salinity										
Nutrient deficiency	0.004	0.008	0.004	0.013	0.000	0.002	0.000	0.003	0.009	0.026
Iron toxicity										
Total	0.034	0.096	0.062	0.131	0.015	0.044	0.039	0.087	0.149	0.357
Other pests										
Weeds	0.017	0.029			0.013	0.032	0.010	0.015	0.041	0.075
Crabs	0.005	0.007			0.002	0.008	0.000	0.002	0.007	0.017
Rodents	0.009	0.014			0.005	0.012	0.000	0.003	0.015	0.028
Birds	0.000	0.000			0.003	0.008	0.000	0.000	0.003	0.008
Total	0.031	0.050			0.023	0.060	0.011	0.019	0.065	0.129
Bioefficiency										
Plant design			0.039	0.068	0.009	0.017	0.018	0.039	0.066	0.124
Photosynthetic efficiency			0.010	0.019	0.003	0.014	0.014	0.031	0.027	0.064
Growth duration			0.028	0.055	0.010	0.024	0.016	0.055	0.055	0.133
Grain quality			0.033	0.055	0.006	0.019	0.020	0.060	0.059	0.134
Total			0.110	0.198	0.029	0.074	0.068	0.184	0.207	0.456
Grand total	0.116	0.253	0.207	0.446	0.097	0.278	0.191	0.460	0.611	1.438

Appendix Table 22.9. Priority portfolios: upland rice, eastern India.

Research problem area	Management research Low	Management research High	Conventional breeding Low	Conventional breeding High	Wide crossing Low	Wide crossing High	Transgenic breeding Low	Transgenic breeding High	All techniques Low	All techniques High
Insects										
Stemborer	0.003	0.008	0.002	0.004	0.001	0.002	0.008	0.015	0.014	0.029
Brown planthopper	0.001	0.004	0.001	0.002	0.000	0.002	0.001	0.004	0.003	0.012
White-backed planthopper	0.002	0.005	0.001	0.003	0.000	0.001	0.000	0.002	0.004	0.011
Leaf folder	0.001	0.002	0.000	0.001	0.000	0.000	0.000	0.002	0.002	0.006
Hispa										
Green leafhopper	0.005	0.021	0.010	0.019	0.002	0.006	0.009	0.018	0.026	0.064
Gall midge										
Caseworm	0.001	0.002	0.000	0.001	0.000	0.000	0.000	0.002	0.002	0.006
Armyworm	0.018	0.032	0.007	0.011	0.001	0.004	0.007	0.025	0.032	0.072
Mealy bug										
Rice bug	0.015	0.031	0.006	0.021	0.001	0.004	0.007	0.027	0.029	0.082
Total	0.046	0.106	0.027	0.062	0.006	0.020	0.032	0.093	0.112	0.282
Diseases										
Blast	0.035	0.070	0.020	0.094	0.013	0.035	0.044	0.091	0.112	0.290
Leaf scald										
Gerlachia leaf spot										
Brown spot	0.010	0.024	0.005	0.007	0.008	0.013	0.006	0.014	0.029	0.058
Sheath rot	0.000	0.001	0.000	0.001	0.000	0.000	0.000	0.000	0.001	0.002
Sheath blight	0.005	0.014	0.000	0.002	0.001	0.003	0.006	0.013	0.013	0.031
Stem rot										
Bacterial blight	0.000	0.001	0.001	0.005	0.002	0.003	0.001	0.003	0.005	0.012
Bacterial leaf streak										
False smut										
Atom blight										
Tungro										
Ragged stunt										
Ufra										
Total	0.051	0.110	0.027	0.108	0.025	0.055	0.057	0.121	0.160	0.394
Abiotic stresses										
Drought	0.052	0.157	0.105	0.203	0.026	0.089	0.073	0.150	0.257	0.600
Submergence	0.001	0.004	0.002	0.005	0.001	0.001	0.001	0.003	0.006	0.013
Cold	0.000	0.001	0.001	0.001	0.000	0.001	0.001	0.002	0.002	0.004
Heat										
Acidity	0.012	0.036	0.039	0.091	0.005	0.026	0.018	0.035	0.075	0.189
Alkalinity	0.003	0.006	−0.002	0.009	0.001	0.003	0.003	0.006	0.005	0.024
Salinity	0.003	0.007	0.008	0.017	0.002	0.005	0.005	0.011	0.019	0.040
Nutrient deficiency	0.009	0.018	0.008	0.029	0.001	0.003	0.001	0.007	0.019	0.058
Iron toxicity	0.001	0.002	0.002	0.004	0.001	0.002	0.001	0.002	0.004	0.011
Total	0.083	0.232	0.164	0.360	0.037	0.131	0.103	0.216	0.386	0.940

Research problem area	Management research Low	Management research High	Conventional breeding Low	Conventional breeding High	Wide crossing Low	Wide crossing High	Transgenic breeding Low	Transgenic breeding High	All techniques Low	All techniques High
Other pests										
Weeds	0.120	0.200			0.093	0.221	0.070	0.101	0.282	0.522
Crabs										
Rodents	0.009	0.014			0.005	0.012	0.000	0.003	0.015	0.028
Birds	0.000	0.000			0.015	0.041	0.000	0.000	0.015	0.041
Total	0.129	0.213			0.112	0.274	0.070	0.104	0.311	0.591
Bioefficiency										
Plant design			0.033	0.059	0.008	0.015	0.016	0.033	0.057	0.107
Photosynthetic efficiency			0.005	0.010	0.001	0.007	0.007	0.015	0.013	0.032
Growth duration			0.014	0.027	0.005	0.012	0.008	0.027	0.027	0.067
Grain quality			0.022	0.037	0.004	0.013	0.013	0.040	0.040	0.089
Total			0.074	0.133	0.019	0.046	0.044	0.115	0.137	0.294
Grand total	0.309	0.662	0.292	0.663	0.199	0.525	0.306	0.651	1.106	2.501

Appendix Table 22.10. Priority portfolios: upland rice, Nepal.

Research problem area	Management research Low	Management research High	Conventional breeding Low	Conventional breeding High	Wide crossing Low	Wide crossing High	Transgenic breeding Low	Transgenic breeding High	All techniques Low	All techniques High
Insects										
Stemborer	0.011	0.026	0.006	0.012	0.003	0.007	0.026	0.048	0.045	0.092
Brown planthopper	0.001	0.006	0.001	0.003	0.001	0.002	0.002	0.005	0.005	0.017
White-backed planthopper										
Leaf folder	0.009	0.016	0.003	0.008	0.001	0.002	0.000	0.011	0.013	0.038
Hispa	0.007	0.018	0.006	0.010	0.006	0.013	0.008	0.036	0.027	0.076
Green leafhopper										
Gall midge	0.005	0.011	0.005	0.010	0.001	0.003	0.002	0.007	0.014	0.030
Caseworm										
Armyworm	0.014	0.024	0.005	0.009	0.001	0.003	0.005	0.019	0.025	0.055
Mealy bug	0.005	0.009	0.002	0.005	0.000	0.001	0.003	0.016	0.010	0.031
Rice bug	0.021	0.044	0.009	0.030	0.001	0.005	0.009	0.038	0.041	0.117
Total	0.073	0.153	0.036	0.087	0.013	0.037	0.056	0.180	0.179	0.457

	Management research		Conventional breeding		Wide crossing		Transgenic breeding		All techniques	
Research problem area	Low	High	Low	High	Low	High	Low	High	Low	High
Diseases										
Blast	0.053	0.107	0.031	0.144	0.020	0.054	0.067	0.139	0.171	0.444
Leaf scald										
Gerlachia leaf spot										
Brown spot	0.007	0.017	0.003	0.005	0.006	0.009	0.004	0.010	0.021	0.040
Sheath rot										
Sheath blight	0.006	0.015	0.000	0.002	0.001	0.004	0.006	0.014	0.014	0.034
Stem rot	0.003	0.005	0.004	0.007	0.003	0.005	0.007	0.012	0.017	0.028
Bacterial blight	0.007	0.016	0.013	0.057	0.021	0.040	0.017	0.040	0.058	0.153
Bacterial leaf streak										
False smut	0.000	0.000	0.000	0.001	0.001	0.002	0.002	0.003	0.004	0.006
Atom blight										
Tungro										
Ragged stunt										
Ufra										
Total	0.077	0.159	0.051	0.216	0.052	0.112	0.103	0.218	0.284	0.705
Abiotic stresses										
Drought	0.014	0.043	0.028	0.055	0.007	0.024	0.020	0.041	0.069	0.162
Submergence	0.001	0.002	0.001	0.003	0.000	0.001	0.001	0.002	0.003	0.008
Cold	0.001	0.004	0.003	0.006	0.001	0.002	0.004	0.008	0.009	0.020
Heat										
Acidity										
Alkalinity										
Salinity										
Nutrient deficiency	0.007	0.014	0.006	0.021	0.001	0.003	0.001	0.005	0.014	0.042
Iron toxicity										
Total	0.023	0.062	0.039	0.085	0.009	0.030	0.025	0.056	0.095	0.233
Other pests										
Weeds	0.024	0.040			0.019	0.044	0.014	0.021	0.057	0.105
Crabs										
Rodents	0.016	0.023			0.009	0.020	0.001	0.005	0.025	0.048
Birds	0.000	0.000			0.003	0.008	0.000	0.000	0.003	0.008
Total	0.040	0.063			0.031	0.073	0.015	0.025	0.085	0.161
Bioefficiency										
Plant design			0.033	0.059	0.008	0.015	0.016	0.033	0.057	0.107
Photosynthetic efficiency			0.005	0.010	0.001	0.007	0.007	0.015	0.013	0.032
Growth duration			0.014	0.027	0.005	0.012	0.008	0.027	0.027	0.067
Grain quality			0.022	0.037	0.004	0.013	0.013	0.040	0.040	0.089
Total			0.074	0.133	0.019	0.046	0.044	0.115	0.137	0.294
Grand total	0.213	0.438	0.200	0.521	0.124	0.298	0.243	0.594	0.780	1.851

Appendix Table 22.11. Priority portfolios: upland rice, Indonesia.

	Management research		Conventional breeding		Wide crossing		Transgenic breeding		All techniques	
Research problem area	Low	High	Low	High	Low	High	Low	High	Low	High
Insects										
Stemborer	0.002	0.005	0.001	0.002	0.001	0.001	0.005	0.010	0.009	0.019
Brown planthopper	0.014	0.077	0.016	0.044	0.007	0.032	0.027	0.071	0.064	0.224
White-backed planthopper										
Leaf folder	0.002	0.003	0.001	0.002	0.000	0.000	0.000	0.002	0.003	0.008
Hispa										
Green leafhopper										
Gall midge	0.000	0.001	0.000	0.001	0.000	0.000	0.000	0.001	0.001	0.003
Caseworm										
Armyworm	0.014	0.025	0.005	0.009	0.001	0.003	0.006	0.020	0.026	0.058
Mealy bug										
Rice bug	0.007	0.015	0.003	0.010	0.000	0.002	0.003	0.013	0.014	0.040
Total	0.040	0.128	0.027	0.069	0.009	0.039	0.041	0.117	0.117	0.353
Diseases										
Blast	0.002	0.005	0.001	0.006	0.001	0.002	0.003	0.006	0.007	0.019
Leaf scald										
Gerlachia leaf spot	0.000	0.000	0.000	0.001	0.000	0.000	0.000	0.000	0.001	0.002
Brown spot	0.001	0.002	0.000	0.000	0.001	0.001	0.000	0.001	0.002	0.004
Sheath rot										
Sheath blight	0.000	0.000	0.000	0.000	0.000	0.000	0.000	0.000	0.000	0.000
Stem rot										
Bacterial blight	0.004	0.009	0.007	0.033	0.012	0.023	0.010	0.023	0.034	0.088
Bacterial leaf streak										
False smut										
Atom blight										
Tungro	0.000	0.000	0.000	0.000	0.000	0.000	0.000	0.000	0.000	0.000
Ragged stunt										
Ufra										
Total	0.008	0.016	0.009	0.041	0.014	0.027	0.013	0.031	0.044	0.114
Abiotic stresses										
Drought	0.002	0.007	0.005	0.009	0.001	0.004	0.003	0.007	0.012	0.028
Submergence	0.002	0.006	0.004	0.009	0.001	0.002	0.002	0.006	0.009	0.022
Cold										
Heat										
Acidity										
Alkalinity										
Salinity										
Nutrient deficiency										
Iron toxicity										
Total	0.005	0.013	0.009	0.018	0.002	0.006	0.006	0.013	0.021	0.050

	Management research		Conventional breeding		Wide crossing		Transgenic breeding		All techniques	
Research problem area	Low	High	Low	High	Low	High	Low	High	Low	High
Other pests										
Weeds	0.017	0.029			0.013	0.032	0.010	0.015	0.041	0.075
Crabs										
Rodents	0.018	0.027			0.011	0.024	0.001	0.006	0.029	0.056
Birds	0.000	0.000			0.003	0.008	0.000	0.000	0.003	0.008
Total	0.035	0.056			0.027	0.064	0.011	0.021	0.073	0.140
Bioefficiency										
Plant design			0.033	0.059	0.008	0.015	0.016	0.033	0.057	0.107
Photosynthetic efficiency			0.005	0.010	0.001	0.007	0.007	0.015	0.013	0.032
Growth duration			0.014	0.027	0.005	0.012	0.008	0.027	0.027	0.067
Grain quality			0.022	0.037	0.004	0.013	0.013	0.040	0.040	0.089
Total			0.074	0.133	0.019	0.046	0.044	0.115	0.137	0.294
Grand total	0.088	0.212	0.120	0.261	0.070	0.182	0.115	0.297	0.393	0.951

Appendix Table 22.12. Priority portfolios: deepwater rice, eastern India.

	Management research		Conventional breeding		Wide crossing		Transgenic breeding		All techniques	
Research problem area	Low	High	Low	High	Low	High	Low	High	Low	High
Insects										
Stemborer	0.055	0.133	0.029	0.064	0.014	0.033	0.133	0.244	0.230	0.474
Brown planthopper										
White-backed planthopper										
Leaf folder										
Hispa										
Green leafhopper										
Gall midge										
Caseworm										
Armyworm										
Mealy bug										
Rice bug										
Total	0.055	0.133	0.029	0.064	0.014	0.033	0.133	0.244	0.230	0.474

	Management research		Conventional breeding		Wide crossing		Transgenic breeding		All techniques	
Research problem area	Low	High	Low	High	Low	High	Low	High	Low	High
Diseases										
Blast	0.016	0.033	0.009	0.044	0.006	0.016	0.021	0.042	0.052	0.135
Leaf scald										
Gerlachia leaf spot										
Brown spot	0.005	0.011	0.002	0.003	0.004	0.006	0.003	0.006	0.013	0.025
Sheath rot	0.002	0.006	0.002	0.004	0.002	0.003	0.001	0.004	0.006	0.017
Sheath blight	0.006	0.016	0.000	0.002	0.002	0.004	0.007	0.016	0.015	0.038
Stem rot										
Bacterial blight	0.009	0.022	0.018	0.079	0.030	0.055	0.023	0.055	0.080	0.212
Bacterial leaf streak										
False smut										
Atom blight										
Tungro										
Ragged stunt										
Ufra										
Total	0.038	0.088	0.032	0.132	0.043	0.084	0.055	0.123	0.167	0.427
Abiotic stresses										
Drought	0.014	0.042	0.028	0.055	0.007	0.024	0.020	0.041	0.069	0.161
Submergence	0.011	0.034	0.023	0.051	0.005	0.009	0.014	0.032	0.052	0.126
Cold										
Heat										
Acidity										
Alkalinity	0.019	0.037	−0.009	0.056	0.006	0.017	0.019	0.039	0.034	0.150
Salinity										
Nutrient deficiency										
Iron toxicity										
Total	0.044	0.114	0.041	0.162	0.018	0.051	0.052	0.112	0.155	0.438
Other pests										
Weeds	0.055	0.091			0.043	0.101	0.032	0.046	0.130	0.239
Crabs										
Rodents	0.016	0.025			0.009	0.022	0.001	0.005	0.027	0.051
Birds	0.000	0.000			0.011	0.031	0.000	0.000	0.011	0.031
Total	0.071	0.116			0.063	0.154	0.033	0.051	0.167	0.321
Bioefficiency										
Plant design			0.017	0.029	0.004	0.007	0.008	0.017	0.028	0.053
Photosynthetic efficiency			0.003	0.007	0.001	0.005	0.005	0.010	0.009	0.022
Growth duration			0.007	0.014	0.003	0.006	0.004	0.014	0.014	0.033
Grain quality			0.011	0.018	0.002	0.006	0.007	0.020	0.020	0.045
Total			0.038	0.068	0.010	0.024	0.023	0.060	0.071	0.153
Grand total	0.209	0.451	0.140	0.425	0.147	0.347	0.295	0.591	0.791	1.813

23 Summary, Conclusions and Implications

R.W. Herdt

Agricultural Sciences, Rockefeller Foundation, 420 Fifth Avenue, New York, NY 10018-2702, USA

I. Introduction

This book addresses the question: how should rice research investments intended to benefit people in Asia be directed? Research generates benefits through its invention of new technology that reduces the per kilogram cost of rice production. That reduction is shared between producers and consumers (Hayami and Herdt, 1977). Such reductions can be achieved by reducing losses after the crop is harvested, by reducing losses during production, by increasing the amount or quality of output produced with a given input cost or by reducing the cost associated with a given quantity of output. Each outcome will result in greater output per unit of input and hence generate a return to an investment in research.

In the preceding chapters these opportunities are generally termed 'yield losses avoided', but some of the most important opportunities for research are those directed at increasing productivity (output per unit input).

This chapter presents a summary of the approach reported in earlier chapters in section II, and, in section III, a summary of the comprehensive international rice research priorities based on the empirical evidence reported in the book.

II. Methodological Approach

Intuitively, if it is just as difficult for research to solve one problem (A) as to solve another (B), and if problem A is responsible for twice the loss or could generate twice the gain as problem B, then it seems rational to invest twice the effort or cost to solve problem A as problem B. If the yield loss associated with problem A occurs on only a small part of the harvested area, while that associated with problem B occurs on a large part of the planted area, then one must also recognize the difference in area affected when determining the allocation of research resources.

Of course, some problems are more difficult than others for research to address – they require more time or money to investigate, or simply have been less explored previously and so more different aspects must be investigated or there is greater uncertainty associated with conducting research on them.

The methodology used in the book reflects the magnitude of losses or opportunities for gains associated with each problem/opportunity and the extent of area affected. The methodology also reflects differences in difficulty – cost, time and degree of uncertainty – associated with making gains toward solving each problem/opportunity. The information on losses caused by a problem/opportunity and the difficulty of the research needed to solve the problem/opportunity are combined to generate a portfolio of rice research activities that is selected so as to give the largest expected increase in economic value. Chapters 5, 6 and 22 present the details of the methodology.

A. Yield gaps

In most Asian countries, farmers' average yields are significantly lower than maximum yields on experiment stations – 3–5 tons ha^{-1} compared to 6–9 tons ha^{-1}. The yield gap is defined as the difference between the yield observed on experiment stations and actual yields on farmers' fields. This yield gap has two components.

Yield gap I is the difference between the maximum yields on the experiment station and the maximum possible yields on farmers' fields. This gap probably is largely attributable to inherent differences in soil and other difficult-to-change environmental factors between the experiment station and typical farmers' fields, and production research or changes in socioeconomic conditions can do little to exploit it.

Yield gap II is the difference between maximum possible yields and actual yields on farmers' fields. It is a measure of the yield that could be gained through the invention of economic technology to overcome constraints – pests, diseases, temperature extremes, drought, submergence, soil fertility constraints and so forth. This gap could be overcome if practical, profitable means to control constraints and exploit opportunities were invented. But, because such means do not exist, it is difficult to know its exact size. It is the target that applied rice research seeks to exploit.

The authors of the country studies used what each believed to be the most appropriate ways to measure the yield gaps of interest (Table 23.1). These were defined for the main rice ecologies within each country. In most cases the maximum yields were obtained from records of experiments, and in China from records of 11 years of maximum yield trials in over 150 prefectures. In the case of Nepal the highest yield was obtained from a survey asking farmers their highest, normal and lowest yields.

In addition to those yields and gaps, researchers have conceptualized several other yield concepts and yield gaps (see Chapter 1). The other primary

Table 23.1. Sources of raw data reflecting yield gaps and yield constraint composition.

Region	Potential yield	Actual yield	Composition of constraints based on
China	Annual prefectural experiments	Secondary data	Rice researchers in 152 prefectures
Eastern India	Researcher survey	Researcher survey	Rice researchers
West Bengal	Recorded experiments	Secondary data	Farm household survey
Southern India	Recorded experiments	Secondary data	240 researchers and extension specialists
Bangladesh	N/A	Extension farmer survey	Farmers and extension specialists, 120 thanas
Indonesia	N/A	N/A	Survey, 640 farmers
Thailand	Recorded experiments	Farm survey	73 provinces, 1986–1991
Nepal	Farmer survey	Secondary data	Survey, 34 district extension specialists
Philippines	Experiments, 1993	Secondary data	Survey plots in five provinces

Source: Country study chapters in this volume.

gap of interest is the difference between the regularly observed maximum experiment station yield and the conceivable experiment station yield. This gap is attributed to management improvements not yet devised but conceived, and genetic differences between the present best varieties and types of varieties conceived but not yet produced or perfected. By definition it cannot be measured.

B. Data

Biological scientists having experience with rice production problems and knowledge of how to increase productivity identified all the possible constraints or opportunities to prevent losses or make gains in farmers' rice production systems. Each such opportunity was defined as a research problem area (RPA). Rice is grown under a wide range of agroecological conditions, from the most well-controlled irrigation to dryland rainfed situations, on both flat and sloping fields. The opportunity to increase productivity or prevent yield losses from each RPA may be different for each of those various rice ecologies, so information was elicited separately for each.

Social scientists in China, several regions of India, Bangladesh, Nepal, Indonesia, Thailand and the Philippines compiled quantitative judgements by biological scientists and other knowledgeable people on the losses suffered or the opportunities for gains associated with each RPA. Farmers have deep personal knowledge of conditions on their own farms, and are good sources of such information. A well-constructed sample of farmers can generate information on opportunities and losses for all RPAs within their experience. Several country studies reported farmers' estimates of potential RPA gains. However, because of the heterogeneity across the rice ecologies of most countries, the spatial distribution and sample size required for good estimates from farmers is quite large and hence quite costly to obtain. A second group of knowledgeable individuals are extension agents who, because they deal with a large number of farmers and cover a large geographic area relative to individual farmers, are able to reflect judgements across broader geographic areas. Many country studies obtained data from groups of extension workers. Researchers compose the third knowledgeable group who may have views representative of a larger extent of rice area because of their responsibility for even broader geographic regions than extension agents. All of the individual country studies compiled data from rice researchers.

Yield losses from insects and diseases that attack crops are often quantified through field plot experiments. Such experiments give statistically accurate estimates of the losses associated with various, quantified, degrees of infestations (see Chapter 16). However, they do not provide any reflection of the extent to which farmers' fields are attacked by each pest and so cannot, alone, give the aggregate estimates of crop production losses that are needed for priority setting.

Well-designed field surveys or careful record keeping by local agricultural officials can produce statistically accurate estimates of the extent of pest prevalence and degree of infestation, and hence good estimates of crop losses. China has such records based on the opinions of local officials but, to our knowledge, no Asian country conducts pest surveys on a regular, statistically valid basis.

Hence, as indicated in Table 23.2 and discussed in the country chapters, the data on which estimates of the yield gaps are based are different for each country. They are not strictly comparable, but the information available is useful for understanding the general picture of research opportunities.

C. Disaggregation of yield gaps

Considerable effort was expended to understand the relative importance of the many factors responsible for yield gap II. First, knowledgeable respondents were asked to list all possible contributing factors to the yield gaps; these are the RPAs. Then they were asked a series of questions designed to elicit consistent estimates of the quantity by which each factor caused losses or presented opportunities. Data were obtained for each distinct rice ecology separately.

The specific procedures used varied somewhat across the countries, but typically, for pest-caused yield losses, respondents were asked to estimate the proportion of land in each rice ecology damaged by each pest in an average year, and the estimated loss in kilograms per hectare suffered on each damaged hectare of rice land, on average. The comparable information was obtained for each yield constraint/opportunity for which the respondents felt qualified to respond. The hectares of rice in each rice ecology was multiplied by the average per hectare yield loss/opportunity associated with each RPA. That gave a measure of the absolute quantity of rice associated with 'solving' or addressing the constraint/opportunity represented by each RPA.

Research targets must eventually be very sharply focused if opportunities are to be exploited or losses prevented. It is not sufficient to identify cold as a constraint. Rather, the actual temperature and the crop growth stage when it is cold, and even whether the cold is from irrigation water or from the atmosphere, can be important in devising ways to respond. Overcoming a soil deficiency requires an exact description of the problem, and so forth. Detailed information was obtained on the individual constraints, in most cases disaggregated by rice production ecologies within countries.

D. Difficulty of alternative research activities

It is trivial to observe that different research goals entail different degrees of difficulty. For example, the goal of increasing the biomass production efficiency of a cereal grain is acknowledged by most to be more difficult than incorporating the semidwarf trait in a particular line, in part because the latter is a well-recognized trait for which a gene has been identified. Most research goals fall between these two extremes.

A procedure of eliciting the relative difficulty of achieving various research goals was devised and applied. Respondents were asked to estimate how many years of research at current levels would be required to resolve each RPA to a small extent (25% of the loss/opportunity) and to a rather more complete extent (75% of the loss/opportunity). Each RPA might be addressed by any one or all of four research approaches: management, conventional plant breeding, wide crossing and genetic engineering; so estimates were obtained for each of the approaches (see Chapter 22). It was also recognized that considerable progress had already been made in some RPAs so an adjustment was made for research achievement.

The key concepts required to judge research difficulty – time required, alternative possible research approaches and progress already made – are quite different from the losses/opportunities data. Whereas farmers and extension workers have experience on which to base response about losses/opportunities, they have little basis for estimating what research is needed and how long it might take to address the various problems/opportunities. Only individuals with experience in conducting research have such insights, so a set of researchers were persuaded to provide estimates of the key parameters. Hence, while data on constraints/opportunities came from country studies (Tables 23.2 and 23.3) data on difficulty to achieve the gain were based on a sample of knowledgeable individuals – seven IRRI scientists and ten Asian scientists in national programs.

An analytical procedure was devised to determine the amount of investment in each RPA that would generate a 25% internal rate of return from each different research activity (see Chapter 22). This was done on a disaggregated

Table 23.2. Estimated differences between maximum yields on experiment stations and potential farm yields (gap I), and between potential and actual farm yields (gap II) in the main rice seasons

Region	Yield gap I (kg ha^{-1}) in season 1	2	3	Yield gap II (kg ha^{-1}) in season 1	2	3
China[a]	4584	3628	6448	2564	2981	3391
Eastern India[b]	391	533	783	1887	2262	866
West Bengal[b]	284	700	141	1698	786	1561
Southern India[c]	813	939	1416	1147	591	544
Bangladesh[d]	na	na	na	575	762	1380
Thailand[e]	264	1637	1125	401	765	517
Nepal[f]	900	1400	900	300	900	1300
Philippines[g]	460	na	860	3480	2450	3050

[a] 1 is early season, 2 is late season, 3 is single season.
[b] 1 is rainfed lowland, 2 is irrigated wet season, 3 is irrigated dry season.
[c] 1 is Karnartka, 2 is Kerala, 3 is Tamil Nadu; all are rainfed.
[d] 1 is rainfed upland, 2 and 3 are as in West Bengal.
[e] 1 is rainfed upland, 2 is rainfed lowland, 3 is irrigated.
[f] 1 is upland, 2 is rainfed lowland, 3 is irrigated; all are in Tara.
[g] 1 is wet season rainfed, 2 is wet season irrigated, 3 is dry season irrigated.
Source: Country study chapters in this volume.

basis for the five rice ecologies, the four research approaches and the 65 RPAs, giving, for any geographic region, 4 × 5 × 65 or 1300 'optimal' levels of research investment (see tables in Chapter 22). The final part of section III of this chapter summarizes the results.

III. Summary of Results

Different research organizations have different responsibilities. Most Asian countries have a unit at the national level responsible for all rice research, and many have state or provincial units responsible for research at those levels. The IRRI has some responsibility for rice research for all developing nations, and the Rockefeller Foundation has also undertaken to support advanced genetic research directed at solving rice genetics problems of developing nations.

The optimal research portfolio will be different for each national rice research organization because the quantity of rice production expected to be 'saved' or 'gained' by successful research on each RPA is dominated by the area of rice in each agroecology. As a result, not only is the absolute size of the optimal research investment quite different for China and Bangladesh, but the allocation of that investment is also quite different. In China it will make little sense to invest much on deepwater rice because little is grown, while in Bangladesh deepwater rice composes an important proportion of the area. At

Table 23.3. Estimated yield losses by groups of constraints (kg ha^{-1}) in the main rice seasons.

Region	Insects	Diseases	Drought	Soils	Other abiotic[a]	Others
China[b]	92	88	na	635	557	91
Eastern India[c]	104	93	177	74	81	167 (weeds, lodging, birds)
West Bengal[d]	43	213	na	na	772	–
Southern India[e]	215	137	44	72	7	87 (weeds, lodging, nutrients)
Bangladesh[f]	135	111	172	27	106	22 (weeds)
Indonesia[g]	399	24	na	na	na	189 (rats)
Thailand[h]	78	12	na	na	na	5 (rats)
Nepal[i]	810	560	210	150	10	150 (rats, weeds, lodging)
Philippines[j]	250[k]		198	na	407	90 (variety, others)

[a]Cold, heat, submergence, 'weather'.
[b]Single season rice; Table 10.10.
[c]Wet season rainfed lowland; Table 7.5.
[d]Rainfed lowland aman; Table 8.6; submergence is the other abiotic stress.
[e]All ecosystems; Table 9.2.
[f]Rainfed lowland aman; Table 11.7; submergence is the other abiotic stress.
[g]Irrigated ecosystem, from a survey of 640 households in 32 villages; from percentage damage to an assumed 3.5 tons ha^{-1} yield.
[h]All ecosystems, from Table 13.9; assuming 10 million ha of rice area.
[i]All ecosystems, extension workers' views; derived from data in Tables 12.1. and 12.8.

the same time, in Bangladesh, it will make sense to devote a larger fraction of rice research resources to rainfed lowland than to deepwater rice because the gains expected from rainfed lowland rice are much greater than for deepwater rice. This holds true even though few other research organizations are doing deepwater rice research.

Likewise, the optimal portfolio of research for the IRRI will be different from that of any individual country because the IRRI has a mandate to serve all countries. One might think the optimal portfolio for the IRRI and the Rockefeller Foundation would be identical; however, the Rockefeller Foundation has different goals to the IRRI. The Rockefeller Foundation considers its rice biotechnology program to be a device for providing developing rice-dependent countries with the tools of plant biotechnology so that those tools can be applied by developing country scientists to any set of problems they consider of importance. That is, the Rockefeller Foundation has the goal of creating Asian scientific capacity for plant biotechnology in addition to generating new rice technology to improve incomes and food availability. The IRRI, on the other hand, probably gives much more weight to food production and income. As a result, the Rockefeller Foundation's portfolio includes only biotechnology research activities while the portfolio of the IRRI and the national institutions have a mix of all research approaches.

The portfolio of research activities summarized in this chapter reflects rice research needs for the aggregation of all Asian countries.

A. Estimated yield gaps

A summary of the data on yield gaps is reported in Table 23.2. Yield gap II is the target for research: it reflects the difference between the average yield on farms and the maximum possible yield in the same place. Depending on season and location, it ranges from around 3 tons ha^{-1} in China to around 300 kg ha^{-1} in the poorest growing conditions of Nepal and Thailand. In many cases in India it reaches 2 tons ha^{-1} and in Bangladesh as much as 1.3 tons ha^{-1}. It is important for researchers to understand, as well as is possible, what accounts for the difference and how the factors accounting for the difference can be overcome if that is possible.

Yield gap I ranged from over 6 tons ha^{-1} in China, where the maximum experiment station yield for single season rice averaged 16 tons ha^{-1} across the reporting prefectures to 0.3 tons ha^{-1} in Thailand. Other values of yield gap I are not nearly as high, apparently because the organizational structures and rewards for demonstrating extremely high yields are not nearly as great in other countries as in China. Yield gap I, in any case, is of less interest because of the acknowledged differences in soil and other environmental conditions on experiment stations and farmers' fields.

B. Contributions to the yield gap

Relatively few individual factors accounted for a high proportion of the specific factors identified as constraints/opportunities. In China, for example, 18 individual factors were identified, with several of the least important contributing less than 1% of the identified yield gap. In Bangladesh, the least important of the top 20 factors occurred on average in less than 1% of the rice area over 10 years and resulted in an average of about 4 kg ha^{-1} of lost yield.

The most important constraints were typically quite different across countries, seasons and rice ecosystems. In some cases losses from drought, temperature extremes and other abiotic factors were large, while in others losses from the biotic factors of insects and diseases were much larger. Although the richness of the data prevent any but the most aggregate representation in this chapter, some observations on Table 23.3 may be of interest, always keeping in mind that the contributions reported are for one important rice production season or rice ecology and do not fully represent the respective nations.

Insects and diseases cause yield losses estimated at around 100 kg ha^{-1} in many locations. In China, these are relatively small losses compared to the other categories, but China has an effective system for developing resistant varieties, an elaborate system for monitoring rice fields and well-developed methods for controlling pests (see Chapter 10). Most pests are controlled so the remaining losses are small. All three of the studies reporting on India and Bangladesh indicate that insects and diseases are quite important there, in stark contrast to China. Insects and diseases caused important losses for all the other countries, although the farm survey data from the Philippines study made it impossible to separate the two. Losses reported for Nepal appear to be considerably larger than for other locations.

Drought and other abiotic stresses were quite significant as a whole but differed considerably across countries. In China, nutrient supply, cold temperatures and other abiotic stresses are estimated to cause losses of over 1 ton ha^{-1}. Submergence is an important problem in rainfed lowland parts of West Bengal, Bangladesh and other places in eastern India. Drought is an important contribution to losses in India and Bangladesh, even in the wet season lowland conditions reported for eastern India. Although most farmers are using modern varieties that are semidwarf, lodging remains a problem in a number of places.

Other pests including weeds, snails and rats are a problem for many rice farmers in some areas. Researchers, accustomed to thinking in terms of modifying genetics to address constraints, seem to have little to offer farmers other than cultural practices as a control.

C. Optimal research portfolios

Detailed versions of the optimal research portfolios are derived and discussed in Chapter 22. Table 23.4 shows a summary of the results based on scientists' conservative estimates of research success, classified by the four rice agroecologies. It shows the annual research investment in all four categories of research for each major problem area in each rice agroecology that would generate an internal rate of return of 25% over the long run. For example, US$58 million could be spent annually on insect control research for irrigated agroecologies to generate an expected internal rate of return of 25%. This includes research on the management of all insects that damage rice, by cultural control, conventional breeding, wide crossing and transgenic approaches. The corresponding amount for diseases is US$31 million, for other pests it is US$28 million, etc.

In total, US$568 million could be allocated as indicated on Table 23.4 and it would be expected to generate a social internal rate of return of 25% annually. If one assumes the optimistic expectations of scientists are correct, society would be justified in investing US$1304 million annually. To put this number in perspective: Asia produces some 350 million tons of rice annually; at a conservative price of US$200 ton^{-1}, it has a value of US$70 billion. Thus, the low expectations of research success imply investing about 0.8% of output value, while the high estimate of research success implies investing about 1.8%.

Several points stand out from Table 23.4. The bulk of research, about 60%, should be focused on rice problems in irrigated areas. Even if one had

Table 23.4. An optimal portfolio of Asian rice research investment based on conservative expectations of research success, by rice agroecosystems.*

Research problem area category	Annual research investment (US$million) for all research techniques directed at			
	Irrigated	Rainfed	Upland	Deepwater
Insects	58	15	8	6
Diseases	31	14	9	5
Other pests	28	13	17	5
Abiotic stress	76	23	20	4
Bioefficiency	178	31	9	2
Total	388	95	63	22

*Conservative expectations reflect the average number of years until 25% of the remaining potential yield increase is obtained.
Source: Derived from research intensity data in Chapter 22. Rice production by ecosystem is: irrigated, 352 million tons; rainfed, 74 million tons; upland, 34 million tons; deepwater, 14 million tons. Rice is valued at US$200 ton^{-1}.

optimistic expectations about the likely success in the other agroecologies and low expectations for irrigated areas (the opposite of historical experience) the optimal investment would still direct 46% of the total investment toward irrigated agroecologies. Part of this concentration on irrigated areas is because China has nearly half of Asia's total production, so its needs weigh heavily, and virtually all of rice production in China is irrigated. The appropriate investment for the rest of Asia would be somewhat differently balanced, but still would have a heavy proportion devoted to the irrigated ecosystem.

The largest gains – almost 40% – appear possible from improving biological efficiency. This is largely because the greatest per hectare gains are expected from these sources. But this result also derives from the belief that bioefficiency gains can be applicable to all rice agroecologies, even though it is recognized that they will take somewhat longer to achieve than control of insects and diseases. Overcoming abiotic stress should receive the second largest amount of research support. Significant gains can also be made by providing rice protection from insects, diseases and other pests so these areas should also continue to be supported.

Table 23.5 summarizes the findings in a complementary format. It shows how the optimal annual research investment of US$568 million should be allocated among the five major RPA categories and four main research approaches. Conventional breeding is targeted for the largest investment. Transgenic approaches are expected to contribute the most to insect and disease control while (somewhat surprisingly) gains against abiotic stresses and in bioefficiency are expected to come from conventional breeding. Management research is expected to make significant contributions to reducing insect damage and to overcoming abiotic stress. Of course, the total

Table 23.5. An optimal portfolio of Asian rice research investment based on conservative expectations of research success, by rice research technique.*

Research problem area category	Annual research investment (US$million) for all rice agroecologies using			
	Management	Conventional breeding	Wide crossing	Transgenetics
Insects	28	17	6	36
Diseases	15	10	12	20
Other pests	27	0	23	12
Abiotic stress	33	50	11	29
Bio-efficiency	0	126	33	77
Total	104	204	84	176

*Conservative expectations reflect the average number of years until 25% of the remaining potential yield increase is obtained.
Source: Derived from research intensity data in Chapter 22. Rice production by ecosystem is: irrigated, 352 million tons; rainfed, 74 million tons; upland, 34 million tons; deepwater, 14 million tons. Rice is valued at US$200 tons^{-1}.

research investment is the same as for Table 23.4 because the two tables are simply two different arrangements of the same data.

IV. Implications and Conclusions

The methodology used in the studies reported in this book is an advance over earlier methods for setting priorities among alternative rice biotechnology research possibilities (Herdt, 1991). The new methodology considers not only the relative importance of problems or opportunities but also the extent to which apparently appropriate solutions have been devised (although those 'solutions' may still be in the pipeline on the way to being adopted). The new methodology also is superior because it generates an optimal portfolio of research investments while earlier methodology generated a rank ordering of problem areas based on losses/gains. Empirically, this new methodology is also a step beyond earlier work because it includes four research approaches – management, conventional breeding, wide crossing and transgenic methods.

The results are inevitably a reflection of the data that went into the analysis. To the extent that the country studies provided: (i) comprehensive lists of RPAs, (ii) good estimates of the yield losses/opportunities associated with each, and (iii) good estimates of the area over which such losses/opportunities prevailed, they are good indications of the opportunities for research targets. The country studies provide excellent coverage of China and good coverage of South Asia. However, only Indonesia is included from Southeast Asia. The omission of Vietnam, the Philippines and other countries is a serious limitation. The sample included researchers at the IRRI and in the national programs, but few work primarily with the new techniques of biotechnology so the potentials of that kind of work may be underestimated. As in any empirical work, the samples used have limitations, and these affect the quality of the data. Despite these limitations, the work reported in this book is a major step forward in efforts to establish rice research priorities.

The quantitative results are useful pointers for research investment policy. They imply that there are quite large social returns to be made by investing in rice research. Those returns will be shared between rice producers and consumers, and, perhaps, in the case of hybrid rice or where there is strong enforcement of intellectual property rights, seed producers who incorporate attractive traits into seed. Despite the high returns, however, there seems little obvious way to induce private investment because the gains from management innovations or open pollinated varieties cannot be captured by private firms. Therefore, the Asian public, all of whom are rice consumers and many of whom are rice producers, have an interest in ensuring continued significant public investment in rice research.

The allocation of public rice research funds across individual RPAs (see Chapter 22) and across the five main RPA groupings (see Table 23.4) provides useful guidance for those charged with Asian rice research. The analysis suggests that large benefits would result from concentrating research funds on increasing the biological efficiency of the rice plant. This would come from success in pursuing hybrid rices adapted to many varied conditions, apomixis to fix desirable traits, improved plant capacity to generate photosynthates and the ability to store them in the grain, and re-designing the plant's architecture from the currently widely grown semidwarf to a plant of about the same height but with fewer tillers and more grains per tiller (see Chapter 4). Data are not available to indicate the present allocation of research funds to this set of objectives, but indications are that it is relatively modest, so returns from increased investment to this broad area could be large.

The results indicate that scientists clearly expect biotechnology – the modern techniques for genetic transfers (wide crossing and transgenic) – to provide large gains, but that conventional plant breeding still has much to offer. Some of these gains may come from using molecular markers to assist conventional breeding, but, unfortunately, the four-technique classification does not indicate the role of molecular marker techniques. For example, advances in characterizing the blast pathogen using molecular techniques have given plant breeders new strategies for breeding plants with more durable resistance to blast. Similarly, molecular markers that can be associated with the length and penetration ability of rice roots could help breeders develop more drought-resistant rices. It is clear that a mix of research approaches should be used into the future.

Finally, the list of RPAs in the analysis explicitly excluded social science subjects like research on policy and agricultural extension. All acknowledge that policy and extension are important factors contributing to the speed with which biological innovations are adopted by farmers. However, the purpose of the work reported in this book was to determine how allocations should be made among biological research opportunities, so non-biological subjects were omitted from the list of possible research areas.

References

Hayami, Y. and Herdt, R.W. (1977) Market price effects of technical change on income distribution in semi-subsistence agriculture. *AJAE* 59(2), 245–256.

Herdt, R.W. (1991) Research priorities for rice biotechnology. In: Khush, G.S. and Toenissien, G.H. (eds) *Rice Biotechnology*. CAB International/International Rice Research Institute, Wallingford, UK, pp. 19–54.

Index

Note: Page numbers in *italic* refer to figures and/or tables

agroecological zones (AEZ) 36, 111, 293
    definition of terms used in classification *45*
    foodgrain production 49, *50*
    geographical delineations *46*
    incidence of poverty 49
    interface with rice ecosystems *47*
    population pressure 49
    relative importance of foodgrain crops in *47*
    rice production *50*
    socioeconomic characterization by 44–51
aksip na pula 267
allelopathy in weed control 283
allocative efficiency 3, 4, 73, 74
aman 134, 180
Ammannia baccifera 281
Andhra Pradesh *see* India, southern
armyworm 266–7
    eastern India *120*, 122, 124, 126
    Indonesia 254, *255*, 257
    Nepal *200*, 207, 208, *209*, 210, *213*
    priority units *352*
    Thailand 227
Asian riceland soil constraints database *39*, *40*
Assam 109
atom blight *353*
aus 134, 180
average internal rate of return (AIRR) *10*, 11

bacterial leaf blight 268, 269, 275, 276, 312, *313*, 314
    Bangladesh 182, *184*, *185*, *188*, 189, *190*
    China 321–2
    eastern India *120*, 121, 122, 123, 124, 126
    Indonesia 254, *255*, *256*, 257
    Nepal 199, *200*, 207, *208*, *209*, *210*, 211, *212*, *213*
    priority units *352*
    southern India *153*, *157*
    Thailand 226, 227, 233, 234, *235*, 236
    West Bengal 138, 139, 140, *141*
bacterial leaf streak
    Bangladesh 182, *184*, *185*, *188*
    priority units *352*
    Thailand 227
bacterial red stripe 275
bakanae 276
    Bangladesh 182, *184*
    Thailand 227
Bangladesh *31*, 179–80
    agricultural research systems 180–1
    consumption trends 26, *27*
    contribution of rice to national economy 18, *19*
    ecosystems *41*, *42*, 180
    pests and diseases 182–3, *184*, *185*
    priority portfolios 378–9, *384–5*
    priority units 352–4
    production *20*, *29*

Bangladesh *contd*
　research institutions 181–2
　yield 21, *22, 23*
　yield gap 184–90, 293
Bangladesh Agricultural Research Council
　(BARC) 181
Bangladesh Agricultural Research
　Institute (BARI) 181
Bangladesh Institute of Nuclear
　Agriculture (BINA) 181–2
Bangladesh Rice Research Institute
　(BRRI) 181–2
benefit/cost ratio (BCR) 92, 93, 147–8,
　349
Bihar *see* India, eastern
biological weed control 283–4
biomass and yield 60, 64–5
biotechnology 148–9, 151
　tool development 328
　　assessment of time-to-
　　　achievement *335,* 340, *342*
　see also International Program on
　　Rice Biotechnology
birds
　eastern India 115, *120,* 125, *126*
　priority units *353*
black bug 227
black-streaked dwarf virus 311, 312
blast 267, 269, 275, 276, 312, *313*
　Bangladesh 182, *184, 185, 188, 189,
　　190*
　China 176, 317–24
　eastern India *120,* 124, 125, *126*
　Indonesia 254, *255,* 257
　Nepal 199, *200,* 207, *208, 209,* 210,
　　211, 212, *213*
　priority units *352*
　southern India 153, 154–5, 156, *157*
　Thailand 227, 228, 229, 230, 235,
　　236–7
　West Bengal 138, 139, 140, *141*
boro 134, 179, 180, 296, 300
Brazil 29
brown planthopper 269, 275, 276, 308,
　309
　Bangladesh 182, *183, 189,* 190
　eastern India *120,* 123
　Indonesia 252, 254, *255,* 256, 257
　Philippines 248
　priority units *352*
　southern India 153, 154, 156, *157*
　Thailand 226, *227, 228, 229, 230,
　　233, 234, 236*
　West Bengal 138, 139, 140, *141*
brown spot 267, 275
　Bangladesh 182, *185*
　eastern India *120,* 125
　Indonesia 255
　Nepal 207, *208, 209, 210, 211,* 212,
　　213
　priority units *352*
　southern India 153, *157*
　Thailand 226, 227
　West Bengal 138
bulus 66, 67
bunchy stunt virus 311

C4-63 252, 253
Cambodia *31,* 32
　production trends 20, 21
　yield 21, *22, 23*
canopy characteristics 63–4
caseworm
　Bangladesh 183
　eastern India *120,* 124
　priority units *352*
　Thailand 227, 230
caterpillar, West Bengal 139, 140, *141*
CERES/RICE model 265
CH45 204
China *31,* 161–3, 175, 177, 317
　consumption trends 26, 27
　contribution of rice to national
　　economy 18, *19*
　cropping systems 318–19
　cultural practices 319–20, *321*
　ecosystems and natural
　　conditions 163–4
　priority portfolios *380–1*
　priority units *352–4*
　production 29
　　constraints 167–75, *176*
　　trends 20, 21
　research
　　infrastructure 161
　　priorities 164–7
　varietal improvement 162
　yield *22, 23*
　yield gap 164–75, *176,* 292
　yield loss 321–3
Cimandiri 252

Cipunegara 252
Cisadane 252, 254
Citandui 252
Citarum 252
CMS 68
Cnaphalocrocis medinalis see leaf folder/roller
cold weather *see* temperature, low
comparative advantage 73, 84
congruence rule 5, 81
cost efficiency 3, 4, 73–4
cost minimization 74–6
crabs
    eastern India 115
    priority units 353
    Thailand 227, 229, 230, 235
crop loss *see* yield loss
Cropping Systems Program, Nepal 205
cutworm 226, *230*
Cyperus difformis 188, 281
Cyperus iria 281
cytoplasmic male sterility 68

damping off 182, *184*, 276
dead heart 307
Dicladispa armigera see rice hispa
discount rate 9, 11, 94
disease resistance 65
    biotechnology development 328
    assessment of time-to-
        achievement 337, 341, 342
drought 294–5
    anthesis 294
    Bangladesh 188, 189, 190, 295
    China 176, 294, 295
    eastern India 114, *120*, *121*, *122*, *123*, *124*, *125*, *126*, *127*, 294, 295
    Nepal 201, 207, *208*, *209*, *210*, 212, *213*, 294, 295
    Philippines 248
    priority units 353
    seedling stage 294
    southern India *154*, *157*, 294, 295
    Thailand 225–6, 229, 233, 235
    tolerance 158
    vegetative stage 294
    West Bengal 138, 139, 141
    and yield loss 294–5, 300–1, *302*
dwarf virus 310, 312

ear-cutting caterpillar *183*
ear head bug 309
    southern India 153, 154, 155, 156, *157*
Echinochloa colona 281
Echinochloa crus-galli 188, 189, 269, 278, 281
Eclipta prostrata 281
economic efficiency 3–4, 73–4, 77
economic modelling 79–84
    application 84–9
economic theory of the firm 73
ecosystems
    applications of analysis 44–54
    Bangladesh 180
    classification 35, 37, *38*
        agroecological 36–7
        rice ecosystems 37, *38*
    databases
        macro-level analysis 41, *42*
        mega-level analysis 37–41
        meso-level analysis 41, *43*
        micro-level analysis 43–4
    eastern India 111–13
    importance in rice production 13
    interface with agroecological zones 47
    Nepal 196–7
    piority portfolios 363–70
    priority setting 103
    Thailand 219–21
    West Bengal 134–6
evaluation
    ex ante 4, 5–6, 92–5
    ex post 5
    imputation studies 9, 10, 11
    and priority setting 9–11, *12*, 99–100
    statistical hedonic studies 10–11, *12*, 330–4
    statistical studies 10, 11
    and priority setting 5–6
extension workers 396

false smut
    Nepal *208*, *209*, *210*, *213*
    priority units 352
farming competitiveness 24–5
fertilizer use
    China 292

fertilizer use *contd*
   Indonesia 253
   Nepal 201
   and yield 22–4
Fimbristylis miliacea 281

gall dwarf virus 311
gall midge 266, *275*, 306, 308, 309
   Bangladesh *183*
   eastern India *120*, 122, 123
   Indonesia 254, *255*, *256*, 257
   Nepal 199, *200*, *208*, *209*, *210*, 211, 213
   priority units 352
   southern India 153, 154, 155, 156, *157*
   Thailand 227, 234, *235*
General Agreement on Tariffs and Trade (GATT) 28
germplasm
   adaptive 79
   for new plant types 66–7
   pool-expanding 79
golden snail 248
grain
   density 63
   filling percentage 63
   quality 328
   size 63
grain discolor *153*, *157*
grasshopper
   Bangladesh *183*
   Nepal *208*, *210*
   Thailand 227
grassy stunt virus *310*, 311, 312
   Indonesia *255*
green leaf spot 352
green leafhopper *275*, 308, 309
   Bangladesh 182, *183*
   eastern India *120*, 122, 124
   priority units 352
   southern India *153*, *157*
   Thailand 227, 230, 233, *234*
   West Bengal 138
green revolution 17
gundhi bug
   eastern India *120*
   Nepal *200*, *207*, *208*, *209*, *210*, 211, 212, 213

harvest index and yield 60, 64–5
herbicides 287
   cost 279–80
   monitoring 285
   transgenic rice resistance to 281–2
   weed resistance to 281
heterosis 68
high-yielding varieties (HYVs), Thailand 221
host plant resistance 102
hybrids 8, 28, 67–8, 162, 318
   enhanced use 344
   indica/japonica 68–9

IMV1 varieties 252, 253
IMV2 varieties 252–3
India 31
   consumption trends 26, 27
   contribution of rice to national economy 18, *19*
   eastern 109–10
    climate 114
    crop loss estimates 118–27
    diseases 115
    ecosystems 51–4, 111–13
    kharif 110, *111*, 112, 119–22, 296, 300
    priority portfolios *374–5*, *383–4*, *386–7*, *390–1*
    priority units 352–4
    production 110–15
    rabi 110, *111*, 112, 122–3, 296
    research priorities 127–9
    soil fertility 114
    technical constraints 113–15
    yield 110–11
    yield gap 116–18, 293
   see also West Bengal
   production 20, 29
   southern 145–6, 159–60
    biotechnology research 151
    critical temperatures 296
    crop management research 150
    methodology for defining and costing constraints 146–9
    physiology research 150
    plant protection research 151
    priority portfolios *375–6*
    priority units 352–4

India contd
  research institutions and
    projects 149–51
  research priorities 156–8
  varietal improvement 149–50
  yield gap 151–2
  yield loss 152–6
  yield 21, 22, 23
Indian Council of Agricultural Research
  (ICAR) 149
indicas 19, 300
  hybridized with japonica 68–9
Indonesia 31, 251–2, 257
  consumption trends 26, 27
  contribution of rice to national
    economy 18, 19
  imports 30
  priority portfolios 381–2, 389–90
  priority units 352–4
  production 20, 29, 252–3
  varietal improvement 252–3
  yield 22, 23
  yield loss 253–6
insect pests 305–7, 313–14
  eastern India 114–15
  see also specific insects
insect resistance 65
  biotechnology development 328
    assessment of time-to-
      achievement 336, 341, 342
integrated pest management 305
  Indonesia 255, 257
  Thailand 233
internal rate of return 92, 93
International Agricultural Research
  Centres (IARCs) 84
International Network for Genetic
  Evaluation of Rice (INGER) 12–
  13, 85, 88
International Program on Rice
  Biotechnology
  application of priority setting 327–
    45
  assessment of time-to-
    achievement 334–42
  expected potential benefits 342–5
  research objectives 327–8
invention production functions 77–9
IR5 252, 253
IR8 19, 25, 162, 181, 221, 252, 253, 330
IR20 330

IR36 252–3, 254
IR64 252–3, 254
iron deficiency, eastern India 114, 120,
  123, 125
iron toxicity
  eastern India 120
  priority units 353
  southern India 154
IRRI
  comparative advantage 84
  and resource allocation 85, 86–7
irrigation and yield 22–3
Ischaemum rugosum 281

Japan 31–2
  consumption trends 25, 26
  contribution of rice to national
    economy 18, 19
  decline in farming population 24
  production 20, 29
  yield 21, 22, 23, 25
japonicas 300
  hybridized with indica 68–9
  tropical 66–7
javanicas 66–7

Karnataka see India, southern
Kerala see India, southern
kharif 110, 111, 112, 296, 300
  insect pests 307, 308
  yield loss 119–22
Korea
  DPR
    production trends 20
    yield 23
  Republic of 31–2
    consumption trends 25, 26
    contribution of rice to national
      economy 18, 19
    decline in farming population 24
    production trends 20
    yield 21, 22, 23, 25
Krueng Aceh 252, 254

land units 96
Laodelphax striatellus 312
Laos 31, 32
  contribution of rice to national
    economy 18, 19

Laos *contd*
    production trends *20*
    weed control 279
    yield *22*
leaf characteristics 63–4
leaf folder/roller 266, 306, 307–8, 309
    Bangladesh *183*
    eastern India *120*, 124
    Indonesia 254, *255*, 257
    Nepal 208, *209*, 210, 213
    priority units *352*
    southern India 153, 154, 155, 156, *157*
    Thailand 226, 227, 230, 233, 234
    West Bengal 138, 139, 140, *141*, 188, 189
leaf scald 276
    Bangladesh 182, *184*, *185*
    priority units *352*
leaf streak 188, 189
leafhoppers 266, 306
    Nepal 199, *200*, 208, *209*, 210, 211, 213
    see also green leafhopper; orange-headed leafhopper
Leptochloa chinensis 281
Leptocorisa see rice bug
Limnocharis flava 281
lodging
    China 175, *176*
    eastern India 115, *120*, 122, 124, 126
    Nepal 208, *209*, *210*, 211, 212, 213
    southern India 154, 155, 156–7
    tolerance to 158
long-horned cricket 183
Ludwigia hyssopifolia 281

Madyha Pradesh *see* India, eastern
Malaysia 31–2
    consumption trends 25
    contribution of rice to national economy 18, *19*
    production trends *20*
    yield *22*
management 4
Marasmia patnalis see leaf folder/roller
marginal internal rate of return (MIRR) 11
Marsilea minuta 281

Marsilea quadrifolia 281
Masuli (Mahsuri) 204
mealy bug 276
    Bangladesh 183, *188*
    Nepal 199, *200*, 208, *209*, 210, 213
    priority units *352*
    Thailand 227
modern varieties (MVs) 330
    Bangladesh 183, *185*
    China 318
    Indonesia 251, 252–3
    Nepal 193, 204, *205*
Monochoria vaginalis 269, 281
monsoon, eastern India 114
Myanmar *31*, 32
    consumption trends 26
    contribution of rice to national economy *19*
    production 20, 21, *29*
    submergence 300
    yield 21, 22, *23*

narrow brown spot 226, 227
national agricultural research (NAR)
    India 149
    weak 104–5
National Summer Crops Workshop, Nepal 204–5
necrotic mosaic virus 311, 312
nematodes 268
    see also ufra
Nepal *31*, 193–4, 212
    agricultural research intitutions 202–3
    area and production by region *194*
    contribution of rice to national economy *19*
    ecosystems 196–7
    expected research benefits and priorities 209–12, *213*
    methodology for study of production constraints 195–6
    priorities
        portfolios 377–8, *387–8*
        setting process 204–5
        units 352–4
    production
        constraints 199–202
        trends 20, 21
    research

Nepal *contd*
   manpower 203–4
   resource allocation 206
   yield 22, 23
   yield gap 197–9, 293
   yield loss 207–9
Nepal Agricultural Research Council (NARC) 203, 205
Nephotettix see green leafhopper
net present value (NPV) 93, 147–8, 222, 232, 349
nitrogen deficiency 176

orange-headed leafhopper 183
orange leaf virus 311
Orissa see India, eastern
ORYZAE1 model 265

Pachydiplosis oryzae see gall midge
Pajam 181
Pakistan
   production trends 20, 21
   yield 22
panicle size 62–3
Pelita I-1 252
Pelita I-2 252
pest resistance 158
pesticide use, Indonesia 255
Philippines 31, 32, 239, 248–9
   consumption trends 26
   contribution of rice to national economy 19
   import and export 30
   production 29
     long-term trends 240–1
     trends 20, 21
   yield 22
     constraints 246–7
     and fertilizer consumption 23
     losses 247–8
     sources of growth 241–2, 243
   yield gap
     by region 243–5
     estimation 245–6
phosphorus deficiency 176
photoperiod-sensitive genetic male sterility (PGMS) system 68
Pin Khaew 221
plant breeding
   conventional 350, 357–62

   germplasm 66–7
   for new plant types 65–6
   transgenic 350, 357–62
   wide crossing/tissue culture 350, 357–62
plant height modification 64–5
plant type modification 60–2
   canopy and leaf characteristics 63–4
   disease and insect resistance 65
   grain filling percentage 63
   grain size and density 63
   growth duration 64
   height 64–5
   large panicles 62–3
   present state of breeding for 65–6
   reduced tillering 62–3
   root system 65
   stem thickness 64–5
planthoppers 266, 306
   Nepal 199, 200, 208, 209, 210, 211, 213
   see also brown planthopper; white-backed planthopper
Pondicherry see India, southern
population growth 27, 29, 59
potassium deficiency 176
potential benefits (b/u) 93, 94
   subjective probability estimation 355–62
present value of benefits (PVB) 9, 92, 93–4, 328–30, 348–9
present value of costs (PVC) 9, 92, 93, 348–9
priority portfolios 348–50, 363, 370, 372–3, 398–400
   Bangladesh 378–9, 384–5
   China 380–1
   deepwater rice 370, 371–2, 390–1
   eastern India 374–5, 383–4, 386–7, 390–1
   Indonesia 381–2, 389–90
   irrigated rice ecosystems 364–6, 374–82
   Nepal 377–8, 387–8
   optimal 401–3
   rainfed rice ecosystems 366–8, 383–5
   southern India 375–6
   upland rice 368–70, 386–90
priority setting 4–5, 54–6, 91–2, 261, 347–8
   adjustments 101–5

priority setting *contd*
   application to biotechnology
      program 327–45
   demand side 95–8
   and ecosystem favourability 103
   and environmental concerns 105
   and equity 104
   and *ex ante* evaluation 94–5
   and *ex post* evidence 9–11, *12*, 99–100
   guidelines 79–84
   and international activity 12–13
   methods for obtaining priority portfolio 348–50
   Nepal 204–5
   and project evaluation 5–6
   and sustainability 102–3
   Thailand 222–31
   and yield gaps 6–9
profit maximization 76

rabi 110, *111*, 112, 296
   insect pests 307
   yield loss 122–3
ragged stunt virus *310*, 311, 312
   priority units *353*
   Thailand 226, 227, 233, *234*
rainfed lowland rice ecosystem database 39, 40–1
ranking techniques 100
RD1 221
RD7 221
RD15 221
RD23 221
research germplasm 4
research problem/priority areas (RPAs) 4–5, 6, 91
   abiotic stress loss 357, *360*
   biological efficiency 361, *362*
   biotechnology 329, 334
   crop-loss 96, 98–9
   data on 395–6
   definition 350–1
   disease loss 357, *359*
   estimating difficulty in achievement 397–8, *399*
   general pest loss *360*, 361
   insect loss 357, *358*
   orphan 104
   and priority setting 95–8

   units 93, 94, 96, 104, 351, *352–4*
research programmes 4
research projects 4, 91
research systems 4
research techniques (RTs) 5
   categories 350
   managerial 350, 357–62
resource allocation 73–4
   basic economic rules 74–7
   invention production functions 77–9
   multi-RPA, multi-organization model 79–84
Rhizoctonia solani see sheath blight rice
   demand trends 25–7
   importance in national economies 18, *19*
   recent production trends 19–21
   supply trends 27–8, *29*
Rice Almanac 38
rice bug 266, 306, 307
   Bangladesh *183*, *188*, *189*, *190*
   Indonesia *255*
   Philippines 248
   priority units *352*
   Thailand 227
Rice Ecosystem Geographic Information System (REGIS) 41
rice hispa 266, 276, 306, 309
   Bangladesh 182, *183*, *188*, *189*, *190*
   eastern India 125
   Nepal 199, *200*, 207, *208*, *209*, *210*, *211*, 213
   priority units *352*
   Thailand 227
   West Bengal 139, 140, *141*
rice output units 96
rice tungro virus *see* tungro virus
rice water weevil 306
rice weevil 208
Ripersia oryzae see mealy bug
rodents 269
   Bangladesh *189*
   China *176*
   eastern India 115, *120*, 122
   Indonesia 254, *255*, 256, *257*
   Nepal *208*, *209*, *210*, 211
   priority units *353*
   southern India *153*
   Thailand 226, 227, 230, 233, *235*, 236

root knot 182, *184*
root system modification 65

Sadang 252
salinity
    eastern India 114, *120*, 123
    priority units *353*
    southern India 154, 155, 157
    West Bengal 138, 140
Samuhik Bhraman 205
scaling techniques 100–1
sclerotial blight 275, 321–2
Sclerotium oryzae see stem rot
scoring techniques 100–1
seedbed beetle 208
seedling blight 276
    Bangladesh 182, *184*
    Nepal 208
Semeru 252
sheath blight 62, 268, 275, 276, 312–13
    Bangladesh 182, *184*, *185*, *189*, *190*
    China 175, *176*
    eastern India *120*
    Indonesia 255
    Nepal 199, *200*, *208*, *209*, *210*, *211*, *213*
    priority units *352*
    southern India *153*, 155, 156, *157*
    Thailand 227
    West Bengal 139, 140, *141*
sheath rot 276
    Bangladesh 182, *184*, *185*
    eastern India *120*, 124
    Nepal 199, *200*
    priority units *352*
    southern India *153*
    Thailand 227
    West Bengal 138
single point probability estimate 101
soil
    acid
        China *176*
        eastern India *120*, 122, 123, 125, 126
        priority units *353*
        southern India *154*, 155
        West Bengal 138
    alkaline
        eastern India 114, *120*, 126
        priority units *353*
        southern India *154*
    low fertility
        Bangladesh *188*, *189*, *190*
        China 175, *176*
        eastern India 114
        priority units *353*
        southern India *154*, 156
        Thailand 235
Sorghum bicolor 62
Sri Lanka 31
    contribution of rice to national economy 19
    production trends 20
    yield 22, 23
stem, thickness modification 64–5
stem rot 62, 268, 276
    Bangladesh 182, *184*, *185*, *190*
    Nepal 207, *208*, *209*, *210*, *211*, 212, *213*
    priority units *352*
stemborers 266, 269, 275, 276, 306, 307
    Bangladesh 182, *183*, *188*, *189*, *190*
    China 175, *176*
    eastern India *120*, 122, 123, 124, 125, 126
    Indonesia 254, 255, 256, 257, 309
    Nepal 199, *200*, 207, *208*, *209*, *210*, *211*, *213*
    Philippines 248
    priority units *352*
    southern India 153, 154, 155, 156, *157*
    Thailand 226, 227, 229, *230*, 233, 234, 235, 236, 237
    West Bengal 138, 139, 140, *141*
stink bug 254, 255, 256, 257
storage moth, Nepal 208
stress-tolerance biotechnology
    development 328
    assessment of time-to-achievement 338, *341*, *342*
stripe virus *310*, 311, 312
striped borer 199, *200*
subjective probability distribution estimation (SPDE) 100, 101
subjective probability estimation (SPE) 5, 94, 100–1
submergence 300
    Bangladesh *188*, 189, 190, 295, 300, *301*
    China *176*, 295, 300, *301*

submergence *contd*
    eastern India 114, *120*, 121, 124, 126, *295*, 300, *301*
    Myanmar 300
    Nepal 208, *209*, *210*, 211, *295*, 300, *301*
    priority units 353
    southern India *295*, *301*
    Thailand 225–6, *229*, 300
    Vietnam 300
    West Bengal *138*, *139*, 140, *141*
    and yield loss 300–1, *302*
sulphur deficiency
    Bangladesh *188*, *190*
    eastern India 114
sustainability and priority setting 102–3
swarming caterpillar
    Bangladesh *183*
    Thailand 227
Syntha 252

Taichung-176 204
Taiwan 31–2
    decline in farming population 24
Tamil Nadu *see* India, southern
temperature
    critical 296
    high
      Bangladesh *190*
      priority units *353*
      West Bengal *140*
    low 296–300
      Bangladesh *190*, *295*, 296, 299, 300
      China *176*, *295*, 296, 297–8, 299, 300
      eastern India *295*, 296, 297, 299, 300
      Indonesia 299
      Nepal *295*, 299
      Philippines 299
      priority units *353*
      southern India *154*, *295*
      Thailand 299
      West Bengal *140*
      yield loss 299, 300–1, *302*
Terminology of Rice Growing Environments 37, 38
Thailand 31, 32, 217, 237
    consumption 25, 26, 218–19

    contribution of rice to national economy 18, *19*
    crop quality 230–1
    economy 217–21
    ecosystems 219–21
    exports 218–19
    farm survey of yields 225
    information sources 222
    mechanization 228, 230
    output 218–19
    production *20*, 21, *29*
    research 221
      benefit assessment 232–7
      priority setting 222–31
    yield 21, *22*, *23*
    yield gap 223–4
    yield loss 225–8, *229*–30
thermosensitive genetic male sterility (TGMS) system 68
thrips
    Bangladesh 182
    Nepal *208*
    southern India *153*, *157*
    Thailand 226, 227, *229*, 230, 235
tillering, breeding for reduction 62
time-to-achievement of benefits (W_t) 93, 94
    assessment in biotechnology research 334–42
    subjective probability estimation 355–62
trade liberalization, possible effects on rice market 28–9
trait values, hedonic *ex post* evidence of 330–4
transitory yellowing virus 311, 312
tungro virus 267, *275*, *276*, 310, 311, 312, 314
    Bangladesh 182, *184*, *185*, *188*, *189*
    Indonesia 254, *255*, *257*
    Philippines 248
    priority units *353*
    southern India *153*, *157*
    Thailand 226, 233, *234*
    West Bengal *139*, 140, *141*
typhoons 248

ufra 268, 276
    Bangladesh 182, *184*, *185*, *189*, *190*
    priority units *353*

upland rice ecosystem geographic
  database 38–9
Uttar Pradesh *see* India, eastern

value of marginal product (VMP) 76, 77,
  80–4, 89
varieties
  origins of 85, *86–7*
  short duration, eastern India 125
Vietnam *31, 32,* 218, 221
  contribution of rice to national
    economy 18, *19*
  production *20, 29*
  yield *22, 23*
virus diseases 310–12, 314
  see also specific viruses

Waika virus 312
weeding 279
weeds 269, 277, 286–7
  China 176
  control
    allelopathy 283
    biological 283–4
    cost of 279–80
    information dissemination 286
    and rice culture 285–6
    training in 286
  cultivar competition with 282–3
  in direct seeded rice 280–1
  eastern India 115, *120,* 122, 123,
    124, 125, 126, 127
  ecological studies 284–5
  economic importance 285
  herbicide resistance 281
  integrated management systems 284,
    286
  Nepal 200, 207–8, *209, 210,* 211
  priority units *353*
  research priorities 284–6
  resistance to 158
  southern India 154, 157
  surveys 284
  Thailand 227, 233, 234, *235,* 236
  West Bengal 138, 140–1
  and yield loss 278–9
West Bengal 131–2, 142–3
  criteria for cultivar selection 141,
    142

ecosystems 134–6
research infrastructure and
  programs 132–3
rice area *135*
yield gaps and yield loss 136–42
see also India, eastern
white-backed planthopper 275
  Bangladesh *183*
  eastern India *120,* 123
  priority units *352*
  Thailand 226, 227
white head 307
white tip *184,* 268, 276
whorl maggot 227
wild pig 255

yellow dwarf virus 311
yellow orange leaf virus 227, 311
yellow stunt virus *310*
yield
  actual (A) 6, 223, 262
  attainable 262
  and biomass 60, 64–5
  D 8
  decline in growth of 27–8, *29*
  E 8
  economic 262
  enhancement, biotechnology
    development 328, *339, 341, 342*
  and harvest index 60, 64–5
  increased 20
  no-loss/best-practice (B) 6, 223, 228
  potential (C) 6, 8, 223–4, 262
  reference 261–2
  role of agroclimatic factors and
    modern technologies 21–4
  strategies for increasing genetic
    potential 59–69
yield gap 115–16, 292–3, 394–5
  Bangladesh 184–90, 293
  China 164–75, *176,* 292
  consistency 98–9
  contributions to 400–1
  disaggregation 396–7
  eastern India 116–18, 293
  I 7, 99, 116, 136–7, 151, 165, 184,
    223, 245, 263, 292, 394
  II 7, 8–9, 99, 116, 137, 151, 165,
    184, 223, 245, 263, 292, 394

yield gap *contd*
  III 8, 184
  Nepal 197–9, 293
  Philippines
    by region 243–5
    estimation 245–6
  and priority setting 6–9
  as research target 400
  southern India 151–2
  Thailand 223–4
  West Bengal 136–42
yield loss 275–8
  actual farm-level 266–9, 332
  assessment 116, 261
  attainable 263
  calculation 262–3
  China 321–3
  data
    conceptual framework for use 261–3
    interpretation 261–4
  due to bacterial and fungal diseases 312–13
  due to cold 299, 300–1, *302*
  due to drought 294–5, 300–1, *302*
  due to insect pests 305–9
  due to submergence 300–1, *302*
  due to virus diseases 310–12
  eastern India 118–27
  economic 263
  Indonesia 253–6
  measurement 98–9
  Nepal 207–9
  Philippines 247–8
  potential 263, 332
    estimation 264–5
  problems with estimations 332
  southern India 152–6
  status 264–9
  Thailand 225–8, *229–30*
  and weeds 278–9
  West Bengal 136–42

Zea mays 62
zigzag leafhopper *227*
zinc deficiency
  Bangladesh *189*, *190*
  eastern India 114, *120*, 123, 126
  Nepal 201, 207, *208*, *209*, *210*, 211
  southern India *154*, *155*, *157*